Springer Series in Optical Sciences Volume 31
Edited by Jay M. Enoch

Springer Series in Optical Sciences

1 **Solid-State Laser Engineering**
By. W. Koechner

2 **Table of Laser Lines in Gases and Vapors**
3rd Edition
By R. Beck, W. Englisch, and K. Gürs

3 **Tunable Lasers and Applications**
Editors: A. Mooradian, T. Jaeger, and P. Stokseth

4 **Nonlinear Laser Spectroscopy**
By V. S. Letokhov and V. P. Chebotayev

5 **Optics and Lasers**
An Engineering Physics Approach
By M. Young

6 **Photoelectron Statistics**
With Applications to Spectroscopy and Optical Communication
By B. Saleh

7 **Laser Spectroscopy III**
Editors: J. L. Hall and J. L. Carlsten

8 **Frontiers in Visual Science**
Editors: S. J. Cool and E. J. Smith III

9 **High-Power Lasers and Applications**
2nd Printing
Editors: K.-L. Kompa and H. Walther

10 **Detection of Optical and Infrared Radiation**
2nd Printing
By R. H. Kingston

11 **Matrix Theory of Photoelasticity**
By P. S. Theocaris and E. E. Gdoutos

12 **The Monte Carlo Method in Atmospheric Optics**
By G. I. Marchuk, G. A. Mikhailov, M. A. Nazaraliev, R. A. Darbinian, B. A. Kargin, and B. S. Elepov

13 **Physiological Optics**
By Y. Le Grand and S. G. El Hage

14 **Laser Crystals** Physics and Properties
By A. A. Kaminskii

15 **X-Ray Spectroscopy**
By B. K. Agarwal

16 **Holographic Interferometry**
From the Scope of Deformation Analysis of Opaque Bodies
By W. Schumann and M. Dubas

17 **Nonlinear Optics of Free Atoms and Molecules**
By D. C. Hanna, M. A. Yuratich, D. Cotter

18 **Holography in Medicine and Biology**
Editor: G. von Bally

19 **Color Theory and Its Application in Art and Design**
By G. A. Agoston

20 **Interferometry by Holography**
By Yu. I. Ostrovsky, M. M. Butusov, G. V. Ostrovskaya

21 **Laser Spectroscopy IV**
Editors: H. Walther, K. W. Rothe

22 **Lasers in Photomedicine and Photobiology**
Editors: R. Pratesi and C. A. Sacchi

23 **Vertebrate Photoreceptor Optics**
Editors: J. M. Enoch and F. L. Tobey, Jr.

24 **Optical Fiber Systems and Their Components**
An Introduction
By A. B. Sharma, S. J. Halme, and M. M. Butusov

25 **High Peak Power Nd : Glass Laser Systems**
By D. C. Brown

26 **Lasers and Applications**
Editors: W. O. N. Guimaraes, C. T. Lin, and A. Mooradian

27 **Color Measurement**
Theme and Variations
By D. L. MacAdam

28 **Modular Optical Design**
By O. N. Stavroudis

29 **Inverse Problems in Laser Sounding of the Atmosphere**
By V. E. Zuev and I. E. Naats

30 **Laser Spectroscopy V**
Editors: A. R. W. McKellar, T. Oka, and B. P. Stoicheff

31 **Optics in Biomedical Sciences**
Editors: G. von Bally and P. Greguss

32 **Fiberoptic Rotation Sensors**
Editors: S. Ezekiel and H.J. Arditty

33 **Integrated-Optics Technology**
By R.G. Hunsperger

Optics in Biomedical Sciences

Proceedings of the International Conference,
Graz, Austria, September 7–11, 1981

Editors
G. von Bally and P. Greguss

With 212 Figures

Springer-Verlag Berlin Heidelberg GmbH 1982

Dipl.-Phys. GERT VON BALLY
University of Münster, Ear-Nose-Throat Clinic,
Medical Acoustics and Biophysics Laboratory,
D-4400 Münster, Fed. Rep. of Germany

Professor Dr. PAL GREGUSS
Technical University Budapest, Applied Biophysics Laboratory,
H-1111 Budapest, Hungary

DOI 10.1007/978-3-540-39455-6

Originally published by Springer-Verlag Berlin Heidelberg New York in 1982
MyCopy version of the original edition 1982

2153/3130-543210
www.springer.com/mycopy

Preface

As there recently has been increased interest in the applications of optical techniques in biomedical research and clinical diagnostics, it seemed to be appropriate to organize a comprehensive international conference on optics in medicine and biology. Such a broad international meeting had not been held before. An international conference on *Optics in Biomedical Sciences* was organized and took place in Graz, Austria, September 7th through 11th, 1981, sponsored by the International Commission for Optics (ICO) in cooperation with the European Optical Committee, the Austrian Association on Biomedical Engineering, and the German Society for Applied Optics.

It seemed timely to establish a forum for communication among specialists on an international level. This book, presenting the papers given at this conference, demonstrates the state of the art of this increasingly expanding field of applications of optics. Furthermore, the interested reader will find an extended list of references in the various contributions. This book helps to overcome the difficulty, inherent in all interdisciplinary research fields, of gathering widely scattered literature.

The contributions to this book are focussed on the following topics:

Biomedical applications of

- *unconventional imaging in microscopy,*
- *image processing,*
- *interferometry and holography,*
- *speckle-techniques and spectroscopy,*
- *optometry, and*
- *Moiré methods.*

In addition, the brilliant and humorous closing remarks of Nils Abramson from the Royal Institute of Technology, Stockholm, have been included.

The idea of promoting optics in the biomedical sciences has stimulated worldwide interest. Specialists from 18 countries submitted manuscripts to these proceedings. It is hoped that this book may contribute to the necessary interdisciplinary dialogue among clinicians and scientists in optics.

In their capacity as conference chairmen, the editors would like to thank Dr. K. Schindl, C. Reichert Optische Werke, Vienna, general chairman of the organizing committee, Professor F.R. Aussenegg, University of Graz, Institute of Experimental Physics, chairman for local arrangements in Graz, and all the numerous helping hands in Graz and Vienna for the excellent organization. Furthermore, our thanks are directed to Professor A. Lohmann, former president of ICO, Professor J. Tsujiuchi, president of ICO, and Professor H.F. Frankena, secretary general of ICO, for promoting the idea of the conference.

The editors also wish to express their thanks to Professor R. Röhler, University of Munich, Institute of Medical Optics, and Professor S. Boseck, University of Bremen, Dept. of Physics, for their assistance in reviewing the contributions given in their respective scientific fields. Finally, the editors gratefully acknowledge the helpful support of Mrs. H. Janutta, Mrs. A. Sütfeld, and Mrs. R. Dargatz for patiently and carefully retyping most of the submitted manuscripts, without which this book could not have been published.

Münster, Fed. Rep. of Germany and *Gert von Bally*

Budapest, Hungary, March 1982 *Pal Greguss*

Contents

Part I. **Unconventional Imaging in Microscopy**

Part II. **Image Processing**

Part III. Interferometry and Holography

Part I

Unconventional Imaging in Microscopy

Electron Microscopy in Medicine and Biology: Methods and Recent Developments

P.W. Hawkes

Laboratoire d'Optique Electronique du C.N.R.S., B.P. 4347
F-31055 Toulouse Cedex, France

1. Introduction

"The history of biomedical science is in large measure a chronicle of the invention of new instruments and ingenious new analytical methods. The development of the transmission electron microscope was clearly one of the most momentous events in the long history of biology... The introduction of the electron microscope... has progressively obliterated the boundary between the structurally and functionally oriented disciplines" (34). These remarks by FAWCETT, supported as they are by a wealth of experimental evidence, show convincingly the importance of the electron microscope to biologists from the 1950s onwards. In the remainder of this introduction, we draw attention to some other historical aspects of electron microscopy after which subsequent sections are devoted to recent developments in image interpretation. The history of biological electron microscopy is at once the history of instrumental development and the history of specimen preparation techniques, and for technical and even commercial reasons, these have always been intimately related. Thus the vast majority of commercial electron microscopes nowadays operate at around 100 kV ($\lambda \simeq 4$ pm) or less, since microtomes are available to cut very thin sections ; electrons with energies of the order of 100 keV can penetrate these sections with little loss of energy. Before such microtomes were on the market, however, microscope manufacturers were actively developing much higher voltage machines, generating electrons that could penetrate the thick specimens then in use ; this effort was abandoned for many years with the advent of the new "ultramicrotomes" (100). More recently, commercial high-voltage instruments have been put on the market and a few biological institutions are equipped with them, but the results have been uneven, sometimes disappointing, sometimes invaluable (33, 46, 48, 77, 101). This mention of high-voltage devices recalls another problem of specimen preparation, namely the very drastic treatment that the specimen must undergo to enable it to tolerate the very low pressure within the microscope (typically of the order of mPa). For many years (2, 25, 69, 96, 97), attempts have been made to observe living material in electron microscopes, encased in special wet cells. High voltage is essential if the electrons are to make their way through wet-cell walls, water vapour and specimen but only indistinct images have so far been obtained in this way.

Fortunately, wet cells are not the only answer to the problem. After dehydration, embedding,section-cutting and staining, one may - indeed should - suspect that the microscope image bears at best an imperfect resemblance to the structure of the living specimen. Chemical methods of specimen preparation were therefore supplemented by a physical method, freezing (7, 26, 37, 92, 133) and this in turn was radically improved by the introduction of high-speed freezing (17, 64, 65), for which the freezing-time is of the order of milliseconds. These techniques have in turn had repercussions on microscope design and modern high-performance instruments offer special accessories for the transfer and insertion of frozen material into the specimen chamber.

We might say that the object of all these efforts has been to render morphological studies more reliable - in the early days, credible would not have been too strong a term. Electron beams can, however, be used not merely to form an image but also to perform local elemental analysis, that is, to present a chart showing the localization of very small quantities of given chemical elements within the specimen : this is the task of the microanalyser, in which a very fine electron probe explores the specimen, generating X-rays as it moves. These are characteristic of the element struck by the probe and can be used to indicate the presence of a particular element in a given cell organelle, for example. This field is in active development (5, 9, 11-14, 47, 54, 78, 84-89, 111, 115-116, 129). We refer in particular to the biological parts of the annual *Scanning Electron Microscopy* (e.g., 6, 39, 102, 104, 112) and *Microbeam Analysis* (93) for evidence of the usefulness of the technique throughout the biomedical spectrum : thus ABRAHAM (1), for example, has invoked it to study toxic dust in lungs and particulate material in liver, kidney, salivary glands and lymph nodes, and CHANDLER, to study reproductive physiology (13).

Before leaving these topics, we should emphasize first, that we have by no means exhausted the list of instrumental modifications inspired by or associated with improvements and novelties in specimen preparation ; and second, that alongside the more spectacular innovations, such as the introduction of freezing, innumerable small but valuable improvements are constantly being made to the more standard methods of fixing, embedding and staining specimens for electron microscopy. The series edited by GLAUERT (45) and HAYAT (52, 53) are useful sources of information here.

2. Recent Developments in Instrumentation : the STEM

We have mentioned briefly in section 1 the high-voltage microscope as a tool that is occasionally useful in biology, when thick specimens must be penetrated. A long survey listing each of its potential uses has been prepared by KING *et al.* (77), and we shall therefore not consider it further here. A new type of instrument that is likely to become much more instantly attractive as it finds its way into the biological laboratory is the scanning transmission electron microscope (STEM), which combines the resolution of a traditional electron microscope with the X-ray and other analytic tools of a scanning electron microscope and the flexibility and power of a simple computer interface. To counterbalance these attractions, it has two disadvantages : it can only operate successfully with a field-emission electron gun, which requires an extremely high vacuum (of the order of 10 nPa) and it is extremely expensive.

The principle is straightforward : in the image-forming mode, a fine probe (a few tenths of a nm in the smallest case) methodically scans the specimen, which is a thin section. The electrons are deflected within the latter and collected by a pair of collectors, one in the form of a plate with a central hole, the other a second plate behind the hole. In a more refined scheme, the latter detector is replaced by an electron prism, which enables us to separate electrons that have lost different amounts of energy in the specimen, which is of diagnostic value (most of the electrons that lose energy in the specimen pass through the hole). The first two detectors provide information, from which we can generate bright field (central) and dark-field (peripheral) images : the current collected by each is used to modulate the brightness of a TV monitor synchronized with the beam scanning the object, and the corresponding images are thus built up simultaneously.

The currents collected by the detectors need not, however, be sent directly to the TV screen ; they can be modified or combined in any way that seems useful or they can be despatched to a computer for more sophisticated processing before being used to form an image. This means, for example, that the contrast range of the image can be adjusted for visual comfort, and linear combinations or the ratio of the bright and dark-field signals can be displayed immediately. The ratio is interesting because it generates so-called Z-contrast, that is, contrast that varies with the atomic weight of the atoms of the specimen. For an account of analogue processing of STEM images, see (71) and for discussion of the STEM as a biomedical tool, see (28-30, 32, 68, 95, 120, 127).

The STEM is still a comparative newcomer to the biological laboratory (3a, 28, 41, 73, 110, 120, 130, 137) but has already begun to provide information that would have been difficult to obtain any other way : mass determination of medium to large biological molecules is not easy and the STEM is being successfully used for this task (27, 31, 38, 83, 128). It seems likely that the fact that the STEM can so straightforwardly be interfaced to a computer will become particularly valuable as simplified software is made available to the non-specialist user of the microscope (57, 73, 75). Furthermore, there is reason to believe that interpretable images can be obtained from unstained or lightly stained specimens with the STEM (eg. 74), which is in itself highly desirable and should also shed light on the relation between stain and structure in the images of stained sections. The use of low temperatures is as vital in STEM as in the conventional microscope, and specialized accessories are being developed to facilitate this (e.g., 135).

Finally, before leaving the STEM, we wish to draw attention to an intrinsic feature of scanning microscope optics that is creating much excitement among instrument designers : we are referring to the fact that the detector geometry can be chosen freely. This is attractive because each small zone of the object, as it is momentarily illuminated by the probe, generates a far-field diffraction pattern of this same zone, which is spread out over the detector plane. To a first approximation, the central detector collects the central order of this diffraction pattern, while the outer (dark-field) detector collects the remainder. We can, however, show that it can be very informative to subdivide the detector further (51) ; into rings, for example (103) or into semicircles (19, 20) or quadrants (49), a geometry that has already been found useful in practice (125), though not yet in biology. We return to this point in Section 5.

We have already mentioned energy analysis very briefly and we conclude this limited selection of recent instrumental developments by mentioning the increasing popularity of electron energy loss spectroscopy (EELS). Energy losses can in principle be as powerful diagnostic tools as X-rays, for in passing through the specimen some electrons lose amounts of energy that are characteristic of the specimen composition. Spectrometers of various kinds have therefore been attached to electron microscopes, and are routinely incorporated in scanning transmission instruments. We refer to the annual volumes of *Scanning Electron Microscopy* for many examples of the technique and to the following general papers (16, 72, 76).

3. Image Processing

For a variety of reasons to which we return below, it is not possible to extract all the information coded in an electron micrograph (or set of micrographs) with the unaided eye. Attempts to remedy this range from major and difficult modifications to the microscope itself, through light-optical processing of the micrograph, to microdensitometry and digital computer processing of the latter. One of the attractions of the STEM, as we have seen, is that microdensitometry is unnecessary and the processed image can in principle be examined on the microscope monitor ; ordinary micrographs must first be measured with a microdensitometer (a very expensive piece of equipment) and, after treatment, converted back into pictorial form, ideally with a filmwriter (yet another very expensive machine).

Why are we driven to such lengths to extract useful information from micrographs ? There are numerous reasons : we may be interested in the three-dimensional structure of the specimen whereas the micrograph gives only a two-dimensional projection ; or we may be interested in very high resolution (< 2 nm, say), in which case the microscope itself cannot be relied upon to transmit information faithfully from object to image - some detail may be suppressed and some contrast reversed ; or the image may be virtually invisible - this occurs when we try to study very fragile specimens, which are destroyed by the electron dose needed to form a visible image, or unstained material. This does not exhaust the list - the solution of the non-linear problem of extracting the phase and amplitude of the specimen transparency in particular has been intensively studied (94, 106, 107) but has not yet generated new knowledge in the biomedical sciences.

Each of these needs has produced its own body of image processing lore, which we now survey briefly. Some of the techniques can be implemented either digitally or optically. Although there are enthusiastic supporters of the optical methods, most processing is nowadays performed digitally for the increased flexibility of the computer usually outweighs the main advantage of optical processing, namely, the ease with which large images can be manipulated. It is only fair to say that the situation may possibly alter as "optical computing" (8, 10) gains ground and that the rather few experimental comparisons of the two procedures show that each can be brought to a successful conclusion (82, 118, 119).

3.1 Three-Dimensional Reconstruction

With the successful introduction of computer-aided tomography into medical radiology, the idea that a three-dimensional image can be built up from a series of two-dimensional projections has become commonplace and the various methods of performing the reconstruction are well-understood (61). In the late 1960s, when the first attempts to produce three-dimensional reconstructions from electron micrographs were made, this was by no means the case and it was the success of the technique that helped to establish the computer as a natural "accessory" of an electron microscope. Most reconstructions are made from structures that are known to possess some internal symmetry, which is exploited in the reconstruction in the sense that the effective number of views of the structure is greater than the number actually recorded. For a full account of work of this kind, with many examples, see (18, 90, 91), to which we may add recent work, such as that on gap junctions (124, 134) by UNWIN and ZAMPIGHI.

The lower the natural symmetry of the specimen, the more distinct views are necessary to achieve a given resolution and the greater are the practical difficulties of the procedure. In the limiting case in which the specimen possesses no useful symmetry properties, it may even be worth redesigning the electron microscope to facilitate the task of data collection - a project with this end in view is progressing in the laboratory of W. HOPPE (81, 98, 99, 121, 122). There are many problems to be overcome when the specimen to be reconstructed has no internal symmetry : the range of tilt-angles within which the different views can be obtained is rather narrow ; the axis of tilt is most unlikely to be sufficiently stable during the process of collecting the set of projections. Each of these produces characteristic defects. Conversely, it has been persuasively argued by HOPPE (59, 66) that the information needed for a three-dimensional reconstruction can be collected without subjecting the specimen to a higher electron dose than is needed for two-dimensional image formation : "Two-dimensional microscopy is in fact a waste of information" (66a, p.128). Although this and related comparisons (66) need careful interpretation (105, 108, 67), their essential correctness is not in doubt.

A problem common to all three-dimensional reconstruction is the presentation of the results. At present, most practitioners resort to model building or to synthesis of a three-dimensional picture by superimposing transparent sheets on which adjacent cross-sections of the structure have been traced, the resulting stack "containing" the structure. A more elegant solution, which unfortunately requires display software that is not always readily available, consists in presenting the structure on an oscilloscope screen connected to the computer in which full details of the structure are stored. It is then possible to study different external views of the structure, look at cross-sections at different levels and even examine the internal arrangement by "cutting out" a segment (like removing a quarter of an apple).

3.2 Linear Image Processing

At high resolution, the contrast seen in an electron microscope image is essentially phase contrast, the "phase-plate" being replaced by a phase shift due to spherical aberration and controlled defocus. This phase shift is not constant, with the result that considerable care is needed when interpreting image contrast created in this way. A body of theory has grown up around this problem, from which we extract one very useful result : for a certain class of specimens, the image contrast (C) is related *linearly* to the amplitude (s) and phase (ϕ) components of the specimen transparency. Furthermore, the spatial frequency spectra or Fourier transforms of these functions are even more simply related :

$$\hat{C} = -2a\,\hat{s}\cos\gamma + 2a\,\hat{\phi}\sin\gamma \qquad (1)$$

in which $\hat{C}$, $\hat{s}$ and $\hat{\phi}$ denote the two-dimensional Fourier transforms of C, s and ϕ, a is a cut-off determined by the size of the objective aperture of the microscope and γ is a measure of the phase shift introduced by spherical aberration (C_s) and defocus (Δ) :

$$\gamma = \pi\lambda^3 C_s p^4/2 - \pi\lambda\Delta p^2 \qquad (2)$$

with p a measure of spatial frequency. A clear physical meaning can be attached to Eq. (1) by considering an amorphous phase specimen, such as a thin carbon film. We may neglect any amplitude contribution (s) and assume that $\hat{\phi}$ is constant over a wide range of spatial frequencies, so that :

$$\hat{C} \propto a \sin \gamma ;$$

$\hat{C}$ is easily obtained either optically or numerically, and in the corresponding diffractogram the rings corresponding to the cycles and zeros of sin γ are clearly seen.

The question that arises therefore is how can we obtain $\hat{s}$ and $\hat{\phi}$, and hence s and ϕ, given that we can only record C ? Clearly, if both s and ϕ vary considerably, a single picture cannot yield much information about them ; even if s is negligible (pure phase object), we can extract only an imperfect estimate of $\hat{\phi}$ from a single micrograph owing to the zeros of sin γ, which suppress some information about the structure or reduce it to too small a value to be usable. The answer is to record a focal series, that is, a series of micrographs taken at different values of defocus and hence corresponding to different forms of sin γ. For such series, filters have been calculated which enable us to obtain optimum estimates (in the least-squares sense) of $\hat{s}$ and $\hat{\phi}$. This in turn generates further problems, of both a practical and a more fundamental kind. From a practical point of view, obtaining a focal series implies submitting the specimen to a considerable dose of radiation, with the result that the last image of the series may correspond to a much more seriously damaged specimen than the first. We return to this in section 3.3. Equally serious is the difficulty of relating the forms of s and ϕ obtained to the specimen structure. Although a great deal is now known about electron scattering in thin films, the inverse problem of deducing specimen composition from s and ϕ is far from solved. For a review of these questions see (50) and (106).

3.3 The Fragile Specimen

The radiation resistance of electron microscope specimens - their ability to endure the onslaught of the electron beam unscathed - varies very considerably (15, 42 - 44, 70, 78, 113, 117). Many types of biological specimen are among the more vulnerable to damage of this kind and even those that are reasonably resistant frequently suffer damage to their fine detail ; needless to say, it is often this fine detail that particularly interests the microscopist.

The risk of radiation damage was recognized even before the first electron microscopes were built : in 1928, Leo SZILARD suggested to Dennis GABOR that he should make an electron microscope. "To this suggestion I (GABOR) gave the answer which, I believe, would have been given by almost all physicists : 'What is the use of it ? Everything under the electron beam would burn to a cinder !' (40).

Various means of avoiding or reducing beam damage have been investigated over the years. An important landmark was an early attempt at "minimum exposure" by WILLIAMS and FISHER (131, 132), in which new detail was seen by arranging that the specimen area actually used to form the recorded image was not irradiated beforehand. Since then, many possible ways of protecting fragile structures have been proposed and tested, two of which we mention here. It has long seemed plausible that cooling the specimen should be advantageous but although the experimental evidence provided some support for this view, it was neither entirely conclusive nor very impressive. In 1979, all this was abruptly changed with the publication of a letter by DIETRICH, FORMANEK, FOX, KNAPEK and WEYL (21) stating that an immense improvement was found if the specimen and its surroundings were at liquid helium temperature ($\sim$ 4 K). Their experiments differed from those of earlier investigators in that they used a superconducting lens rather than just a liquid helium stage in a conventional lens and they could thus be confident of the specimen temperature. This preliminary result has subsequently been confirmed (22 - 24, 80) and it seems reasonable to hope that, for many types of specimen, the radiation damage problem need no longer worry those who have access to a microscope equipped with an appropriate superconducting lens.

If this hope is indeed borne out, then clearly this experimental solution is ideal for the image obtained should represent the undamaged specimen : all structure that survives the process of specimen preparation should be present in the image. The alternative is to use very low exposures indeed and combine several images of similar structures in such a way that a single legible image is obtained. This idea has been exploited in two ways : UNWIN and HENDERSON (60, 123) used it to obtain first two-dimensional images and later a three-dimensional reconstruction of a regular array : the low-dose image of the unstained specimen contained no detail visible to the unaided eye but by averaging over the cells of the array, a good image was obtained. So impressive was this result (which made UNWIN and HENDERSON the first winners of the RUSKA Prize) that KLUG has subsequently written "It would seem that, for the time being, at least, high resolution in the electron microscopy of biological molecules can only be achieved if they are in the form of a thin crystalline specimen" (79).

Alternatively, a set of sub-images of the same structure may be present in a single low-dose image and computer manipulation may be used to superimpose all these individual representations. The computer is needed not only to align the latter with an accuracy at least as good as the resolution desired (35, 35a) but has also been used recently to classify the individual sub-images. It is usually necessary to select good sub-images, for some may be obviously unsuitable : broken or grossly distorted for example. It would clearly be desirable to make this selection on a less subjective basis and a statistical procedure known as "correspondence analysis" has recently been introduced by VAN HEEL and FRANK for this purpose (55, 56). So effective is this technique that it not only allows us to select only particles that bear a strong resemblance to one other but also groups the sub-images into clusters if more than one category is present, even though the difference between members of different clusters may be very slight. In (56), for example, VAN HEEL and FRANK show that images of haemocyanin half-molecules from the horseshoe crab *Limulus polyphemus* fall into distinct categories and propose an explanation for this.

4. Processing Languages

Digital methods of processing images cannot hope to be tolerated, let alone widely accepted in biology and medicine if a substantial computing effort is required for every image. Happily this is not the case, for a number of simple programming languages have now been developed that require very little familiarity with computing. We mention in particular SEMPER, developed in Cambridge by M.F. HORNER, W.O. SAXTON and A.J.PITT (109) ; SPIDER, developed in Albany by J. FRANK and B. SHIMKIN (36) ; IMAGIC, developed in Groningen by VAN HEEL (55, 57) ; EM developed in Münich by R. HEGERL (58);and a program suite developed in Basel by P.R. SMITH (114). Each of these languages consists of a set of relatively simple commands, the use of which can be mastered quite rapidly, and which are used to perform complex sequences of operations on images. The various languages reflect the particular interests of their authors, though SEMPER, SPIDER and IMAGIC have many features in common. The SMITH programs are particularly suitable for three-dimensional reconstruction of periodic structures while EM is to be used for reconstruction of non-symmetric specimens. Further comparisons of these various languages are to be found in (57). A number of less ambitious processing systems have also been produced, such as that described in (126).

5. Image Processing and the STEM

The types of processing described in Section 3, which the languages listed in Section 4 are designed to put into practice, were all devised with the conventional electron microscope in mind, and it is interesting to return to the STEM to see what the latter has to offer. We have already seen that simple contrast-stretching can be used to make the image easier to see and that elementary combinations of the bright-field and dark-field images can be created straightforwardly. Is the STEM superior in any way for the types of manipulation discussed in Sections 3.1 - 3.3 ? So far as I am aware, there has been no discussion of its potential merits for three-dimensional reconstruction. Nevertheless, one of the great attractions of the STEM, as we have already mentioned, is its capacity for direct data collection, and it seems therefore that it should play an intermediate role between the conventional instrument, in which each image is recorded photographically, measured and then submitted to the computer, and such novel designs as that proposed by HOPPE and colleagues (98). One can imagine STEM modifications permitting systematic collection of image data from a tilt series, using the same or different specimens, depending on the damage rate ; it might be preferable to tilt the axis of the probe, which could easily be achieved electronically, and then straighten up the beam beyond the specimen (though this could probably not give very wide angular coverage). Such proposals as these will no doubt be considered in more detail as the STEM becomes a more familiar instrument in the biosciences.

The question of linear processing has already been considered by STEM designers, and a most interesting solution has emerged. The optics of the STEM are such that the detector space is in some respects analogous to the source space of a conventional microscope, and it is known that illumination schemes that are difficult to implement in the latter are potentially very informative. Since the detector is easily accessible in the STEM, it is worth considering whether more complicated geometries than the simple bright-field dark-field pair could be useful and a number of such designs have been discussed. ROSE (103), for example, has examined in detail the signals collected by introducing detectors in the form of concentric rings in the place of the single bright-field detector ; DEKKERS and DE LANG (19, 20) have shown that division of the detector into semi-circular segments provides partial information about the specimen transparency functions s and ϕ and the simple extension to four quadrants (49) completes this. These by no means exhaust the list, which is reviewed in some detail in (51).

Finally, we consider the STEM and radiation damage. When this type of microscope first appeared, there was considerable discussion of the damage rate to be expected and hopes were raised that the STEM would prove superior to the conventional instrument in this respect. Claim and counter-claim alternated in the literature but it gradually became clear that no general conclusion could be drawn : if it could be argued that the STEM would sometimes be superior, it is by no means a universal panacea. Once again, however, the straightforward data collection of the STEM eliminates a tiresome stage if correspondence analysis, for example, is to be performed on a set of sub-images. Thus ZINGSHEIM *et al.* used STEM images in their study of membrane-bound acetylcholine receptor from *Torpedo marmorata* (136).

Having stressed the fact that STEM signals can so easily be despatched to a computer, it is only fair to point out that direct links between the conventional microscope and the computer have been built in a number of research laboratories, though none is on the market. We draw attention in particular to two very different systems, the Electrotitus developed in LABERRIGUE'S laboratory at the University of Reims (4) and the elaborate chain built in the Fritz-Haber-Institut in Berlin by HERRMANN and colleagues (62, 63). For a more complete list, see (50).

6. Concluding Remarks

The electron microscope is only one tool among many used in the biosciences to furnish information in the range from nanometres or smaller up to the resolution of the light microscope. Nevertheless, it has provided - and continues to provide - an immense amount of biological information that could have been obtained in no other way and, as we have attempted to show in the foregoing account, new arrows are continually being added to its quiver.

What directions are future developments likely to take ? No doubt the next few years will be largely taken up with establishing the usefulness or otherwise of the techniques that are at present in active growth : electron energy loss spectroscopy, for example, the various methods of examining fragile specimens, biological microanalysis, reconstruction of specimens having little or no internal symmetry. The computer can be expected to be more and more invasive but, with the development of simple languages, less and less obtrusive.

References

1. J.L. Abraham : Scanning Electron Microscopy Part IV, 171-178 and 188 (1980).
2. D.L. Allison : In (52), vol. 5, 62-113 (1975).
3. W. Baumeister and W. Vogell (eds) : *Electron Microscopy at Molecular Dimensions* Proc. in Life Sciences (Springer, Berlin, Heidelberg, New York 1980).

3a. M. Beer, J.W. Wiggins, D. Tunkel, C.J. Stoeckert : Chemica Scripta 14, 263-266 (1978/79).

4. A. Beorchia, P. Bonhomme, N. Bonnet : Optik 55, 11-22 (1980).
5. A. Boekestein, A.L.H. Stols, A.M. Stadhouders : Scanning Electron Microscopy Part II, 321-334 (1980).
6. I.D. Bowen, G.H.J. Lewis : Scanning Electron Microscopy Part IV, 179-187 (1980).
7. S. Bullivant : In *Advanced Techniques in Biological Electron Microscopy*, ed. by J.K. Koehler (Springer, Berlin, Heidelberg, New York, 1973) pp.67-112.
8. H.J. Butterweck : Prog. Opt. 19, 211-280 (1981).
9. I.L. Cameron, N.K.R. Smith : Scanning Electron Microscopy Part II, 463-474 (1980).
10. D. Casasent (ed.): *Optical Data Processing*, Topics in Applied Physics, Vol.2 Vol. 23 (Springer, Berlin, Heidelberg, New York, 1978).

11. J.A. Chandler : In *Electron Probe Microanalysis in Biology*, ed. by D.A. Erasmus (Chapman and Hall, London, 1978), pp. 37-93.
12. J.A. Chandler : Scanning Electron Microscopy Part II, 595-606 (1979).
13. J.A. Chandler : Scanning Electron Microscopy Part II, 475-484 (1980).
14. J.R. Coleman : Scanning Electron Microscopy Part II, 911-926 (1978).
15. V.E. Cosslett : J. Microscopy 113, 113-129 (1978).
16. V.E. Cosslett : Scanning Electron Microscopy Part II, 575-582 and 534 (1980).
17. M.J. Costello : Scanning Electron Microscopy Part II, 361-370 (1980).
18. R.A. Crowther, A. Klug : Ann. Rev. Biochem. 44, 161-182 (1975).
19. N.H. Dekkers, H. de Lang : Optik 41, 452-456 (1974).
20. N.H. Dekkers, H. de Lang : Philips Tech. Rev. 37, 1-10 (1977).
21. I. Dietrich, H. Formanek, F. Fox, E. Knapek, R. Weyl : Nature 277, 380-381 (1979).
22. I. Dietrich, J. Dubochet, F. Fox, E. Knapek, R. Weyl : In (3), pp. 234-244.
23. J. Dubochet, E. Knapek : Chemica Scripta 14, 267-269 (1978/79).
24. J. Dubochet, E. Knapek, I. Dietrich : Ultramicroscopy 6, 77-80 (1981).
25. G. Dupouy : Adv. Opt. Electron Micr. 2, 167-250 (1968).
26. P. Echlin : In *Advanced Techniques in Biological Electron Microscopy II*, ed. by J.K. Koehler (Springer,Berlin, Heidelberg, New York, 1978), pp. 89-112.
27. A. Engel : Ultramicroscopy 3, 273-281 (1978).
28. A. Engel : Ultramicroscopy 3, 355-357 (1978).
29. A. Engel : In (3), pp. 170-178.
30. A. Engel : J. Microsc. Spectrosc. Electron 5, 581-594 (1980).
31. A. Engel, J. Meyer : J. Ultrastruct. Res. 72, 212-222 (1980).
32. A. Engel, J. Dubochet, E. Kellenberger : J. Ultrastruct. Res. 57, 322-330 (1976).
33. P. Favard : In *Proc. 7th Eur. Cong. Electron Microscopy, The Hague*, vol. 4, 414-419 (1980).
34. D.W. Fawcett : J. Electron Microsc. 28 Suppl., S-73 - S-78 (1979).
35. J. Frank : In *Computer Processing of Electron Microscope Images*, ed. by P.W. Hawkes, Topics in Current Physics, Vol. 13 (Springer,Berlin, Heidelberg, New York, 1980) pp. 187-222.
35a. J. Frank, W. Goldfarb : In (3), pp. 261-269.
36. J. Frank, B. Shimkin : In *Proc. 9th Int. Cong. Electron Microscopy, Toronto*, vol. 1, 210-211 (1978).
37. F. Franks : Scanning Electron Microscopy Part II, 349-360 (1980).
38. R. Freeman, K.R. Leonard : J. Microscopy 122, 275-286 (1981).
39. W. Fuchs, H. Fuchs : Scanning Electron Microscopy Part II, 371-382 and 296 (1980).
40. D. Gabor : Preface to L. Marton : *The Early History of the Electron Microscope* (San Francisco Press, San Francisco, 1968).
41. L.T. Germinario, R. Reed, M.D. Cole, S.D. Rose, J.W. Wiggins, M. Beer : Scanning Electron Microscopy Part I, 69-76 (1978).
42. R.M. Glaeser : J. Ultrastruct. Res. 36, 466-482 (1971).
43. R.M. Glaeser : In *Physical Aspects of Electron Microscopy and Microbeam Analysis*, ed. by B.M. Siegel and D.R. Beaman (Wiley, New York and London, 1975) pp. 205-229.
44. R.M. Glaeser : In (68), pp. 423-436.
45. A. Glauert (ed.) : *Practical Methods in Electron Microscopy* (North Holland, Amsterdam, New York and Oxford, 1973-).
46. A. Glauert : J. Cell. Biol. 63, 717-748 (1974).
47. T.A. Hall : In *Microbeam Analysis in Biology*, ed. by C. Lechene and R. Warner (Academic Press, New York and London, 1979) pp. 185-208.

48. K. Hama : In *Advanced Techniques in Biological Electron Microscopy*, ed. by J.K. Koehler (Springer, Berlin, Heidelberg, New York, 1973), pp. 275-297.
49. P.W. Hawkes : J. Optics (Paris) 9, 235-241 (1978).
50. P.W. Hawkes : In *Computer Processing of Electron Microscope Images*, ed. by P.W. Hawkes, Topics in Current Physics, Vol. 13, (Springer, Berlin, Heidelberg, New York, 1980), pp. 1-33.
51. P.W. Hawkes : Scanning Electron Microscopy Part I, 93-98 (1980).
52. M.A. Hayat (ed.) : *Principles and Techniques of Electron Microscopy, Biological Applications* (Van Nostrand Reinhold, New York and London, 1970-).
53. M.A. Hayat (ed.) : *Principles and Techniques of Scanning Electron Microscopy* (Van Nostrand Reinhold, New York and London, 1974-).
54. M.A. Hayat (ed.) : *X-ray Microanalysis in Biology* (University Park Press, Baltimore, 1980).
55. M. van Heel : *Image Formation and Image Analysis in Electron Microscopy* (Proefschrift, University of Groningen, 1981).
56. M. van Heel, J. Frank : Ultramicroscopy 6, 187-194 (1981).
57. M. van Heel, W. Keegstra : IMAGIC, a fast, flexible and friendly image analysis software system, to be published ; included in (55).
58. R. Hegerl : In *Proc. 7th Eur. Cong. Electron Microscopy, The Hague*, vol. 2, 700-701 (1980).
59. R. Hegerl, W. Hoppe : Z. Naturforsch. 31a, 1717-1721 (1976).
60. R. Henderson, P.N.T. Unwin : Nature 257, 28-32 (1975).
61. G.T. Herman : *Image Reconstruction from Projections* (Academic, New York and London, 1980).
62. K.H. Herrmann, D. Krahl, H.P. Rust : Ultramicroscopy 3, 227-235 (1978).
63. K.H. Herrmann, D. Krahl, H.P. Rust: In (3), pp. 186-193.
64. J.E. Heuser, T.S. Reese, D.M.D. Landis : Cold Spring Harbor Symposia on Quantitative Biology 40, 17-24 (1976).
65. J.E. Heuser, T.S. Reese, M.J. Dennis, Y. Jan, L. Jan, L. Evans : J. Cell. Biol. 81, 275-300 (1979).
66. W. Hoppe : In (3), pp. 278-287.
66a. W. Hoppe and R. Hegerl : In *Computer Processing of Electron Microscope Images*, ed. by P.W. Hawkes, Topics in Current Physics, Vol. 13 (Springer, Berlin, Heidelberg, New York, 1980), pp. 127-185.
67. W. Hoppe, R. Hegerl : Ultramicroscopy 6, 205-206 (1981).
68. J.J. Hren, J.I. Goldstein, D.C. Joy (eds.) : *Introduction to Analytical Electron Microscopy* (Plenum, New York and London, 1979).
69. S.W. Hui, D.F. Parsons : In *Advanced Techniques in Biological Electron Microscopy II*, ed by J.K. Koehler (Springer, Berlin and New York, 1978), pp. 213-235.
70. M.S. Isaacson : In (52), Vol. 7, 1-78 (1977).
71. M.S. Isaacson, M. Utlaut, D. Kopf : In *Computer Processing of Electron Microscope Images*, ed. by P.W. Hawkes, Topics in Current Physics, Vol. 13 (Springer, Berlin, Heidelberg, New York, 1980), pp. 257-283.
72. D.E. Johnson : In (68), pp. 245-258.
73. A.V. Jones : J. Microsc. Spectrosc. Electron. 5, 595-609 (1980).
74. A.V. Jones, K.R. Leonard : Nature 271, 659-660 (1978).
75. A.V. Jones, K.C.A. Smith : Scanning Electron Microscopy Part I, 13-26 (1978).
76. D.C. Joy : In (68), pp. 223-244.
77. M.V. King, D.F. Parsons, J.N. Turner, B.B. Chang, A.J. Ratkowski : Cell Biophysics 2, 1-95 (1980).
78. J. Kirz : Scanning Electron Microscopy Part II, 239-249 (1980).
79. A. Klug : Chemica Scripta 14, 245-256 and 291-293 (1978/79).
80. E. Knapek, J. Dubochet : J. Mol. Biol. 141, 147-161 (1980).

81. V. Knauer, W. Hoppe : In *Proc. 7th Eur. Cong. Electron Microscopy, The Hague,* vol. 2, 702-703 (1980).
82. O. Kübler, M. Hahn, J. Seredynski : Optik 51, 171-188 and 235-256 (1978).
83. M.K. Lamvik : J. Mol. Biol. 122, 55-68 (1978).
84. C. Lechene : In *Microbeam Analysis,* ed. by D.E. Newbury (San Francisco Press, San Francisco, 1979) pp. 59-60.
85. C. Lechene : Fed. Proc. 39, 2871-2880 (1980).
86. C. Lechene, R. Warner (eds.) : *Microbeam Analysis in Biology* (Academic Press, New York and London, 1979).
87. C. Lechene, J.V. Bonventre, R.R. Warner : In (86), pp. 409-426.
88. A.T. Marshall : Scanning Electron Microscopy Part II, 335-348 (1980).
89. A.T. Marshall : Scanning Electron Microscopy Part II, 395-408 (1980).
90. J.E. Mellema : J. Microsc. Spectrosc. Electron. 5, 611-629 (1980).
91. J.E. Mellema : In *Computer Processing of Electron Microscope Images,* ed. by P.W. Hawkes, Topics in Current Physics, Vol. 13 (Springer, Berlin, Heidelberg, New York, 1980), pp. 89-126.
92. H.W. Meyer (ed.) : *Freeze-etching, Methods and Applications in Membrane Research* (Fischer, Jena, 1981) ; Supplement XXIII of Acta Histochemica.
93. *Microbeam Analysis* ed. by D.E. Newbury (1979) and by D.B. Wittry (1980) ; (San Francisco Press, San Francisco, 1979-, appears annually).
94. D.L. Misell : Adv. Opt. Electron Micr. 7, 185-279 (1978).
95. D.L. Misell : *Image Analysis, Enhancement and Interpretation* (North-Holland, Amsterdam, New York and Oxford, 1978).
96. D.F. Parsons : In *Physical Aspects of Electron Microscopy and Microbeam Analysis* (Wiley, New York and London, 1975), pp. 267-278.
97. D.F. Parsons, V.R. Matricardi, R.C. Moretz, J.N. Turner : Adv. Biol. Med. Phys. 15, 161-270 (1974).
98. E. Plies, D. Typke : Z. Naturforsch. 33a, 1361-1377 (1978).
99. E. Plies : Nucl. Instrum. Meth. to be published (1981).
100. N. Reid : *Ultramicrotomy,* vol. 3, Part II of (45), 1975.
101. H. Ris : In *Proc. 9th Int. Cong. Electron Microscopy, Toronto,* vol. 3, 545-556 (1978).
102. G.R. Roomans : Scanning Electron Microscopy Part II, 309-320 (1980).
103. H. Rose : Optik 39, 416-436 (1974) and 42, 217-244 (1975).
104. A.J. Saubermann : Scanning Electron Microscopy Part II, 421-430 and 420 (1980).
105. B.E.H. Saxberg, W.O. Saxton : Ultramicroscopy 6, 85-89 (1981).
106. W.O. Saxton : *Computer Techniques for Image Processing in Electron Microscopy* (Academic, New York and London, 1978).
107. W.O. Saxton : In *Computer Processing of Electron Microscope Images,* ed. by P.W. Hawkes, Topics in Current Physics, Vol. 13 (Springer, Berlin, Heidelberg, New York, 1980), pp. 35-87.
108. W.O. Saxton : In *Proc. 7th Eur. Cong. Electron Microscopy, The Hague,* vol. 1, 486-493 (1980).
109. W.O. Saxton, T.J. Pitt, M. Horner : Ultramicroscopy 4, 343-354 (1979).
110. A.M.H. Schepman, J.A.P. van der Voort, J. Kramer, J.E. Mellema : Ultramicroscopy 3, 265-269 (1978).
111. H. Shuman, A.V. Somlyo, A.P. Somlyo : Ultramicroscopy 1, 317-339 (1976).
112. H. Shuman, A.V. Somlyo, A.P. Somlyo : Scanning Electron Microscopy Part I, 663-672 (1977).
113. B.M. Siegel, D.R. Beaman (eds.) : *Physical Aspects of Electron Microscopy and Microbeam Analysis* (Wiley, New York and London, 1975). Section C, "Biophysical : radiation damage", pp. 205-285.
114. P.R. Smith : Ultramicroscopy 3, 153-160 (1978).
115. A. Somlyo, A. Somlyo, H. Shuman : J. Cell Biol. 81, 316-335 (1979).
116. A.P. Somlyo, A.V. Somlyo, H. Shuman, M. Stewart : Scanning Electron Microscopy Part II, 711-722 (1979).

117. K. Stenn, G.F. Bahr : J. Ultrastruct. Res. 31, 526-550 (1970).
118. G.W. Stroke, M. Halioua, F. Thon, D. Willasch : Optik 41, 319-343 (1974).
119. G.W. Stroke, M. Halioua, F. Thon, D.H. Willasch : Proc. IEEE 65, 39-62 (1977).
120. W. Tichelaar, G.T. Oostergetel, J. Haker, M.G. van Heel, E.F.J. van Bruggen : Ultramicroscopy 5, 27-33 (1980).
121. D. Typke : Nucl. Instrum. Meth. to be published (1981).
122. D. Typke, M. Burger, N. Lemke, G. Lefranc : In *Proc. 7th Eur. Cong. Electron Microscopy, The Hague,* vol. 1, 82-83 (1980).
123. P.N.T. Unwin, R. Henderson : J. Mol. Biol. 94, 425-440 (1975).
124. P.N.T. Unwin, G. Zampighi : Nature 283, 545-549 (1980).
125. E.M. Waddell, J.N. Chapman : Optik 54, 83-96 (1979).
126. R. H. Wade, A. Brisson, L. Tranqui : J. Microsc. Spectrosc. Electron. 5, 699-715 (1980).
127. J. Wall : In (68), pp. 333-342.
128. J.S. Wall : Scanning Electron Microscopy Part II, 291-302 (1979).
129. R.R. Warner, C. Lechene : In *Microbeam Analysis in Biology,* ed. by C. Lechene and R.R. Warner (Academic Press, New York and London, 1979), pp. 161-169.
130. J.W. Wiggins, J.A. Zubin, M. Beer : Rev. Sci. Instrum. 50, 403-410 (1979).
131. R.C. Williams, H.W. Fisher : J. Mol. Biol. 52, 121-123 (1970).
132. R.C. Williams, H.W. Fisher : In *Proc. 28th Ann. Meeting EMSA,* 304-305 (1970).
133. J.H.M. Willison, A.J. Rowe : *Replica, Shadowing and Freeze-etching Techniques* (North-Holland, Amsterdam, New York and Oxford, 1980), vol. 8 of (45).
134. G. Zampighi, P.N.T. Unwin : J. Mol. Biol. 135, 451-464 (1979).
135. K. Zierold, R. König, K.H. Olech, D. Schäfer, D.W. Lübbers, K.H. Müller and H. Winter : Ultramicroscopy 6, 181-186 (1981).
136. H.P. Zingsheim, D-Ch. Neugebauer, F.J. Barrantes, J. Frank : In (3), pp. 161-169.
137. J.A. Zubin, J.W. Wiggins : Rev. Sci. Instrum. 51, 123-131 (1980).

Electron Microscopy in Medicine and Biology: Applications

S. Boseck

Fachbereich Physik der Universität, Bremen, Postfach 330440
D-2800 Bremen 33, Fed. Rep. of Germany

Electron microscopy has developed into a standard research method in medicine and biology. Because of the possibilities to represent structures in the range of macromolecular dimensions it is used in routine diagnostic work and pathology as well as high-resolution structure research [49]. Today electron microscopy has a permanent place in diagnostics and biopsies of the liver, kidney, muscle, nerves, intestine, in hygiene and virology, and particularly in molecular biology.

This article describes the special physical and methodical working conditions which are typical for the application of electron microscopy in biology and medicine [41,42].

First we have to draw attention to the fundamental difference between conventional transmission electron microscopy (CTEM) and scanning (transmission) electron microscopy S(T)EM, which affects the selection of preparations, the method of analysis, the obtainable resolution and the definite analytical statement. In CTEM the whole preparation is "illuminated" at the same time by a beam of electrons, sent from the crossover of the electron source by means of the condensor system. The electrons, partly scattered, diffracted and absorbed in the probe, are focussed on the screen by the objective, the intermediate lens and the projective. In S(T)EM the crossover is focussed in a single spot, which is scanned over the area of the preparation synchronously with the writing beam of monitor tube. The different interaction signals of the electrons with the probe can be collected by detectors at both sides of the preparation and can be used for image construction [14], especially after suitable data processing. While in the S(T)EM a resolution of 70-130 Å (30 keV, 80.000x) is obtained; in the CTEM a magnification of 1 million (100 kV) and a lattice resolution of 2.5 Å may be possible [6]. With super-conducting lens systems the resolution limit may be increased to 1.5 Å, at 220 kV, because astigmatism can be reduced and the low temperature of 4.2 K gives a negligible rate of contamination and radiation damage [3].

Which Phenomena Are Important for Electron Microscopy of Biological Specimen [2,41,42]?

1. In biological ultrathin sections (300-1500 Å) the absorption of electrons does not play an important role. Electrons are scattered and diffracted elastically on the amorphous and crystalline parts of structures, which affects scattering absorption contrasts as well as diffraction contrasts in the image. The presented objects have to be interpreted as phase structures with weak interaction, which are reproduced by suitable underfocussing of the objective lens in connection with its spherical aberration. Imagine that the periodical parts in the object can be shown by a certain scattering angle of electrons. By image formation at certain values of

defocussing not all components of spatial frequencies are transmitted with uniform contrast. For practical use with the electron microscope such a defocussing value of the objective lens is important where a maximum bandwidth of spatial frequencies is transmitted into the image at maximum contrast in order to avoid a misinterpretation of particular structures by change of sign. This is done by slight underfocussing, depending on the object.

2. The chromatic aberration of the objective and the bandwidth of electrons as a consequence of alternations in the beam, variations of accelerating voltage or energy loss in the object leads to focussing differences in the image and a chromatic error disc. Using a field emission gun and energy analyzer (electrostatic or magnetic), a superior image quality at the STEM is obtainable by reducing the chromatic aberration and taking into account for image formation the amount of energy lost due to inelastic scattering processes in the probe. This offers the possibility of analyzing the amplitude contrast in the object [53].
3. The error of distortion can be corrected nearly completely in modern electron microscopes.
4. The diffraction image in the focal plane of the objective contains the Fourier transform of the structures included in the object. By methods derived from X-ray diffractometry and crystallography, the 3rd dimension of the object and its periodical character can be found. Principly in the same way one can succeed in 3d reconstruction of aperiodic objects like proteins. One strict assumption is that the object has no or only little radiation damage, which can be controlled.
5. The SE and RE electrons, released by the interaction of the electron beam with the preparation, have different velocities depending on origin, depth of the probe, topography and Z numbers of the material. Therefore, each effect belongs to a different resolution limit [42,6]. The same is valid for measurement of probe current. The Auger electrons, coming from the uppermost layers (~ 10 Å) may be registered at UHV (10^{-9} Torr). Their application biological analysis is aspired.
6. The emitted electromagnetic radiation consists of gamma quanta of the inner shells of the elements and of light generated by interaction with the molecules. In the first case the energy dispersive X-ray analysis (EDX) or angle dispersive X-ray analysis (WDX) is derived, in the course of which the quanta is absorbed by a Si-Li diode and analyzed by electric pulse-density distribution or by turning an analyzing crystal and counting the quanta [6]. The lateral resolution limit is ~ 1 µm with a sensitivity limit of 0.1% for $Z > 11$, which typically corresponds to a mass of 10^{-17} to 10^{-21}g [36,46]. The light quanta (cathode luminescence) are used to measure the distribution of molecules and especially fluorochromes in the object by glassfibres, spectrometer and photomultiplier.
7. The excessive ions and fragments of molecules do not yet play a role in electron microscopy of biological specimen since the pressure of the rest gases in the column normally is too high.
8. The biologist has to be informed of the following phenomena: the dependence of the image contrast on the acceleration voltage, the variation of the resolution limit as a consequence of the variable penetration of electrons, but also of the origin of radiation damage in the object. Therefore, the loss of structure in the molecular range begins at a dose level under 60 e/Å ^{2}. For imaging purposes the 100-1000x of this value is needed [22].

9. For correct observation of biological and medical objects the reduction of contamination of the preparation by crack products of pump oil is important. This may be achieved by decreasing the object temperature with the help of cooling devices or by replacing the diffusion pumps by oil-free systems like ion-getter pumps and turbo-molecular pumps.

Which Preparation Methods Are Important in the Electron Microscopy of Biological Objects [15]?

For submicroscopic histopathology and histochemistry at ultrathin section and for X-ray microanalysis those methods are of importance which do not change the fine structure of the tissue and keep the distribution of chemical elements undestroyed. Both requirements can hardly be fulfilled at the same time. Concerning the first requirement, in the routine embedding methods for the material after dehydration the inclusion of heavy metal atoms along the membranes is used. In the other case deep freezing, drying and substitution processes are involved. Concerning then purposes of molecular biology and virology, negative staining methods (by adding contrast materials on the surface of the object) or direct mounting methods without artificial contrast for dark field illumination and strioscopy are found. Membrane research is done by freeze etching and freeze fracturing.

In ultramicroscopical autoradiography the distribution of developed grains of silver is measured in the area surrounding radioactive particles, which are tracers for the destination of biochemical activities in a section. With the help of statistics on the distribution of reaction products in an extended volume a resolution limit of 500-1000 Å is possible. For purposes of S(T)EM the evaporation and sputtering with Au, C and Ag on the predried material is proved.

Is It Possible to Determine 3d Structures of Biomacromolecules Directly?

As already mentioned, the square of the Fourier transform is constructed in the back focal plane of the electron microscope objective. Certain spots can be analyzed directly in CTEM by their diffractograms by imaging the diaphragm for selected area diffraction (SAD) backwards into the object area. The signal distribution also allows one to index crystallographic planes and zone axes for biological crystals [8,37,45]. As the objective aperture diaphragm is moved away from the beam, by this indirect method, structures under 1 Å are found. In methods similar to X-ray crystallography, especially by Patterson synthesis and Fourier transform [33], the unit cell can be found and recondstructed. This method will be of use if you intend to construct the real 3d structure of aperiodic objects like protein molecules. In this case, so many sections in the Fourier-transform space are made as necessary for total 3d reconstruction by tilting the electron beam in the electron microscope. It is necessary to have reduced radiation damage by averaging low-dose images, to correct faults of the electron microscope, and to have solved the sampling and interpolation problems [7,20,21,24,26-29, 32].

Which Methods Are Important for Optical and Digital Image Processing on Electron Micrographs of Biological Specimen?

In addition to the automatic image analyzing computer for revealing stereometric, morphometric and statistical values, digital and optical image processing plays a special and important role in biological research, especially in molecular biophysics [9,19,27,39,43].

Coherent optical and digital image processing is done to study basic problems of weak and strong phase contrast and to evaluate principles of image manipulation and 3d reconstruction [17,18,30,35,38,47,48] as well as for ultrastructure research on biomacromolecules and virology [1,9-12,24,25,43].

As far as linear systems are concerned, only Fourier-transform and related convolution methods are used [5]. In this case, coherent optics lead to a better resolution than computer methods, whereas computing is more convenient and of more practical use for varying problems [39,50,51]. Nonlinear methods of image restauration and special strategies (on the basis of maximum entropy or maximum likelihood) can be applied only digitally [4,13,34,52].

Coherent optical image processing leads to special apparatus for reconstruction enhancement or suppression of details and noise by holographic and diffractometric filtering with the help of physical and hybride masks [35]. In the latter case - especially for correcting the electron microscope by inverse filtering [22,23,35] - light optical diffraction allows one to find the zeros and absolute amplitudes as well as the phases of the MTF by using C foils as a spatial frequency generator. Optical diffraction methods also make it possible to measure image and electron microscopes defects such as astigmatism, drift, elastical and magnetic fields, defocussing, and noise directly [30].

By adding the information content of differently defocussed micrographs in the Fourier space, the whole area of transmission can be corrected. Similar efforts are made in order to add different aspects of the same object. By incoherent optical methods image details can also be revealed, especially by periodical integration of densities in a line or rotational superposition over a periodic angle [40]. Periodics can be averaged and symmetrized.

Computer methods achieve principally the same as coherent optical methods [16]. Normally algorithms for the reconstruction of specimen boundaries can be handled better digitally than optically, in spite of the limited resolution with normally 256×256 pixels (8 bits). Colour coding, easily done by computer, leads to modified display modes for image contents, which are more convenient to the human eye. Histogram and grey-level modifications as well as filtering methods allow reducing backgrounds and undesired structures. By using cross-correlation [44], one can exactly measure the status of correlation and the MTF of the electron microscope. Therefore, it is possible to compare different exposed micrographs at the same focus to study radiation damage. At the same level of correction one can reconstruct the 3d structure similar to coherent optical methods by superposition of exactly defocussed images.

Computer image processing also allows direct methods for 3d reconstruction by superpositioning corrected 2d projections in a similar way as in X-ray tomography. It is also possible to reconstruct highly complicated aperiodic molecule structures in 3d [9].

References

1 U. Aebi, P.R. Smith: In *Signal Processing*, ed. by M. Kunt, F. de Coulon (North Holland, Amsterdam 1980) pp.219-228
2 A.W. Agar, R.H. Alderson, D. Chescoe: In *Practical Methods in Electron Microscopy*, Vol.2, ed. by M. Glauert (North Holland, Amsterdam 1974) pp.143-296
3 See Ref.2, p.87
4 N. Baba, K. Murata, K. Okada, Y. Fujimoto: Optik *58*, 233-239 (1981)
5 S. Boseck, R. Hilgendorf, G. Hoffmann, J. Schmidt, A. Wasiljeff: Bedo *10*, 707-712 (1977)

6 J.A. Chandler: In *Practical Methods in Electron Microscopy*, Vol.5 (II), ed. by M. Glauert (North Holland, Amsterdam 1977) pp.327-518
7 R.A. Crowther, D.J. De Rosier, A. Klug: Proc. R. Soc. Lond. A*317*, 319-340 (1970)
8 H.-G. Fehn, H. Hartwig, J. Kirchmeier, A. Wasiljeff, S. Boseck: Biotechnische Umschau *3*, 354-358 (1979)
9 J. Frank: In *Advanced Techniques in Biological Electron Microscopy*, ed. by J.K. Koehler (Springer, Berlin, Heidelberg, New York 1979)
10 J. Frank, W. Goldfarb, D. Eisenberg, T.S. Baker: Ultramicroscopy *3*, 283-290 (1978)
11 J. Frank, W. Goldfarb, M. Kessel: Proc. 9th Int. Congr. Electron Microscopy, Toronto 1978
12 J. Frank, B. Shimkim: Proc. 9th Int. Congr. Electron Microscopy, Toronto 1978, Vol.1, pp.210-211
13 B.R. Frieden: J. Opt. Soc. Am. *62*, 511-518 (1972)
14 J. Frosien, H.-M. Thieringer: Siemens Analysentechnische Mitteilungen 187
15 A.M. Glauert, N. Reid (eds.): *Practical Methods in Electron Microscopy*, Vol.3 (North Holland, Amsterdam 1974)
16 W. Goldfarb, J. Frank: Proc. 9th Int. Congr. Electron Microscopy, Toronto 1978, Vol.2, pp.22-23
17 K.-J. Hanßen: PTB Bericht, Braunschweig (1977)
18 K.-J. Hanßen, R. Lauer, G. Ade: PTB Bericht Aph-15, Braunschweig (1980)
19 P.W. Hawkes (ed.): *Computer Processing of Electron Microscope*, Topics in Current Physics, Vol.13 (Springer, Berlin, Heidelberg, New York 1980)
20 R. Hegerl, W. Hoppe: Z. Naturforsch. *31a*, 1717-1721 (1976)
21 R. Henderson, P.N.T. Unwin: Nature *257*, 28 (1975)
22 K.-H. Herrmann, D. Krahl, V. Rindfleisch: Siemens Forsch. Entwickl. Ber. *1* (1972)
23 K.-H. Herrmann, E. Reuber, P. Schiske: Proc. 9th Int. Congr. on Electron Microscopy, Toronto 1978, Vol.1, pp.226-227
24 W. Hoppe: Proc. R. Soc. London B*261*, 71-94 (1971)
25 W. Hoppe: Annal. N.Y. Acad. Sci. *306*, 121-144 (1978)
26 W. Hoppe, B. Grill: Ultramicroscopy *2*, 153-168 (1977)
27 W. Hoppe, R. Mason (eds.): *Unconventional Electron Microscopy for Molecular Structure Determination* (Vieweg, Braunschweig 1978)
28 W. Hoppe, H.J. Schramm, M. Sturm, N. Hunsmann, J. Gaßmann: Z. Naturforsch. *31a*, 1370-1379, 1380-1390 (1976)
29 W. Hoppe, H. Wenzl, H.J. Schramm: Hoppe-Seyler's Z. Physiol. Chem. *358*, 1069-1076
30 R.W. Horne, R. Markham: In *Practical Methods in Electron Microscopy*,
31 E. Knapek: Dissertation, Technische Universität Berlin (1981)
32 V. Knauer, H.J. Schramm, W. Hoppe: Proc. 9th Int. Congr. Electron Microscopy, Toronto 1978, Vol.2, pp.4-5
33 T. Kobayashi, L. Reimer: Optik *43*, 237-248 (1975)
34 O. Kübler: Bildqualitätsanalyse in der Elektronenmikroskopie; Lehrgang 4493/46.22 der Tech. Akad. Esslingen (1980)
35 O. Kübler, M. Hahn, J. Seredynski: Optik *51*, 171-188 (1978); *3*, 235-256 (1978)
36 D. Kuschek: Material- und Strukturanalyse (Kontron) *10*, 29-35 (1981)
37 R.H. Lange, H.P. Richter: J. Mol. Biol. *148*, 487- 491 (1981)
38 K.R. Leonard, A.K. Kleinschmidt, N. Agabian-Keshishian, L. Shapiro, J.V. Maizel, Jr.: J. Mol. Biol. *71*, 201-216 (1972)
39 D.L. Misell: In *Practical Methods in Electron Microscopy*, Vol.7, ed. by M. Glauert (North Holland, Amsterdam 1978)
40 T. Mulvey: J. Microsc. *98*, 232-250 (1973)

41 L. Reimer: *Elektronenmikroskopische Untersuchungs- und Präparationsmethoden*, 2nd ed. (Springer, Berlin, Heidelberg, New York 1967)
42 L. Reimer, G. Pfefferkorn: *Rasterelektronenmikroskopie*, 2nd ed. (Springer, Berlin, Heidelberg, New York 1973)
43 W.O. Saxton: *Computer Techniques for Image Processing in Electron Microscopy* (Academic, New York 1978)
44 W.O. Saxton, J. Frank: Ultramicroscopy *2*, 219-227 (1977)
45 G. Schimmel: *Elektronenmikroskopische Methodik* (Springer, Berlin, Heidelberg, New York 1969)
46 M. Schulz-Baldes, R. Lasch, S. Boseck: The EDAX-EDITor *8*, 2, 17-18 (1978)
47 P. Sieber: Dissertation, Technische Universität Berlin (1978)
48 G.W. Stroke, M. Halioua, F. Thon, D.H. Willasch: Proc. IEEE *65*, 39-62 (1977)
49 M. Themann: Mikroskopie *36*, 276- 318 (1980)
50 A. Toepfer, W. Jung, P. Liebherr, S. Boseck, A. Wasiljeff: Poster, 82. Jahrestg. Dt. Ges. angew. Optik, Bremen 1981
51 A. Wasiljeff, S. Boseck, R. Hilgendorf, G. Hoffmann, J. Pickelhan, J. Schmidt: "Design of Two-Dimensional Digital Filters for Electron Microscopy", in *Underwater Acoustics and Signal Processing*, ed. by L. Bjørnø (Reidel, Dordrecht 1981) pp.551-557
52 S.T. Wernecke, L.R. D'Addario: IEEE Trans. C-*26*, 351-364 (1177)
53 D. Willasch: Siemens Analysentechnische Mitteilungen No. 204

Laser Projection Microscope: Principle and Applications in Biology

V.V. Savransky, G.A. Sitnikov and P.N. Lebedev

Physical Institute of the Academy of Sciences of the USSR
Moscow, Leninsky prospect, 53, USSR

Introduction

One of the technical problems in modern biology is the display of microobjects on large screens. Generally, for this purpose "passive" microprojectors are used, but in many cases these systems provide only an insufficient screen illumination. In order to achieve a good visibility the screen illumination should be, e.g. in the green spectral range not less than $0.1 W/cm^2$. For example, if we want to have a 1.000 x linear magnification of the image, the light flux per unit area of the object must be 10^6 x as intensive as that per unit area of the screen. This means that due to absorption and scattering losses an energy density in the object plane 10^7 x higher than the minimum energy intensity on the screen is needed. Although modern light sources can provide the theoretically needed object illumination, the resulting very high light fluxes may have degradation or even destroying effects on the object, especially when living systems are investigated. One of the ways to overcome these difficulties is the use of a laser amplification system.

Laser projection microscope

In spite of the fact that the idea to obtain bright images of microobjects by using an intensity amplifying laser medium and the realization of this technique have been known for quite a while, nevertheless, the laser projection microscope (LPM) is not yet widely used in technology and biology. The basic feature of a LPM is that (low energy) laser light containing object information is amplified by a laser medium. In the simplest version of the LPM, presented by PETRASH et al. [1,2,3], the object is illuminated by light emanating from a laser medium (copper vapor) and the light backscattered from the object is amplified by the same laser medium. In addition to its imaging feature, such an arrangement provides the possibility of intermediately releasing high power laser pulses for microsurgical operations or studies on laser effects in living organisms. Since the principle and technical realization of this prototype is described in detail in [1], we can confine this article to the advanced version we are using now (Fig. 1).

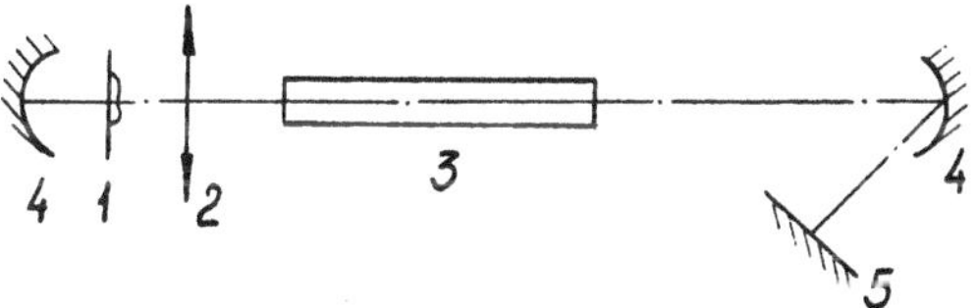

Fig. 1: Schematic diagram of the laser projection microscope: 1 - object, 2 - objective aperture, 3 - laser medium, 4 - mirrors, 5 - screen.

This type of a LPM differs from the previous one mainly by a spherical mirror positioned behind the object in the laser beam path, which reflects the light passing through the object. In this way the transmitted light contributes to the amplification process and the object can be investigated in transmission mode. It could be proved experimentally that this version of LPM is preferable to the first one, especially when low contrasted objects are investigated.

High-speed recording of structure changes in biological objects using the laser projection microscope

Investigations of structure changes in living cells in the submillisecond range are of actual interest in biological research. Conventional microscopes are not applicable in such experiments. Optical intensity amplifiers using capper vapor as laser medium, like the LPM, operate in a pulsed mode with pulse frequencies up to 10 kHz and pulse durations of 10-15 ns. Such frequencies are by far high enough to provide a continous and stable image for the naked eye, due to its time averaging feature. On the other hand, each single light pulse of the LPM contains an energy sufficient for recordings on usual photoemulsions [5].

A combination of the LPM and an appropriately triggered high-speed camera provides the possibility of recording film sequences. Thus, movements and structure changes of transparent microscopic (phase-) objects, occuring in the millisecond and submillisecond range, can be followed up. Biological objects, we investigated up to now, were different isolated cells of animals and plants, protozoa, and unicellulars. Examples recorded with the de-

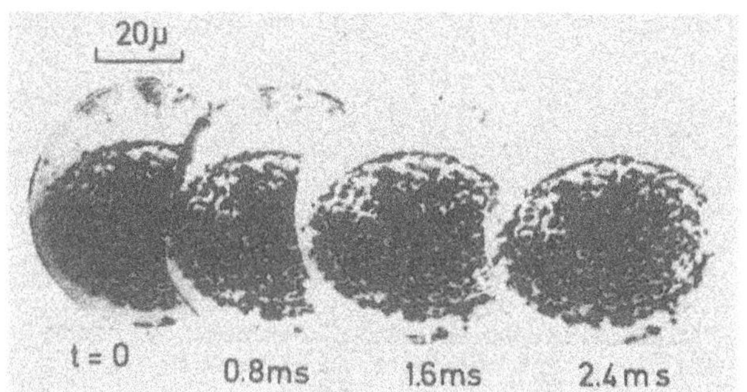

Fig. 2: Image sequence of a neuron taken from a grape snail brain; (laser pulse separation: 0.8 ms).

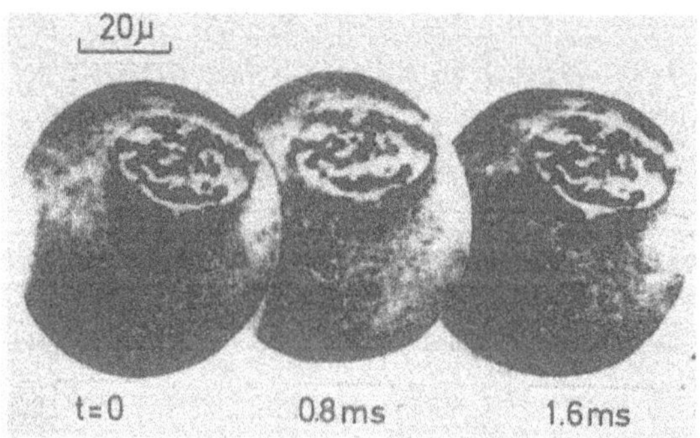

Fig. 3: Image sequence of an infusorian; (laser pulse separation: 0.8 ms).

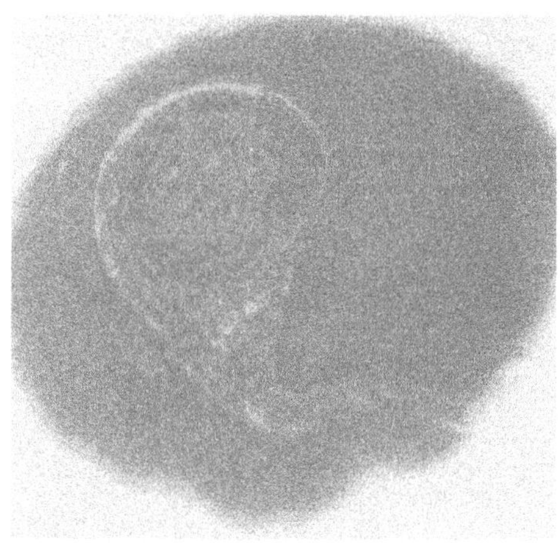

Fig. 4: Isolated neuron with axon; Glia cells are on the left (magnification: 2.000 x).

scribed technique are presented in Fig. 2 and 3, demonstrating image sequences of a neuron taken from a grape snail brain and of an infusorian, respectively. The pictures were recorded with a pulse separation of 0.8 ms.

Further investigations were carried out on isolated viable neurons of the perigullet nervous plexus, prepared according to the method described in [6]. Observing such an isolated neuron (Fig. 4), a folding of the cytoplasma membrane could be detected [7].

After 3 to 4 hours of operation of the laser projection microscope we did not observe any structure changes of the neuron. Relèasing intermediately some high power output pulses of the laser medium to investigate ser beam effects,vacuolization as well as changes in shape and size of the nucleus could be detected using the imaging feature of the LPM.

Conclusion

The presented experiments prove the applicability of the LPM to studies on living cells. The combination of LPM and high-speed photography opens a new field of microscopic investigations on structure changes in living microorganisms occuring in the millisecond and submillisecond range by recording film sequences. Thus, it may be possible to correlate such (short-time) structure changes with the electric activity in nervous cells. Furthermore, by (intermediately) releasing high power output pulses of the laser medium of the LPM the effect of laser radiation on living microorganisms can be investigated.

References

1 RABINOVITCH, P., R. CHIMENTI: J. Opt. Soc. Amer. 60, (1970), 1577.
2 ZEMSKOV, K.I., A.A. ISAEY, M.A. KAZARYAN, G.G. PETRASH: Kvantovaya elektronika 1, (1974), 14.
3 ZEMSKOV, K.I., A.A. ISAEV, M.A. KAZARYAN, G.G. PETRASH: Kvantovaya elektronika 3, (1976), 35.
4 BUNKIN, F.V., M.A. KAZARYAN, M.A. KONDRAT'ev, G.G. PETRASH, A.M. PROKHOROV, V.V. SAVRANSKY, I.N. SISAKYAN, G.A. SITNIKOV, K.I. ZEMSKOV: Microscopica Acta 82, (1979), 229.

5 ZEMSKOV, K.I., M.A. KAZARYAN, G.G. PETRASH, V.V. SAVRANSKY: Paper held at the 2-nd USSR-France Symposium on Optical Instrument Manufacturing, March, 1981, Moscow-Tallin.
E.A. MOROZOVA, A.M. PROKHOROV, V.V. SAVRANSKY, G.A. SHAFEEV: DAN USSR, to be published.
6 KOSTENKO, M.A.: Cytologics 10, (1974).
7 KULAGINA, E.N., S.A. OSIPOVSKY, V.V. SAVRANSKY, A.P. SHURYGINA: Cytologics (USSR), in print.

Ultrasonic Microscopy in Medicine and Biology

C.F. Quate
Department of Applied Physics and Electrical Engineering
Edward L. Ginzton Laboratory, Stanford University
Stanford, CA 94305, USA

E.A. Ash
Department of Electrical Engineering
University College London, Torrington Place
London WCIE 7JE, U.K.

1. Introduction

The acoustic microscope, [1, 2], has something in common with the optical and the electron microscope. The wavelength of the acoustic radiation is equal to that of the optical instrument and the image is acquired through scanning as in the SEM. But, the acoustic instrument differs significantly from the conventional instruments in that it can be used to monitor the elastic properties of biological material. That is the phase of the work that we will emphasise in this presentation. It is hardly surprising that we should be sensitive to elastic parameters since the reflectivity of acoustic waves is determined almost completely by the elasticity of the specimen under study.

The primary elements of the new microscope are illustrated in Fig. 1. There, we find an array of electronic circuits feeding energy into the system marked as the acoustic lens. There is a second system designed to receive pulses of radiation from the lens and process these signals in a form that is appropriate for display on the "CRT display". The information appears as an intensity modulated image on this TV monitor. The display system is similar to that used with the Scanning Electron Microscope. The scan time for a single frame can be as low as one second and this is suitable for direct presentation on a long persistence monitor.

2. Distinctive Characteristics of the Acoustic Microscope

An important difference between optical imaging and acoustic imaging comes from the large differences in acoustic velocities that are encountered between liquids and solids. This permits us - in a way to be described below - to monitor with great sensitivity the changes in the elastic parameters of the surface being examined, [3].

It comes about as a result of the reflection properties of an acoustic beam at the interface between a liquid and a solid. The sketch of Fig. 2 will be used to explain this properly. There we show an acoustic beam converging from the lens to the interface between medium 1 and medium 2. The plane transducer (see Fig.1) is most responsive to those reflected waves that appear to come from the focal point of the beam in the plane. When the specimen is at the focal plane, all of the incident rays are returned and the reflection signal is maximised. When the reflector is displaced from the focal plane toward the lens (as shown in Fig.2), there are only two rays which appear to come from the focal point upon reflection. The first is the normal ray. Its return is unaffected by the position of the reflector. The second ray is shown with an angle of incidence θ_c. It denotes the critical angle for "total internal reflection". The ray at the angle excites a wave that travels along the interface and re-emits energy

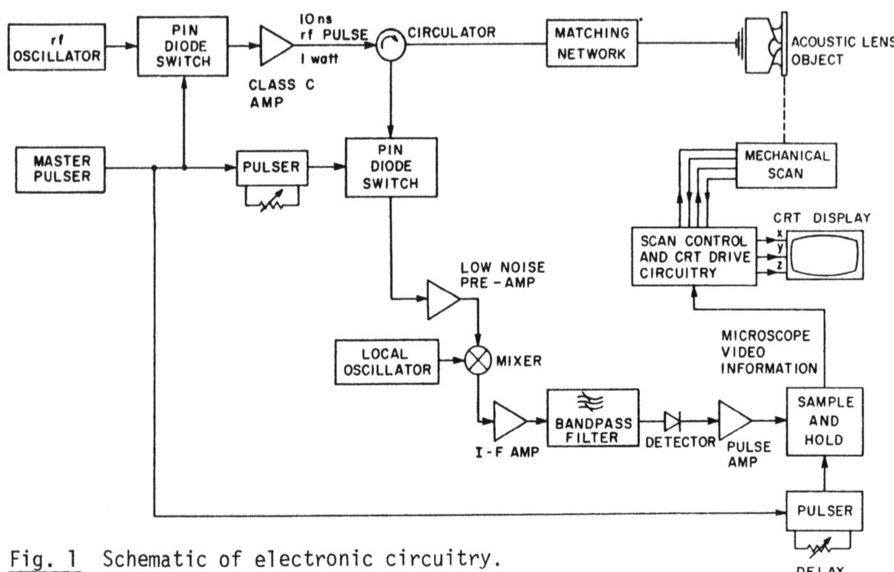

Fig. 1 Schematic of electronic circuitry.

Fig. 2 Ray diagram for acoustic reflection.

into the liquid at an angle equal to the angle of incidence, θ_c. After this ray has traveled to the right by the amount indicated in Fig.2, it re-emits a wave that appears to come from the focal point. This ray after traversing the lens produces a response at the transducer just as does the ray at normal incidence. The two rays, however, are shifted in phase by a different amount. This phase difference depends, among other things, on the velocity of the wave traveling along the interface. If this velocity changes as the probing beam is moved transversely to another part of the specimen, the interface pattern is altered and this produces a change of contrast in the image.

We will recapitulate. In order to monitor small variations in the elastic properties of the specimen's surface, we move the reflecting surface away from the focal plane toward the lens. At this point, the incoming ray at the critical angle will interfere with the ray at normal incidence. The interference parameter will change as the velocity of the wave along the interface changes. The velocity of the wave along the liquid-solid interface is dependent on the elastic properties of the solid and this dependence produces the contrast changes in the final image.

This model works well with a single surface of a solid. With biological cells, we must consider two interfaces, one at the top of the cell where it is in contact with the surrounding liquid, and the other at the bottom of the cell where it is in contact with the substrate. There is an acoustic reflection at both of these points, and the two can interfere constructively or destructively according to the elastic properties of the substrate. The remarkable thing about this system is that we get both. In the outer regions of the cell where the thickness tapers there are regions where the reflection from top and bottom add and other regions where they cancel. The dark regions, or rings, will move and distort as the internal elastic properties of the cell changes. The contours of dark rings give us a precise measuring system that allows us to monitor not only the elastic properties of a given point in the cell - but to record the changes in these properties as the cell moves around and about on the substrates. The images of Fig. 3 and Fig. 4 illustrate the appearance of these dark contours.

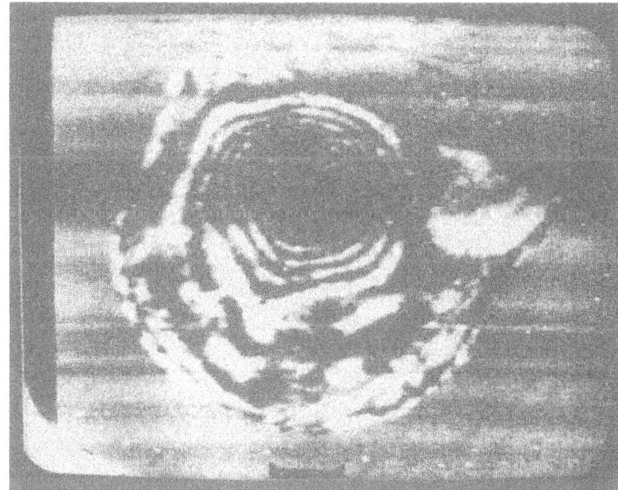

Fig.3 Acoustic reflection image of a chick embryo fibroblast.

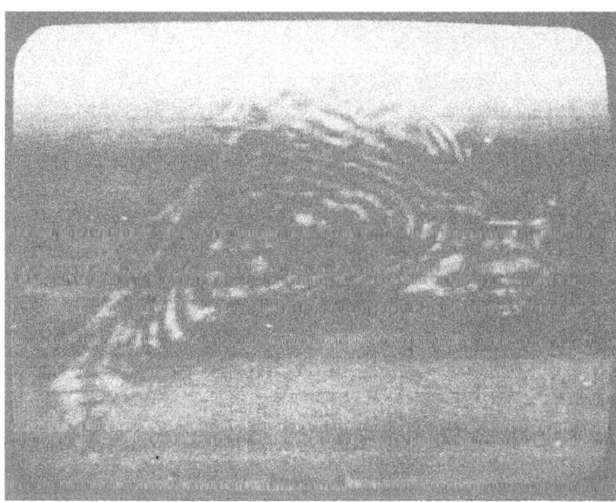

Fig.4 Acoustic reflection image of a chick embryo fibroblast.

3. Imaging in Cryogenic Liquids

As we push for greater and greater resolution, we must search for liquids with low sound velocities for there the wavelength is decreased for a given frequency. Cryogenic liquids offer possibilities for their velocity is much less than water [4]. There are two of interest: (1) liquid nitrogen and (2) liquid helium. The velocity of sound in nitrogen is decreased by a factor of 2 over water, and in helium, it is down by a factor of 8.

To record our images, we have rebuilt the microscope in a form that would fit inside a dewar 75 mm. diameter and immersed the entire unit in the liquid.

In helium at 1.95°K, we now have preliminary images and it looks as though it will soon be possible to push the resolving power to the point where we can resolve periodicities that are in the neighbourhood of 1000 Å. To date, we have succeeded in resolving aluminium lines of silicon wafer with a periodicity of 2500 Å. The width of the individual line was 1000 Å.

In liquid nitrogen, we have imaged fixed frozen cells with detail that is somewhat more distinct than that found in the phase contrast images from the optical microscope. Chick fibroblast cells imaged in this liquid are shown in Figures 5 and 6.

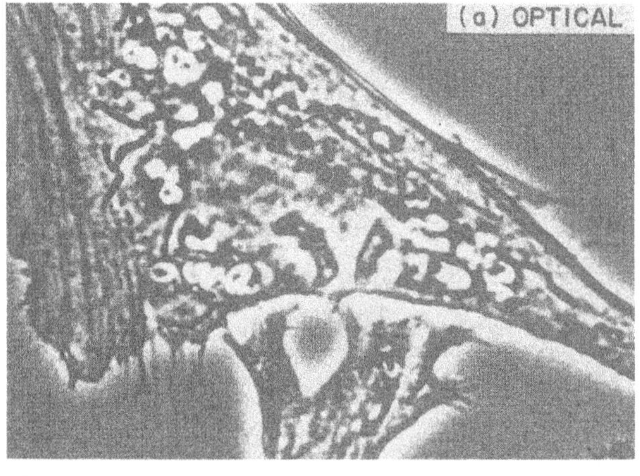

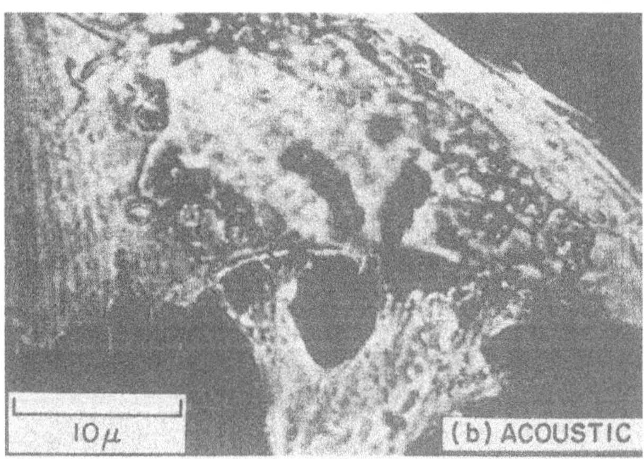

Fig. 5 Cryogenic acoustic image of a chick embryo fibroblast.

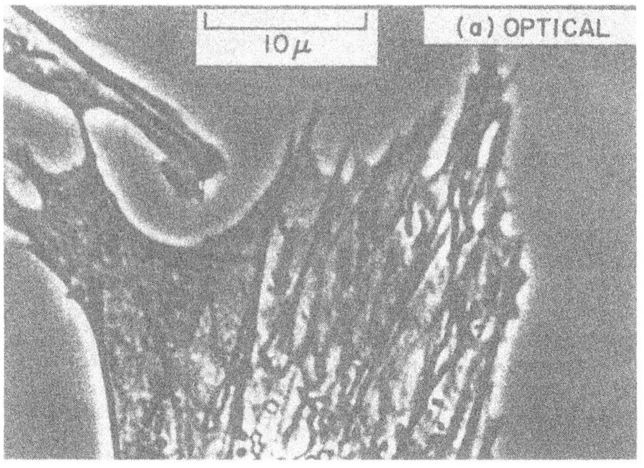

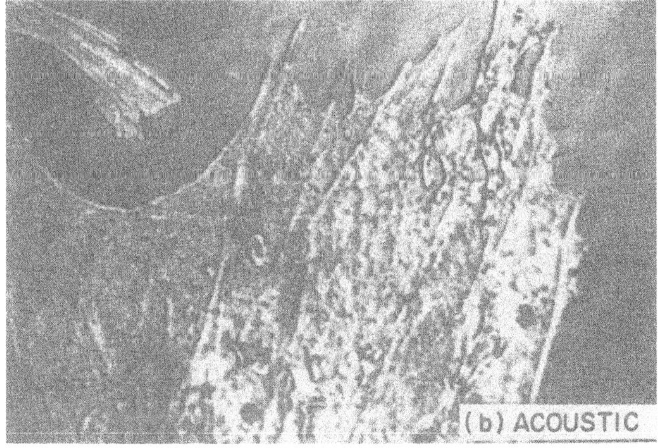

Fig. 6 Comparison of cellular detail for an acoustic image in liquid nitrogen.

4. Low Resolution Acoustic Imaging

The acoustic microscope has a demonstrated resolution capability rivalling that of optical microscopy. The prospects for achieving a radically higher resolution, by resorting to cryogenic liquids described in the last section, or by the use of high pressure gas as a coupling medium [5], look promising for the immediate future. However, the fact that the contrast mechanisms involved are quite different from those exploited in optical microscopy is an attribute which can be of direct value even at low resolutions. In this section, we will give two examples of such low frequency and, hence, low resolution applications.

Clinical ultrasound B-scan imaging is normally conducted at a frequency in the range 2-5 MHz, the resolution being of the order of one millimetre. The contrast mechanisms which enable the clinician to detect, for example, quite small focal and also diffuse liver tumours are not wholly understood.

With the help of acoustic microscopy, it should be possible to determine the specific tissue properties giving rise to contrast. To this end, a resolution of a few tenths of a millimetre is adequate. The need is to analyse the variations in density of the image into separate causal factors - changes in acoustic impedance, velocity or attenuation.

In order to achieve this discrimination, we must make use of the phase as well as the amplitude information in the received acoustic beam. It is also more convenient to work with a transmission microscope rather than the reflection version shown in Figure 1. The phase information is derived by utilising a reference channel in which the phase can be switched by $\pi/2$. From two measurements, the amplitude and phase can, therefore, be computed.

The impedance and velocity of the sample can, in principle, be derived from this data, [6,7]. There are certain difficulties of ambiguity, and accuracy, even when one is dealing with plane waves. For a focussed beam, one can still derive the impedance and velocity in a direct way from the data, provided the angular aperture of the illuminating beam is not too large. For the largest apertures, corresponding to the highest attainable resolutions, it is necessary to derive the information from an optimisation procedure.

Figure 7 shows an example of a ½ mm. section of fixed human liver. An optical image shows very little contrast. The acoustic amplitude and phase, obtained at 11 MHz in transmission, is shown in (a) and (b) respectively. Figures 7 (c) and (d) show the derived impedance and velocity. The numerical

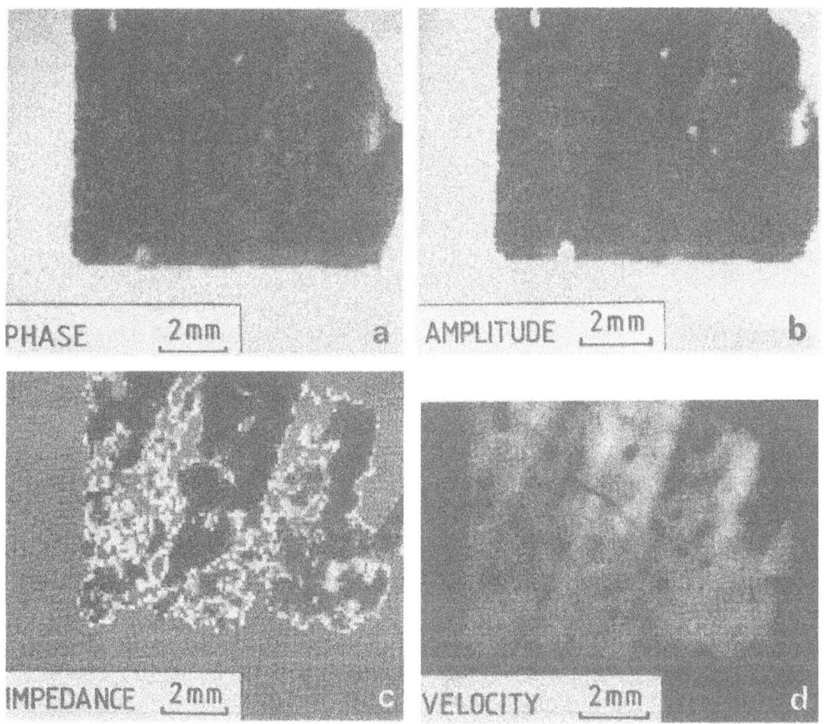

Fig. 7 Human liver imaged at 11 MHz.

values can be read off by reference to a calibrating grey scale, or alternatively, presented on a screen, by placing a cursor on the required portion of the image. A histogram of the velocity shows a peak value of 1.05 relative to the velocity in water, and a range extending from 1.04 to 1.08 - reasonable in the light of known tissue velocities. The impedance values implied by Fig. 7 (c) show a variation below and above that of water which is somewhat wider than one would have expected; an error analysis indicates that a higher level of experimental accuracy is probably needed to ensure quantitative reliability.

A second example of low resolution acoustic imaging is its application to chick embryology, [8]. Fig. 8 shows amplitude and phase images of a 25 somite chick embryo, fixed in ethanol and imaged in methanol, taken at 50 MHz. It is important to emphasise that no staining was used - yet clearly there is no lack of contrast. Indeed, these highly preliminary results suggest evidence that the acoustic imaging permits a more sensitive detection of somite formation than is achieved with staining and optical imaging.

Use of a temperature controlled microscope stage, together with a dialysis unit, would enable the embryo to be kept alive for extended periods. This should allow the observation of growth at a time when it is proceeding with great rapidity.

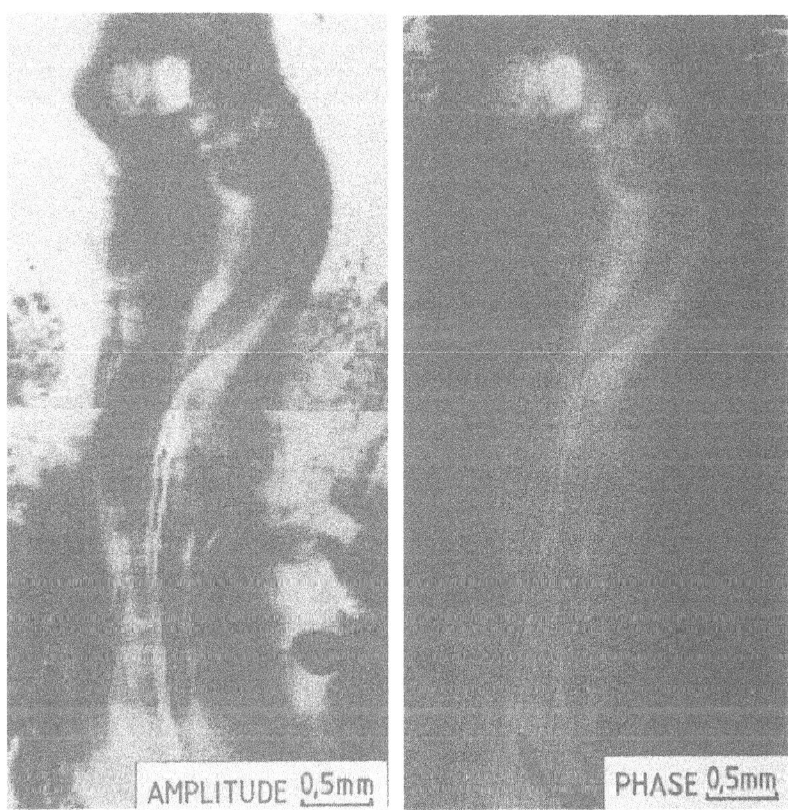

Fig. 8 Chick embryo

5. Future Developments

The basic acoustic microscope, working in liquids and at room temperature, is now reaching a stage of developmental maturity. However, the applications, both to cellular biology as to tissue characterisation are, as yet, in an early stage. The ability to image without staining is just beginning to be exploited. The technique for deriving materials information, on a resolution cell basis, by varying the lens-object spacing or by direct phase measurements is a unique and novel microscopical technique. Further, the fact that the information is derived on a point by point basis, coupled with the fact that one obtains a complex (amplitude and phase) signal, offers opportunities in subsequent image processing which may prove extremely valuable. A simple example is presented by the ability to subtract two complex, successive images, [9], which allows one to display the differences - e.g. growth patterns in a biological specimen.

The higher resolution which one can obtain using cryogenic liquids has been discussed; one can anticipate rapid development of this technique during the next few years. A complementary approach which promises the attainment of very high resolutions is the use of a gas at high pressure as the coupling medium. Resolutions below 1000 Å are anticipated, [5]; if the expectations are realised, and since gas medium microscopy takes place at room temperature, it may be possible to image some classes of living specimens.

Finally, we should mention the rapid development of the application of photo-acoustic spectroscopy, [10], to microscopy, [11]. In this technique, a modulated laser beam incident on the specimen, thermally excites acoustic waves which can be detected by a variety of means, some directly related to those used in the acoustic microscope. This hybrid optical-acoustic approach looks exciting in a number of different respects - notably the fact that it permits spectroscopically selective imaging. It is just reaching the stage where useful images of biological specimens are emerging.

ACKNOWLEDGEMENTS

This work was supported in part by a Research Grant 1R01 GM-25826- from the National Institute of Health in the USA, and by grants from the Medical Research Council and the Science and Engineering Research Council in the UK.

The authors would like to express their gratitude to their colleagues for their generous permission to include results which in some cases are as yet unpublished.

REFERENCES

[1] R.A. Lemons and C.F. Quate, "Acoustic Microscopy" in Physical Acoustics, Ed. W.P. Mason and R.N. Thurston, (Academic Press), V.XIV, p.1, 1979.

[2] C.A. Quate, A. Atalar and H.K. Wickramasinghe, "Acoustic Microscopy with Mechanical Scanning - A Review", Proc. IEEE, V.67, pp.1092-1114, 1979.

[3] C.F. Quate, "Microwaves Acoustics and Scanning Microscopy", Scanned Image Microscopy, Ac. Press, 1980, pp. 23-56.

[4] J. Heiserman, "Cryogenic Acoustic Microscopy", loc cit, pp. 71-96.

[5] H.K. Wickramasinghe, "Gas Medium Acoustic Microscopy", loc cit, pp. 57-70.

[6] D.A. Sinclair, I.R. Smith, "Tissue Characterisation using the Acoustic Microscope", to be published in Ultrasonic Imaging.

[7] S.D. Bennett, "The Determination of Elastic Constants from Acoustic Microscopy Measurements", submitted to IEEE Trans. on Sonics and Ultrasonics, September 1981.

[8] I.R. Smith, "Contrast Analysis in Scanned Image Microscopy", Ph.D. Thesis, Univ. of London, July 1981.

[9] S.D. Bennett and E.A. Ash, "Differential Imaging with the Acoustic Microscope", IEEE Trans. on Sonics and Ultrasonics, V.SU-28, No. 2, pp. 59-64, March 1981.

[10] A. Rosencwaig, "Photoacoustics and Photoacoustic Spectroscopy", John Wiley & Sons, 1980.

[11] Y.H. Wong, "Scanning Photo-acoustic Microscopy", Scanned Image Microscopy, Academic Press, 1980, pp. 247-271.

Imaging and Microscopy with Optoacoustic and Photothermal Methods

G. Busse

Hochschule der Bundeswehr München, Zentrale Wiss. Einrichtung Physik
D-8014 Neubiberg, Fed. Rep. of Germany

1. Introduction

Optoacoustic (also called photoacoustic) and photothermal imaging is based on the propagation of thermal waves [1] and their interaction with thermal structures in a solid [2]. These waves are locally generated in the surface of the sample by the absorption of an intensity modulated focused laser beam which is scanned across the sample. They are detected via the correlated temperature modulation. The classical method is optoacoustic detection where the sample is kept in an airtight cell. The temperature modulation, which depends on optical and thermal sample properties, results in a gas pressure modulation which can be heard as sound [3]. A more modern method uses piezoceramics for detection of modulated thermal expansion [4].

While these two methods still require physical contact with the sample, photothermal detection of thermal waves is based on detection of modulated thermal infrared radiation [5]. It is therefore a remote method which is quite different from conventional thermal imaging.

2. Optoacoustic Imaging and Microscopy

Imaging means mapping of local sample properties. The arrangement used here is shown in Fig. 1. A laser beam is modulated and then deflected in such a way that it scans across the sample which is glued to piezoceramic material (PZT). The AC-output voltage of this detector is analysed by a lock-in amplifier and then shown in an imaging system as a function of the sample coordinate where the signal was obtained.

An example of an optoacoustic image is shown in Fig. 2 where two pieces of metal (thickness 0.5 mm) were welded and then carefully ground to remove optical surface structure. The subsurface structure is clearly revealed with the signal magnitude at 20 Hz in halftone or signal surface representation. As this kind of imaging is sensitive to boundaries, it is a valuable tool for nondestructive testing of opaque materials far beyond the penetration depth of the optical radiation [7-9]. The optoacoustic signal depends on how much heat is produced initially and on how well it is dissi-

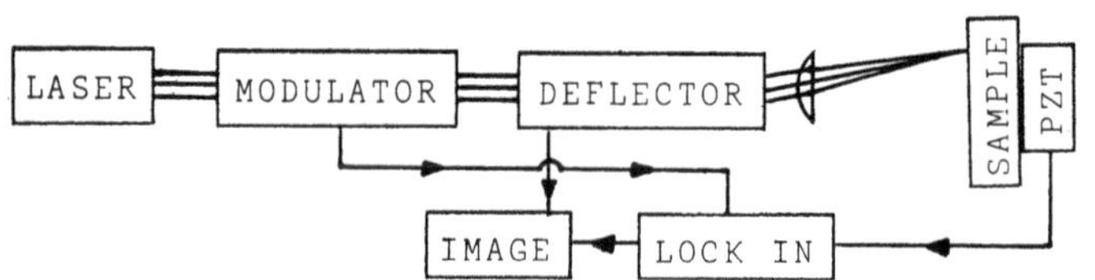

Fig.1. Experimental arrangement [6]

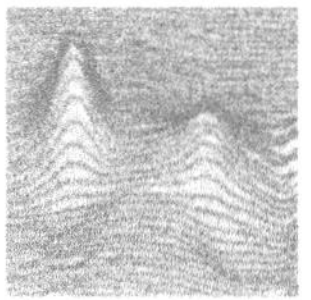

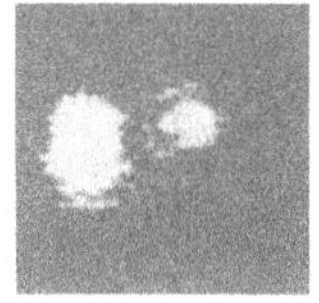

Fig.2. Optoacoustic imaging of welded steel (0.5 mm thickness)

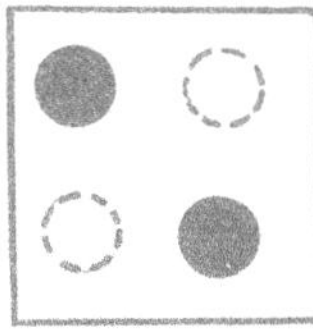

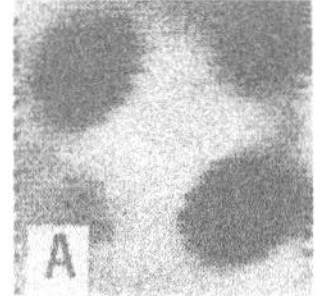

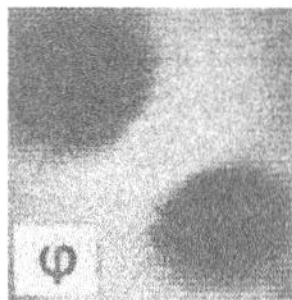

Fig.3. Optical and optoacoustic images of an aluminium sample provided with black spots and subsurface holes (*dashed*)

pated into the interior of the sample [3]. If a sample has optical structure, the optoacoustic image is also affected because the amount of generated heat changes.

On the other hand, temperature modulation as a thermal response is characterized by its magnitude A and its phase lag φ with respect to the optical modulation. As phase can be interpreted as a normalized delay time, it does not depend on how much heat is produced. Therefore phase angle φ shows only thermal structures [10] as is demonstrated in Fig. 3, where an aluminium sample is provided with subsurface holes and black spots on the surface [11]. The optical image can only show the surface structure while the signal phase reveals only the subsurface structure and the magnitude A. Therefore phase φ has the advantage of providing information which does not depend on optical surface absorption. Also it has been shown previously that the depth range is larger [12]. By varying the modulation frequency, depth resolved analysis is possible because the signal has its origin mostly in a surface layer the thickness of which is about the thermal diffusion length [3,6]. Optoacoustic microscopy of thermal structures requires modulation frequencies of 10^5 Hz or more [2,10]. Inspection of biological material is possible with a liquid coupling to the piezoceramic transducer [13]. Microscopy of a hair is shown in Fig. 4. For contrast enhancement electronic filtering has been used. The phase image does not show all the surface details of the magnitude image but mostly lines along the direction of the hair. Laser power can be decreased to about 1 mW in these experiments. Though liquid coupling provides some cooling, high power density ($\approx 10^5$ W/mm^2) is still a problem.

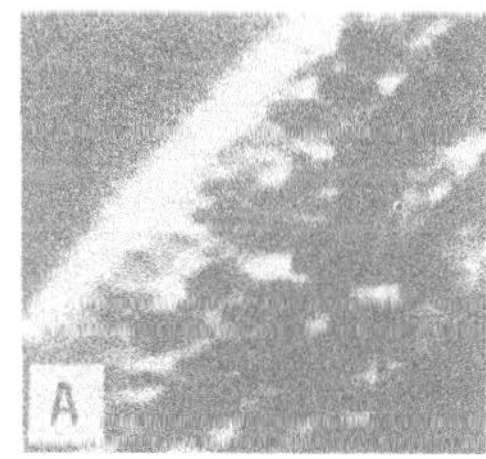

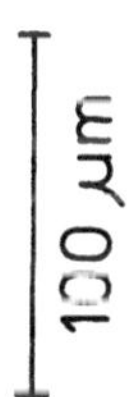

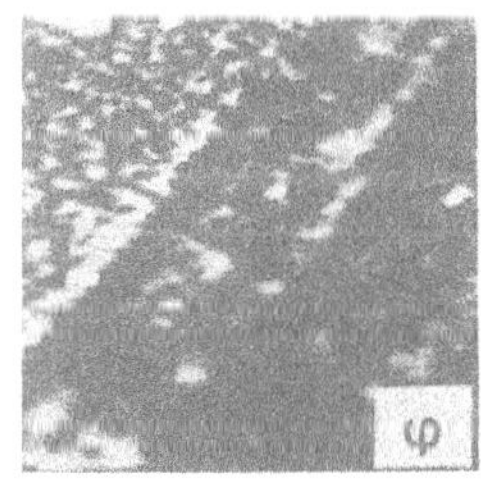

Fig.4. Optoacoustic images of a hair at 110 kHz

3. Photothermal Imaging

Optoacoustic detection requires physical contact with the sample. Remote observation of temperature as it is known in thermal imaging can also be used for thermal wave analysis (photothermal detection). The main difference is that one observes temperature modulation only [5]. Therefore photothermal imaging differs from thermal imaging by the fact that it is a dynamical method where a phase shift allows distinguishing between optical and thermal structures. Photothermal front surface detection has been previously used for material probing [14]. A comparison with optoacoustic detection is shown in Fig. 5a,b where the laser beam scans across 1 mm aluminium provided with a 0.4 mm subsurface hole. Resolution seems to be about the same because it is determined by thermal diffusion length [2].

In a photothermal transmission arrangement (Fig.5c) the thermal wave is detected after it has propagated through the sample. The results (which show a diffraction-like pattern) indicate that at the same modulation frequency 20 Hz resolution is better. As the signal is not dominated by near-surface parts of the sample, one observes mostly the projection of internal structures [15]. Therefore this method should be suited for remote nondestructive measurement or inspection of layers up to several mm of aluminium at 20 Hz [16]. One possible field of applications is testing laser welded thin metals used for medical devices. Such an example is presented in Fig. 6. Photothermal transmission imaging with the phase angle φ shows a discontinuity which is not so well seen with signal magnitude A. From the few results obtained till now it seems that the required optical power exceeds the one needed for optoacoustic detection. Therefore imaging biological substances is more difficult in photothermal transmission.

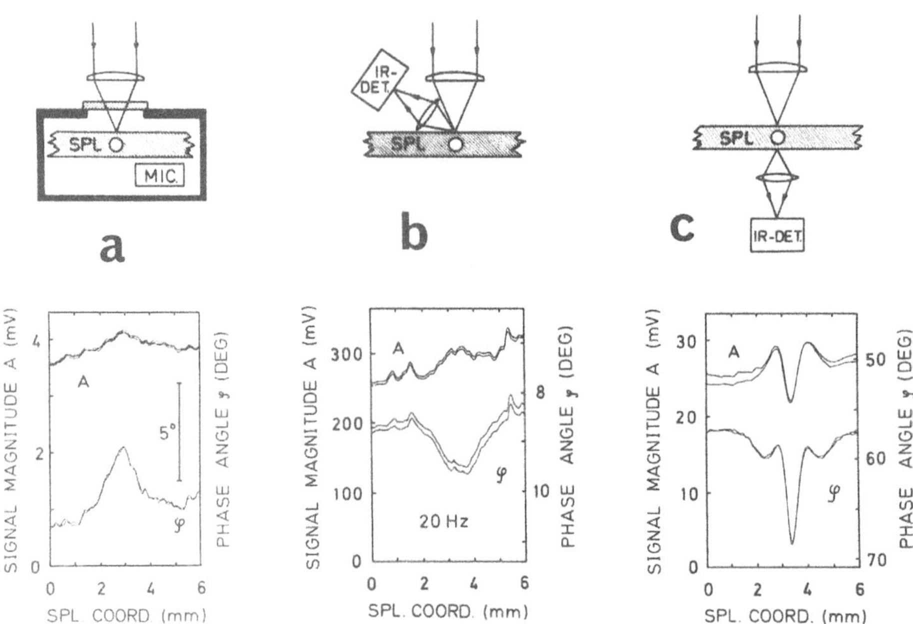

Fig.5a-c. Resolution in different methods: optoacoustic (a), photothermal front surface (b), and photothermal transmission (c) [15]

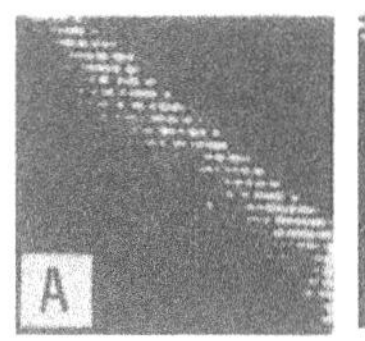

Fig.6. Bright diagonal is seam of laser welded metal. Field of view is 2.3×2.3 mm^2

4. Conclusion

There are several ways of thermal wave imaging and microscopy, and the information obtained can be different. Optoacoustic imaging shows mainly the near surface region, structures in greater depth appear weaker. Range is variable via modulation frequency but generally smaller than with photothermal transmission imaging which does not directly provide depth information. This method should find more application for the inspection of thicker samples with a higher damage threshold.

Acknowledgement. The author acknowledges helpful discussions with Professor B. Bullemer (ZWE Physik) and with Professor G.J. Müller and Professor W. Schmidt (Zeiss, Oberkochen). He is also grateful to Professor M. Schaldach and Professor R. Thull (Zentrales Institut für Biomedizinische Technik der Universität Erlangen-Nürnberg) for providing the sample used in Fig. 6.

References

1 M.A.J. Ångstrom: Philos. Mag. *25*, 180 (1863)
2 A. Rosencwaig: Am. Lab. *11*, 39 (1979)
3 A. Rosencwaig, A. Gersho: J. Appl. Phys. *47*, 64 (1976)
4 A. Hordvig, H. Schlossberg: Appl. Opt. *16*, 101 (1977)
5 P.E. Nordal, S.O. Kanstad: Phys. Scr. *20*, 659 (1979)
6 G. Busse, A. Rosencwaig: Appl. Phys. Lett. *36*, 815 (1980)
7 Y.H. Wong, R.L. Thomas, J.J. Pouch: Appl. Phys. Lett. *35*, 43 (1979)
8 M. Luukkala, S.G. Askerov: Electron. Lett. *16*, 84 (1980)
9 G. Busse: IEEE Ultrasonics Symposium Proc. 1980, p.622
10 A. Rosencwaig, G. Busse: Appl. Phys. Lett. *36*, 725 (1980)
11 G. Busse: To be published
12 G. Busse: Appl. Phys. Lett. *35*, 759 (1979)
13 G. Busse: In *Scanned Image Microscopy*, ed. by E.A. Ash (Academic, New York 1980)
14 P.E. Nordal, S.O. Kanstad: In Ref.13
15 G. Busse: IInd Int. Top. Meeting on Photoacoustic Spectr., Optical Soc. of America, Berkely 1981, paper Th A3
16 G. Busse: Infrared Phys. *20*, 419 (1980)

Part II

Image Processing

Biomedical Image Classification by Data Reduction

M. Mohajeri, M.A. Fiddy, and R.E. Burge

Physics Department, Queen Elizabeth College, University of London
Campden Hill Road
Kensington, London, W8 7AH, U.K.

Introduction

The aim of this paper is to consider the quantitative classification of medical images for diagnostic purposes. Medical images present many common features from patient to patient but the differences between them may not always be due to some abnormality. This variety of format for such images leads to a complexity that restricts the success of image processing. This paper considers the use of data compression techniques to provide a compact representation of the image and, hopefully, a more efficient means for classification [1,2]. We have also studied pre-processing (histogram equalisation [2] and spectral filtering) in order to aid establishing decision criteria. We consider chest X-rays of pneumoconiosis patients and acoustic images of babies heads. The only compression techniques discussed here are the discrete cosine transform (DCT) and the Karhunen-Loève transform (KLT), the latter being of more interest. Confidence in finding criteria is drawn from applications of these transforms to simulated data.

Compression transforms

The DCT and KLT are well known and will only be briefly outlined here, [3,4]. The DCT of a two dimensional NXN image f(x,y) is given by

$$g(u,v) = \frac{1}{2N^3} \sum_{x=0}^{N-1} \sum_{y=0}^{N-1} f(x,y) \left[\cos(2x+1)u\pi\right] \left[\cos(2y+1)v\pi\right]. \qquad (1)$$

This transform can be obtained directly from the FFT and also lends itself to optical implementation. An ultimate goal in our work is the development of a real time optical processor to facilitate routine mass screening and diagnosis. The DCT results in most of the image energy being concentrated towards lower spatial frequency coefficients.

Although not offering a great improvement over the DCT in compression factors, the KLT is based on a completely different representation of the function. Consider an N^2 dimensional vector $\underline{x}$, having components x_j which might be an image or a sub-image. If there are $i = 1,2...M$ such vectors one can approximate their covariance matrix by the average

$$\underline{C}_{\underline{x}} = \frac{1}{M} \sum_{i=1}^{M} (\underline{x}_i - \underline{m}_{\underline{x}})(\underline{x}_i - \underline{m}_{\underline{x}})^T \qquad (2)$$

where for M similar images $\underline{m}_{\underline{x}} = m_{\underline{x}_j} = \frac{1}{M} \sum_{i=1}^{M} x_{ij}$ or $\underline{m}_{\underline{x}} = \underline{m}_{\underline{x}_i} = \frac{1}{N^2} \sum_{j=1}^{N^2} x_{ij}$ for each sub-image.

The KLT consists of obtaining an image vector $\underline{y}$ given by $\underline{y} = \underline{A}(\underline{x}-\underline{m}_x)$ where $\underline{A}$ is the transformation matrix whose rows are the eigenvectors of $\underline{C}_x$. This transform is optimum in the sense that $\underline{C}_y$ is diagonal, the elements being the eigenvalues of $\underline{C}_x$ and hence the elements of $\underline{y}$ are uncorrelated. Compression arises in the multiple image case because only a few $\underline{y}_i$ are significant. If only one image is available one approximates $\underline{C}_x$ as shown in eq. (2) by taking the average of the covariance matrices from the set of sub-images. The 'sharpness' of the diagonal structure of each covariance matrix will vary from sub-image to sub-image. The greater the similarity between these structures, i.e. the more homogeneous the complete image, the better the averaged $\underline{C}_x$ one would obtain, and hence the higher the possible compression. Compression for one image is achieved by considering only the p principal terms of each $\underline{y}$ giving

$$\underline{x}_i \cong \sum_{j=1}^{p} y_{ij}\underline{A}_j^T + \sum_{j=p+1}^{N^2} b_j\underline{A}_j^T + \underline{m}_{\underline{x}_i} \quad \text{where} \quad b_j = E\{y_{ij}\}. \tag{3}$$

We denote the multispectral situation [4] by the MKLT and that for a single image by the SKLT. A recent publication by CAULFIELD et al. [5], indicates the possibility of using a fast parallel non-coherent optical method for determining eigenvectors.

Image classifcation

The DCT and SKLT were applied, first, to simulated textures in order to assess their efficiency at distinguishing differences. The difference between the normal and abnormal conditions in the images mentioned above tend to be local texture changes. The textures shown below correspond to two views of the same texture (fig. a and b), changes in each of these due to line broadening (fig. c and d) and line lengthening (fig. e and f). The graph below (fig. g) shows

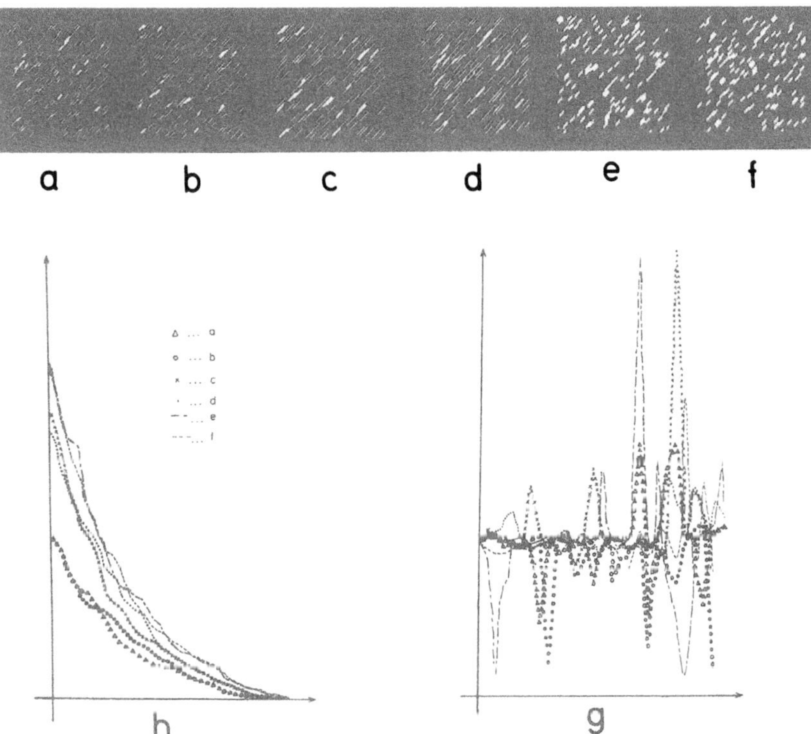

the distribution of the rotationally averaged cosine transform and fig. h shows the distribution of eigenvalues or the diagonal elements of $\underline{C}_y$. The DCT does not show clearly the differences between the six cases. There is a clustering of profiles using SKLT, however. These eigenvalues correspond to the case of an averaged $\underline{C}_x$ derived from subdividing the 64 x 64 images (fig. a to f) into 8 x 8 sub-images.

For a specific classification problem the optimum differences in, for example, eigenvalue profiles would vary. We demonstrate here only that consistent differences do indeed exist, as shown in fig. h. The motivation for preprocessing the data using histogram equalisation is to redistribute the image contrast and thus alter the amplitude of the coefficients or eigenvalues to be monitored. Clearly, the maximum differences that can be introduced will ease the classification criteria.

Using the SKLT the images below show, from left to right, the original image (64 x 64), the averaged covariance matrix, the $\underline{y}$ vectors for each sub-image with maximum to minimum values running from bottom to top, and the cases p = 8,15,40 respectively as defined by eq. (3). The vertical columns correspond to the texture (top), a chest X-ray with and without histogram equalisation followed by an acoustic image with and without histogram equalisation. It can be seen how compact the data may become for image storage or adequate representation.

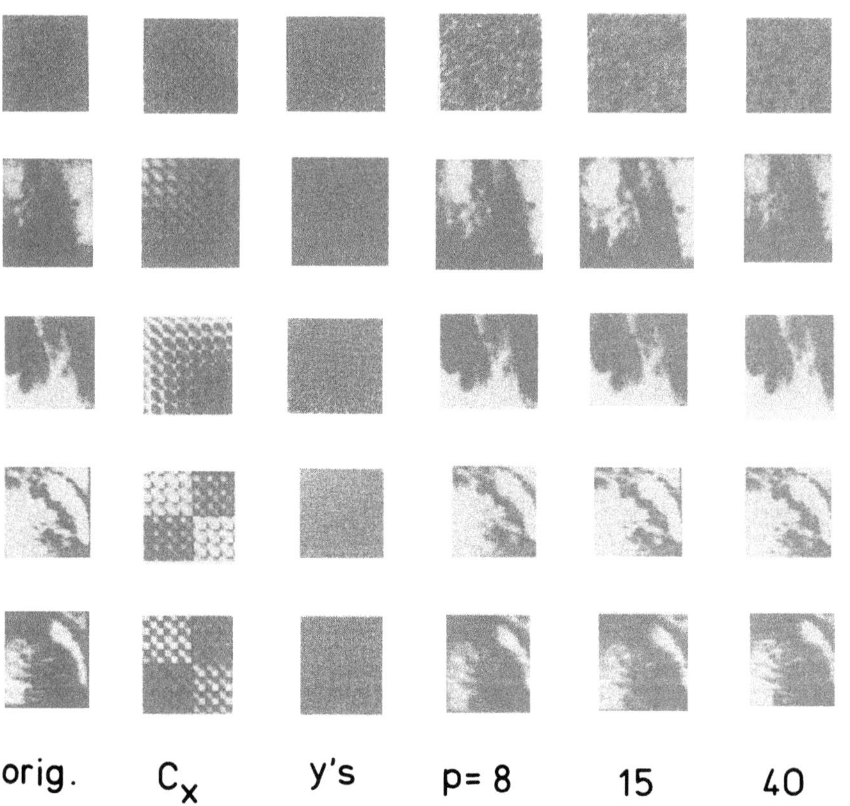

orig. C_x y's p= 8 15 40

While it might be possible in the acoustic case to generate a multispectral set, one is likely to have only one independent image for each patient. In order to generate the equivalent of a multispectral set one can apply a variety of window functions to the transform of an image. Once again, optimum choice of window for the given feature of interest can greatly enhance the separation of information between coefficients or eigenvalues. For specific problems, using the MKLT, one may be able to isolate the relevant information into one principal component and thus monitor only one eigenvalue for diagnosis. This is illustrated below for a set of 5 processed images (on the left) taken from the original and the 5 principal components (on the right). There is a concentration of uninteresting information about, for example, the ribs into the first principal component.

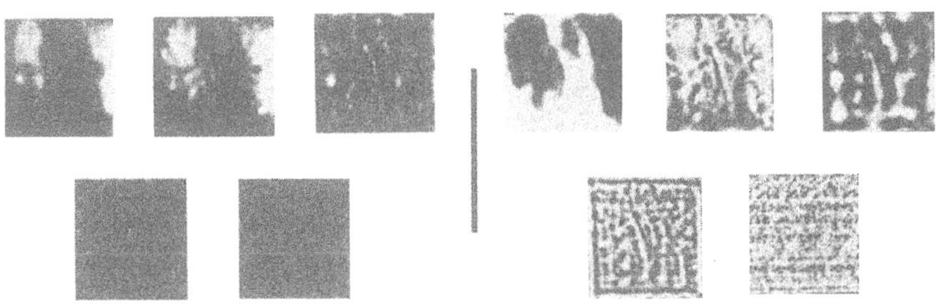

Conclusion

We have demonstrated the potential use of the DCT and KLT, which are amenable to fast optical processing, for classifying medical images. Pre-processing might be required to facilitate some feature classification and two techniques are suggested.

References

[1] G. E. LOWITZ, Pattern Recognition, 10, 359 (1978).

[2] R. C. GONZALEZ and P. WINTZ, Digital Image Processing, Addison-Wesley, (1977).

[3] W. K. PRATT, Digital Image Processing, John Wiley (1978).

[4] E. L. HALL, Computer Image Processing and Recognition, Academic Press (1979).

[5] H. J. CAULFIELD, D. DRONE, J. W. GOODMAN and W. RHODES, Applied Optics, 20, 2263 (1981).

Optimization Criteria for Television-Based Image Digitizers

L.P. Cordella, and A. Pirri

Istituto di Cibernetica del CNR
I-80072 Arco Felice, Naples, Italy

It is well known that digitizing an image implies converting it into an $n \times m$ matrix of numerals, each of which is a function of the integrated average light intensity reflected (or transmitted) by each of the $n \times m$ pixels in which the image is subdivided. At least two main reasons lead one to pay special attention to the problem of medical image digitization. On one hand it is believed that both for preservation and circulation, and even more in the future, storing images will be done by computer memory. Medical images are normally observed in order to reach a decision about the presence and the nature of a pathology. Such a decision is often strongly dependent on the quality of images, so their digital conversion should not deteriorate, but should improve if possible the conditions for an adequate evaluation. On the other hand, automated image analysis already appears as a powerful means of complementing the information attainable by the human observer (often qualitative) with more quantitative information. In this last respect, the digitization system is the eye of the computer: the better the pictorial information we are able to supply the computer with, the greater its possibility of a correct analysis and the smaller the need of preprocessing operations. This is especially true for biomedical images, since correct segmentation and feature extraction are often dependent on the possibility of detecting small density differences and thin lines.

Digitization techniques have greatly improved from Nipkow discs of the 50's, to flying-spot scanners, television scanners, photodiode arrays and more complex, high-speed systems operating with non-coherent light [1]. TV camera based digitizing systems are still widely preferred, essentially because of their speed, the possibility of operating both in transparency and reflection, and the lower cost and size. Among many types, the most commonly used television tubes are vidicon and plumbicon. The quality of a digitized image as obtained with such scanners depends on a number of factors, such as light source characteristics, transfer characteristics of the optics involved (microscope, lenses), and features of tube, the electronics connected to it and analog to digital converter (ADC) [2-6]. Although image quality evaluation could be based on different criteria, according to whether images have to be analyzed by man or computer, a number of optimization techniques can be mentioned.

The source of light should be adequately stable and provide a uniform illumination over the field of interest; its spectral distribution should be also considered according to the task. Both source and optics should be chosen so as to avoid distorsions and shading effects. Attention has to be paid to the relationships between optical characteristics and the actually wanted spatial resolution. Corrections for the modulation transfer function of the system are attainable by nonlinearly shaping the frequency response of the video amplifier [3]. Shading could also originate from tube photosurface; in this respect plumbicon is noticeably better than unselected vidi-

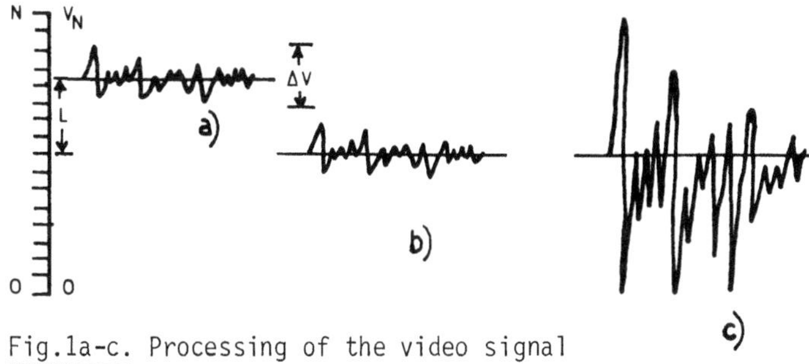

Fig.1a-c. Processing of the video signal

cons, showing a very uniform response over its surface. Use of shading correctors should anyway be considered [7]. As already mentioned, the light transfer function of plumbicon is very linear ($\gamma \cong 1$) while it is rather nonlinear for vidicon [2]. Nonlinearity gives to vidicon a dynamic range that, in reflected light, can be up to 10 times larger than for plumbicon (which is more sensitive), making the former sometimes preferable for images showing a large dynamic range, but whose informative content is confined into a narrower range. However, vidicon suffers from image retention which, besides the mentioned drawbacks, causes a preference for plumbicon in the majority of applications.

Signal to noise ratio is very important for comprehensible reasons. Plumbicon has a better S/N figure than vidicon, so the overall S/N ratio of the system mainly depends on preamplifier and circuitry noise. Figures better than 40 db at 5 MHz should be requested to obtain an adequate number of "meaningful" grey levels. But how many grey levels are necessary for a given application? [3]. Anyway, under the same conditions it appears desirable that the dynamic range of interest for the image at hand be resolved at most, i.e., that any image be digitized using the whole number of available levels. In this connection it should be pointed out that not only the integral linearity of the converter, but even its differential linearity, influences image quality, as channels may be not equally large. A differential linearity better than 5%, with 64 available quantization levels, could be achieved with a 10 bits ADC, by using only the 6 more significant bits.

Let us refer to Fig. 1 and let a) be the video signal corresponding to an image row: it varies in a range ΔV about a DC level V. If the signal would be digitized with a device able to assign, say, 16 levels within the range 0-V_N, it would be resolved in only 4 levels. Moreover, the grey level of every pixel would be represented by a number of bits greater than that actually sufficient for the same resolution. Manually regulating illumination, objective aperture, DC level and video amplification is time consuming and not practical for a number of reasons. Moreover, it would hardly allow one to make "absolute" measurements of optical density. We developed an analog device that seems to overcome such problems. A block scheme is given in Fig. 2.

Suppose that ADC2 converts voltages in the range 0-V_N into numbers in the range 0-N. The video signal is added in S_2 to the DC voltage value ± L necessary to shift its dynamic range about the middle value of the converter input dynamic range (Fig. 1b). The value of L and its sign are automatically evaluated, together with the image dynamic range ΔV, during a raster scan (or a restricted part of it, selected by an analog switch) as shown in the scheme. The signal is then amplified (Fig. 1c) by an amount $A = V_N/\Delta V$ by a digitally programmable gain amplifier (see e.g., [8]). With the components

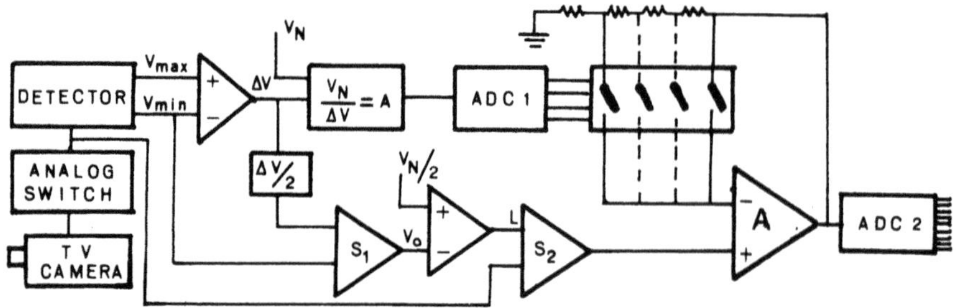

Fig.2. Block scheme of the electronic circuit

used, the noise introduced is negligible with respect to other sources. The noise due to the TV camera and to its preamplifier is, of course, amplified by A as well. While not affecting the overall S/N ratio, this effect gives a further reason for using high S/N ratio cameras. The "true" grey value g of a pixel can be recovered from its actual value G, if the values A and L are stored by computer, since g = (G/A) - L. By suitably shaping the transfer functions of A it is also possible to obtain MTF and γ corrections. In this last case g is given by accordingly modified formulas. Some results are shown in Fig. 3.

Thanks are due to Mrs. A.M. Mazzarella and Mr. S. Piantedosi for their technical assistance.

References

1 R. Schak et al.: J. Histochem. Cytochem. *27*, 1, 153 (1979)
2 G. Palmieri: Pattern Recognition *3*, 157 (1971)
3 K. Preston, Jr.: In *Digital Picture Analysis*, ed. by A. Rosenfeld, Topics in Applied Physics, Vol.11 (Springer, Berlin, Heidelberg, New York 1976)
4 W.F. Schreiber: Proc. IEEE *66*, 12, 1640 (1978)
5 A.N. Netravali, J.O. Limb: Proc. IEEE *68*, 3, 366 (1980)
6 J.F. Brenner et al.: J. Histochem. Cytochem. *24*, 1, 100 (1976)
7 R.M. Haralick: In *Digital Picture Analysis*, ed. by A. Rosenfeld, Topics in Applied Physics, Vol.11 (Springer, Berlin, Heidelberg, New York 1976)
8 B.R. Woo: Proc. IEEE *68*, 7, 934 (1980)

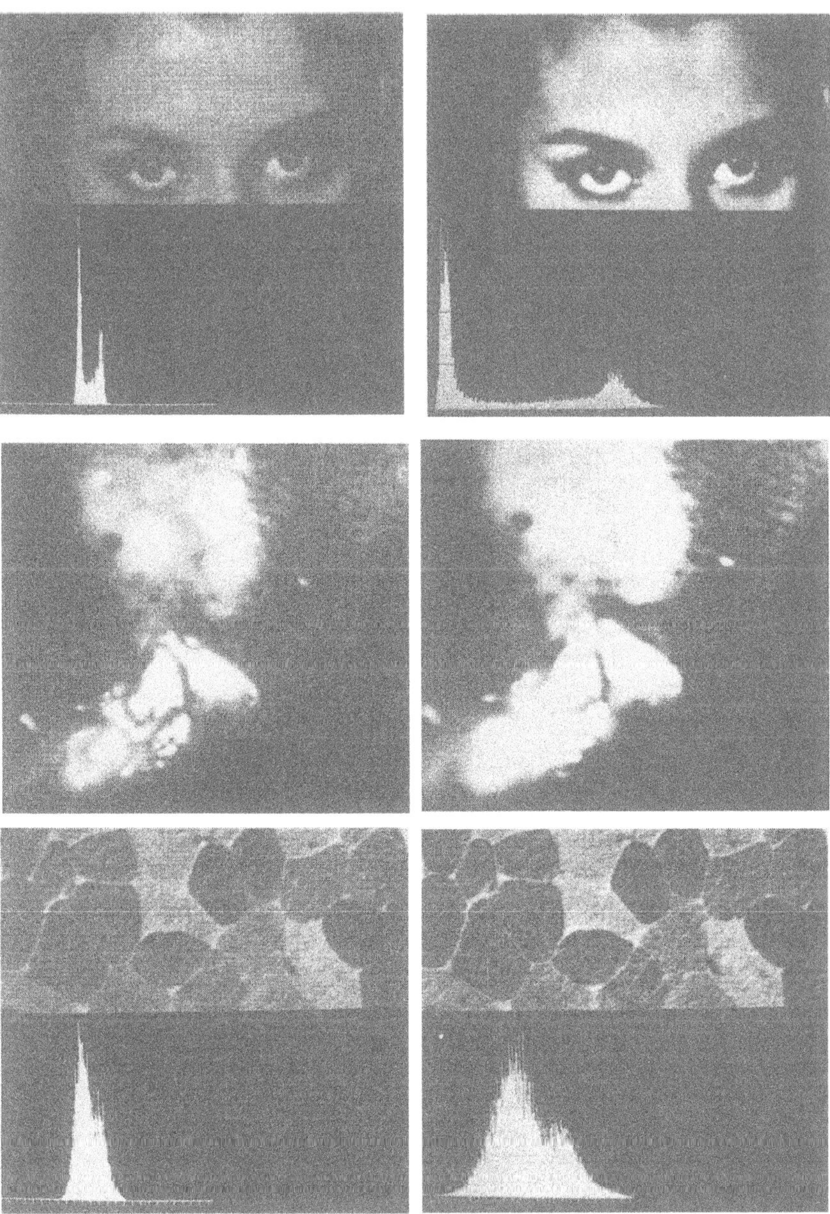

Fig.3a-c. Degraded (*left*) and improved (*right*) digital versions of three photographic images: a) girl, b) colposcopy, c) muscular fiber section. Grey level hystograms are shown. The only processing applied to the images is that illustrated in Fig. 1

Digital Image Analysis of Complex Periodical Viral Structures

A. Santisteban, and N. García

Centro de Investigación UAM-IBM, Universidad Autónoma de Madrid
Madrid 34, Spain

J.L. Carrascosa, and E. Vinuela

Centro de Biología Molecular, C.S.I.C., Universidad Autónoma de Madrid
Madrid 34, Spain

1. Introduction

Structures with two-dimensional translation symmetry are common among viruses. Digital image processing techniques can be used to study their electron micrographs, filtering the periodic structure from the non-periodic noise. We outline in section 2 the method that gives the best results, consisting of the transformation of the image to get its digital Fourier transform (DFT) matched to a pre-defined theoretic lattice-pass filter [1]. In section 3 we extend this technique to the case in which we have two periodic layers producing a moiré pattern that, in the usual presence of heavy noise, is nearly impossible to interpret by simple visual inspection. We apply this method in section 4 to the study of the capsid of the African Swine Fever virus.

2. Single Lattice Analysis

The first step in the analysis of an electron micrograph is digitization. When the structure we are dealing with is a simple two-dimensional periodic array, it is not difficult in general to approximately align the scan direction with one of the lattice axes. Thereafter we display the digital image on an interactive terminal screen, select with the cursor equivalent points in the most clearly distinguishable basic elements of the lattice, and compute with their coordinates the parameters of a geometric transformation that adapts the image to the filter (see below). We use the two-dimensional cubic convolution algorithm [2] for interpolation.

The purpose of the geometric transformation is to have an integer number of repetitions of the basic motif in each direction (apart from minor local distortions), what is equivalent to have an infinite lattice. The Fourier transform in absence of noise is then a lattice of delta functions and the DFT has its samples on the precise locations of these delta functions. Therefore there is no leakage in the DFT, and the non-zero values other than the reciprocal lattice points correspond solely to noise. We apply then a lattice-pass filter that makes zero all DFT values except those in the reciprocal lattice points. The filtered image is obviously pure periodic with the motif identically repeating centered on the direct lattice points. For a more detailed discussion see [1].

3. Multiple Lattice Analysis

When we are dealing with an object composed by several independent periodic structures we can, at least in principle, analyze separately all of them using the procedure outlined in section 2. Nevertheless, this is not straightforward at all. We discuss here the steps of the analysis of a moiré produced by two periodic layers and apply them to a real case in section 4.

When we have a moiré with low signal-to-noise ratio, we cannot, in general, select a relevant direction for scanning; hence we have to digitize an area containing the specimen with an arbitrary axes orientation, using a window size and a sampling distance small enough to ensure that no significant information will be lost in any subsequent resampling.

We display the image on the interactive terminal screen and select an appropriate subimage of the best preserved part of the specimen. We compute the DFT of the subimage using the FFT algorithm [3] and display its modulus on the terminal screen. This Fourier pattern is usually very difficult to interpret due to the unadequate image sampling that introduces a serious leakage in the image frequencies, and to the usual high level of noise. We have to identify the location of the peaks corresponding to the basic frequencies of the lattices, distinguishing those coming from each lattice.

When both lattices are equal though rotated a small angle, it is very difficult to separate the peaks of the DFT that correspond to each one of them. We then have to resample with a lower frequency in order to separate further these peaks. This puts a limit on the resolution which is necessary to evaluate the basic motif of the lattice.

From the approximate coordinates of the peaks of the DFT, we obtain the parameters of a change of coordinates in the direct space that makes one of the lattices to approximately match the filter, defined as outlined in section 2. The leakage of the frequencies of this lattice has been reduced but not completely removed. We obtain a refinement of the change of coordinates and construct a new subimage that achieves a better matching to the filter.

We keep on trying different changes of coordinates until we get a satisfactory matching of the first lattice to the filter. Then we filter the image and obtain the reconstruction of the first layer of the specimen. The final change of coordinates performs, in general, different rotations and scale dilatations for the two basic vectors. The inverse change of coordinates can eventually be applied to transform the filtered image back to the original coordinate frame.

The same iterative procedure is also applied to the second lattice and the filtered image is transformed to the first lattice coordinate frame.

The use of an interactive display system is essential for our trial and error approach to obtain a DFT matching the theoretic lattice-pass filter. We use a Ramtek 9351 display unit attached to an IBM 370/158 computer operating under VMS.

4. Application: Study of the Capsid of the African Swine Fever Virus

As a practical application of the method described in the previous section we study the capsid of the African Swine Fever (ASF) virus. These capsids were obtained after osmotic shock of the virus during 24 h in water, concentrated by sedimentation, resuspended in water and adhered to carbon coated grids previously made hidrophilic by glow discharge. The samples were negatively stained with 2% phosphotungstate pH 7 and micrographs were taken in a

JEOL 100B transmission electron microscope with a magnification x36,600. The micrographs were digitized in a Perkin-Elmer 1010A flat-bed microdensitometer, using a square window 4 μm on a side, sampling distances of 4 μm in both axes and a scale linear in the optical density.

The image that we selected revealed no information in the direct space. Nevertheless, its Fourier transform showed two hexagonal lattices rotated a small angle. In order to achieve a good separation of both lattices (their points at least 2 sampling units apart in the DFT) it was necessary to undersample. By trial and error we finally selected a 256x256 pixel subimage (specimen was not big enough to extract a 512x512 pixel subimage) with 16 rows of 14 motifs each one (notice that $14/16 \simeq \sqrt{3}/2$ the ratio in an exact hexagonal lattice). The axes of the subimage (rotated with respect to those of the original image) were orthogonal and the resampling distances were 3.736 in the horizontal direction and 3.870 in the vertical one (see Fig. 1).

Its DFT (Fig. 2) conformed very accurately to the theoretic lattice-pass filter. Only the 3 first diffraction orders were kept in the filtering provided that the higher orders correspond to frequencies higher than the first zero of the contrast transfer function. Figure 3 represents the reconstruction of this lattice; it is a hexagonal array with a periodicity of 7.6 nm. Again by trial and error we selected another subimage of the same characteristics, for which the second lattice conformed to the theoretic lattice-pass filter. Its horizontal axis formed an angle of -10.8° with the horizontal axis of the first subimage while the vertical axes formed an angle of -11.6°. The resampling distances were the same as before. The peaks of its DFT were less clearly resolved than in the first lattice (probably due to a poorer staining of the corresponding layer). The filtered subimage (Fig. 4) was then geometrically transformed to be in register with the first filtered subimage and added to it to obtain a moiré pattern. This pattern is the reconstruction of the original image having been filtered out the non-periodic noise (compare Fig. 5 and Fig. 1).

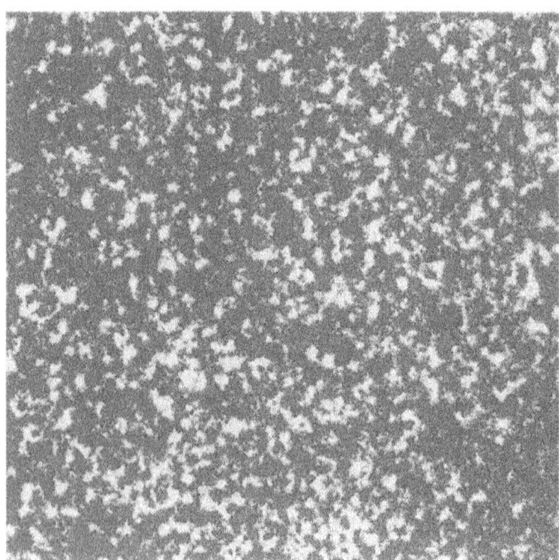

Fig. 1 Subimage of the original micrograph geometrically transformed to match the theoretic filter.

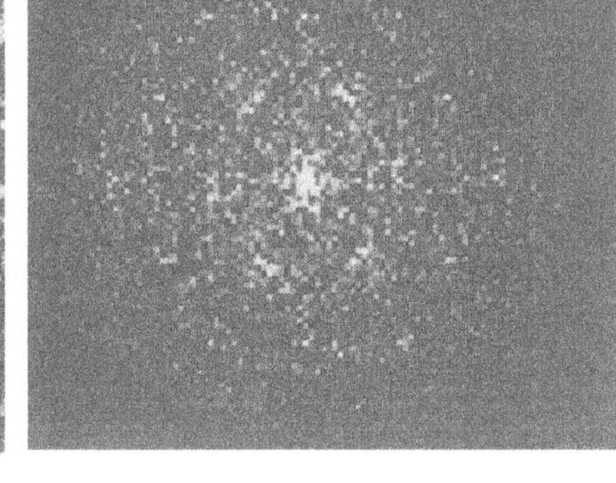

Fig. 2 Modulus of the DFT of Fig. 1 in logarithmic scale (magnified x4).

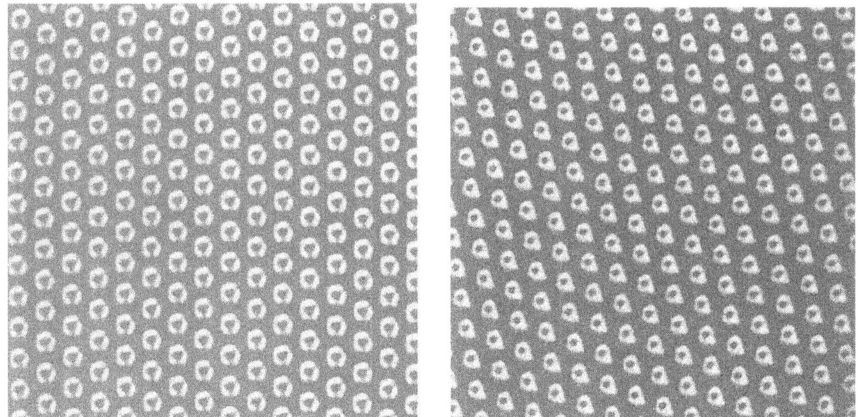

Fig. 3 Reconstruction of the first lattice.

Fig. 4 Reconstruction of the second lattice.

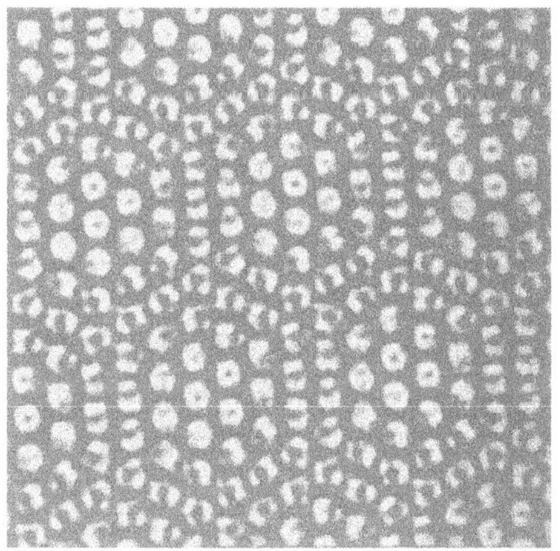

Fig. 5 Reconstruction of Fig. 1 obtained by addition of Fig. 3 and Fig. 4.

References

1. N.García, A.Santisteban and J.L.Carrascosa, Proc. 2nd Scandinavian Conf. on Image Analysis, 444 (1981).
2. S.S.Rifman and D.M.McKinnon, TRW report 20634-6003-TU-00 (1974).
3. J.W.Cooley and J.W.Tukey, Math. Comput. 19, 297 (1965).

Molecular Order of Ceramide Trihexoside (CTH) Inclusions of Fabry's Disease: Study Through Digital Image Processing and Optical Fourier Transformation

F. Delbarre, S. Laoussadi, H. Dupoisot, A. Constans, and G. Daury

Groupe d'Analyse d'Images Biomédicales, Université René Descartes
3, Boulevard Pasteur
F-75015, Paris, France

Since 1979, the structure of various biological membranes has been studied at the G.A.I.B. using optical means. In case the membrane elements are stacked together into lamellar thick inclusions a favourable statistical situation arises. In case of Fabry's disease such inclusions can be found in several tissues and organs, including heart and kidney. This disease is caused by an enzymatic α-galactosidase A deficiency, which is X-chromosome linked, and is quickly mortal to men (45 years). The inclusions consist of ceramide trihexoside molecules and water [1].

Such inclusions were photographed through an Elmiskop I Siemens transmission electron microscope (TEM) with a 40,000 X magnification. Tissue samples were fixed in glutaraldehyde (5%), and O_s O_4 (1%), dehydrated in different gradients of ethanol and embedded in Epon 812. Thin sections (75 nm) were obtained from a LKB III ultrotome, stained with uranyl acetate and lead citrate. Six tissues (mainly muscle and synovium) were examined, derived from nine males and females from five families. The age range was 12-45 for males and 18-70 for females. A thousand photographs were handled, fifty of them being digitalized and processed on an Univac 1110 computer or on a Comtal Vision 1/20 electronic processor in order to increase the visibility and signal-to-noise ratio. Two thousand optical Fourier transforms were performed with a laser set-up. In order to pull out the structural information these Fourier transforms were correlated with family, age, and type of tissue (Figure 1).

The great majority of publications related to Fabry inclusions only illustrate textural aspects (rectilinear or bent crystals, polygons,

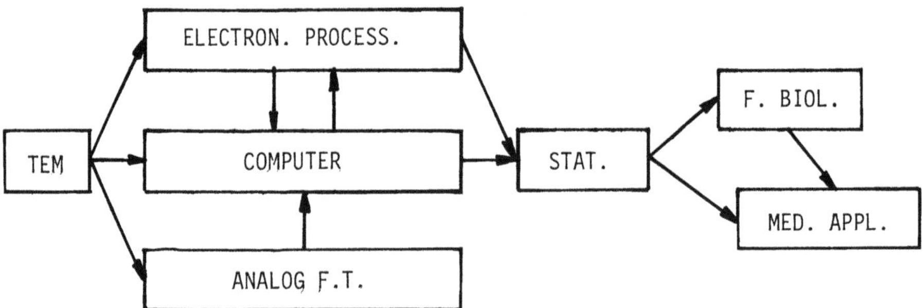

Fig. 1 Schematic diagram of the set-up used for biomedical information extraction

mosaics, etc. ...). Up to now, because of lack of scales along the lamellae, even with a good lamellar visibility, there is a tendency to show only the $L\alpha$ bilayer [2] and $L\alpha$ bilayer stacks (BLS), where each bilayer is supposed to be separated from its neighbours by an intermediate free water layer. The observed differences are associated with cleavage incidence variations while cutting the sample. Since 1979 [3] it has become obvious that, besides the well-known $L\alpha$ structure there are simultaneously (Figure 2) $L\beta$ or $L\beta'$ structures [2], $L\beta$ or $L\beta'$ polybilayers (PBL), $L\gamma$ structures [2], $L\gamma$ double-bilayer stacks (DBLS), and various polybilayer stacks (zebra bodies). Each structure has a characteristic, well-defined repeat distance or longitudinal period. There are various phase transitions depending on tiny chemical or physical variations. Because of their higher compactness, $L\beta$, $L\beta'$ structures and zebra bodies form the inclusion core. The inclusion wall consists of $L\beta$ and $L\gamma$ structures because of their efficient impermeability to water. The core thickness is generally ten times larger than the wall thickness, which justifies that we give preference to the core in our statistics.

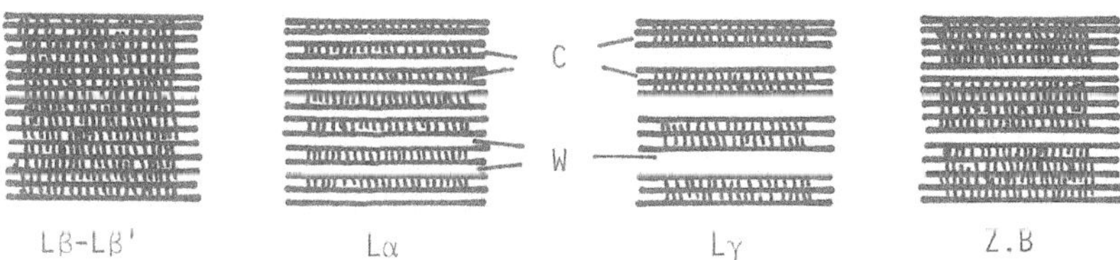

Fig. 2 The various CTH lamellar structures. C : céramide ; W ÷ free water

The image processing techniques used in this study were image smoothing (convolution), filtering (amplitude selection in the reciprocal plane), and density straightening (elimination of density gradients or low periods). Smoothing and filtering help to eliminate the image grain. Density straightening is necessary for display and plotting because of the non-uniform sample thickness and electron illumination. Since $L\beta$ or $L\beta'$ polybilayers appear as grating-like objects of a cosine transparency profile there is no problem for the spectrum interpretation in the Fourier transforms. In order to avoid the lamellar undulations and the non-perpendicularly cut lamellae, caution has to be taken, i.e. to compensate the Young fringes in the reciprocal plane whenever two or more crystals of a given period and orientation are present in a given field. The crystal dimension has to be taken into account in the statistics.

Image processing on one side indicates a possible oblique orientation in the statistical distribution of the ceramides, thus proving the existence of the $L\beta'$ structure in inclusions, and on the other side increases the visibility of transverse periodicities (along the lamellae). The values of these transverse periodicities are multiples (1 to 4) of 1.4 nm, leading to <d> = 1.4 nm for the mean transverse polar diameter [2]. Fourier transform, however, looks more attractive than image processing, particularly if Fourier values are correlated together and if longitudinal period histograms are computed. Tissue histograms generally reveal a faintly visible $L\beta$ - $L\beta'$ multipeak spectrum of repeat distances ranging from 5.8 to 2.6 nm, the main peak of which being located at 5.2 nm. The minimum of the

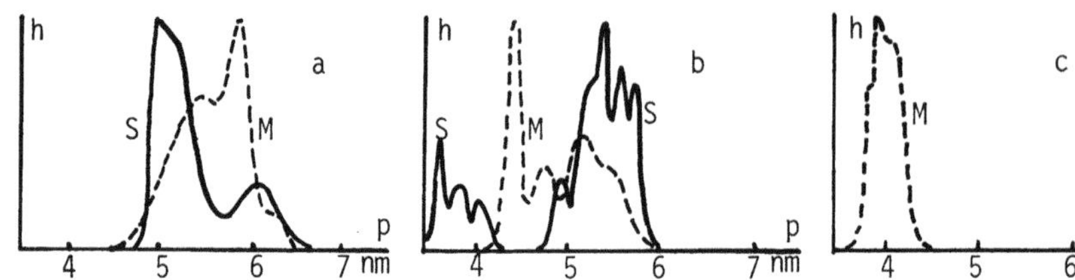

Fig. 3 Histogram of the Lβ or Lβ' repeat distances. a : youth ; b : middle age ; c : end of life. M : muscle. S : synovium

period does not depend on the microscope resolution. Higher values are observed if hydrophobic impurities such as proteins are present in the ceramide lamellae ($p < 7$ nm). Main results of our study are : a) the existence of multiple and well-discriminated periods, b) the influence of age, which induces a step by step reduction of the period (Figure 3), c) the influence of sex, which does not apply to the overall spectrum or to elementary values, but only to the speed of the reduction; this phenomenon is more rapid within males and justifies the statistics of an earlier death in males than in females.

The electron microscope, used so far to detect the presence of Fabry's disease, appears now in combination with image processing techniques as a good tool to follow up the evolution of the disease within a given patient. Since some of our cell membranes may undergo such an evolution, consequently, image analysis could help to understand our ageing process.

REFERENCES

1. Sweeley, C.; Klionsky, B.; Krivit, W.; Desnier, R.; in Wyngaarden, J.; Stanbury, J.; Frederickson, D.: The Metabolic Basis of Inherited Diseases, Third Ed., McGraw Hill, N-Y, (1972) p. 663.

2. Tardieu, A.: Etude cristallographique de systèmes lipides-eau. Thèse de Doctorat ès Sciences, Paris-Sud, (21/3/1972), n 915, Série A, 122 p.

3. Dupoisot, H.; Constans, A.; Daury, G.; Delbarre, F.; Laoussadi, S.: Comptes Rendus Acad. Sc., (Paris), (1979), 288 D, p. 783.

Hierarchical Description of Image Structure

J.J. Koenderink, and A.J. van Doorn

Department of Medical and Physiological Physics
Physics Laboratory, State University Utrecht, Princetonplein 5
3584 CC Utrecht, The Netherlands

1. The Structure of Smooth Halftone Images

Let the image be given as a smooth illuminance distribution $I(\vec{r})$. Then we define the ε-level set $I_\varepsilon(\vec{r})$ as $I_\varepsilon(\vec{r}) = \{\vec{r} | I(\vec{r}) \leq \varepsilon\}$. The boundary of the ε-level set $\partial I_\varepsilon(\vec{r})$ is just the ε isophote. We require that the image is generic, that is, stable against infinitesimal perturbations (any *real* image is generic!). A *regular* point of the image is a point at which $\vec{\nabla} I(\vec{r})$ does not vanish, if $\vec{\nabla} I(\vec{r}_0) = 0$, $\vec{r}_0$ is a *critical* point and $I(\vec{r}_0)$ a critical level. For a generic image the critical points are isolated and all critical levels are distinct. Moreover, at a critical point the Hessian

$$\left[\frac{\partial^2 I}{\partial x^2}\frac{\partial^2 I}{\partial y^2} - \left(\frac{\partial^2 I}{\partial x \partial y}\right)^2\right]$$

does not vanish. Thus the critical points are simple extrema or saddle points. These are basic results of Morse theory.

Now let ε_1, ε_2 be two subsequent critical levels. Then any two isophotes ∂I_μ and $\partial I'_\mu$ with $\varepsilon_1 \leq \mu, \mu' \leq \varepsilon_2$ are homotopic. This means that there is a smooth mapping ψ: $R^2 x[0,1] \rightarrow R^2$, such that

$$\psi(0) = \partial I_{\varepsilon_1} \quad ; \quad \psi(1) = \partial I_{\varepsilon_2} \quad \text{and}$$

$$\psi(\xi) = \partial I_\mu \quad ; \quad \psi(\xi') = \partial I_\mu{}' \quad \text{for some} \quad \xi, \xi' \in [0,1] \ .$$

All these isophotes are topologically equivalent, and can be sketched in qualitatively if only I_{ε_1}, I_{ε_2} are given. The region $I_{\varepsilon_1} \cap I_{\varepsilon_2}$ is completely featureless. By way of the critical levels the whole image can be subdivided into such featureless regions.

The isophotes $\partial I_{\varepsilon_i}$ of the critical levels ε_i consist either of an isolated point (this is the case when the corresponding critical point is an extremum) or a loop with one self-intersection (for saddle-points). There are two types of these loops, so that the canonical foliation induced by the critical levels

generically looks like

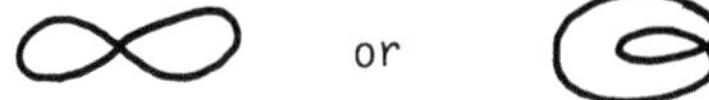

or even more complicated:

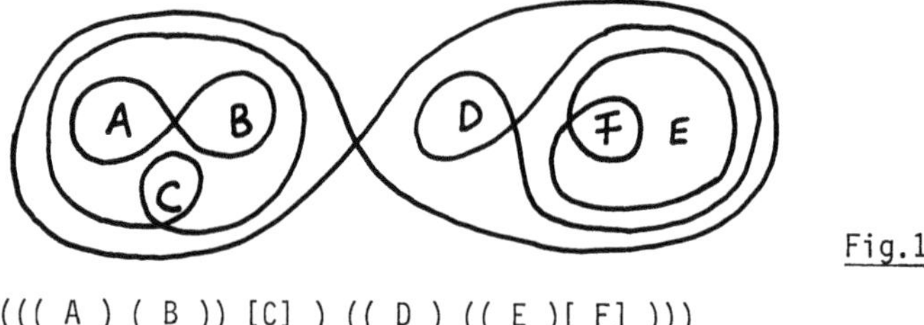

Fig.1

((((A) (B)) [C]) ((D) ((E)[F])))

Every folium can conveniently be represented by a pair of brackets ( ) if the function $I(\vec{r})$ decreases towards the inside (a dark blob) and [ ] if the function increases towards the inside (a light blob).

Then the structure formula gives the structure of the image in terms of a hierarchy of nested and juxtaposed light and dark blobs in an obvious manner. A property table may be added, to specify amplitude, size, shape for every bracket pair. Then *logical* filtering of the image is just a matter of deleting brackets.

2. Results of Perturbations

Because the structure formula is a discrete entity, it can change only in finite steps. Consequently any smooth perturbation (of dimensions, gray scale, resolving power) cannot change the structure unless it exceeds a certain threshold; the description is *stable* against any infinitesimal perturbation. This is a very valuable property of this description.

As one important perturbation you may regard a change of resolving power. The image remains essentially invariant for certain ranges of resolving power, but at certain levels the image changes suddenly through processes like (( )( )) ↔ ( ) or [( )[ ]] ↔ [ ]. It is even feasible to specify the structure formula for *all* resolving powers — in practice this boils down to only a limited number of structure formulas. This set, graded with respect to the critical resolving powers, gives a very complete description of the image at all levels of resolution simultaneously.

3. Images Degraded by Photon Noise

Consider a typical scintigram, which is essentially a finite set of points $\vec{r}_i$ — the points where photon absorptions occurred. At a level of resolution Δ, we sample the image through a circular aperture of diameter Δ, say. It is convenient to tesselate the image plane through the construction of all circles of diameter Δ, centered on the points $\vec{r}_i$. The resulting tessela are regions where a certain constant number of photons is detected if they contain the center of the aperture.

On this discrete structure we can again define ε regions, and we can again obtain the canonical foliation. The only difference is that now not all critical levels need be distinct. This is no problem if we admit the possibility of structures like (( )( )[ ]).

Fig.2

Another problem is that different critical levels need not be different in a statistically significant way. This is because of the photon noise. Critical levels N_1 and N_2 differ only if

$$|N_1 - N_2|^2 > \sigma(N_1 + N_2) \quad .$$

Here σ sets the desired level of significance. Thus we must simplify the structure formula. If bracket pairs are nested and the critical levels are not significantly different, we just delete the innermost pair. Care must be taken because of three nested pairs $(_A(_B(_C\ )))$, pairs A, B and B, C may not be significant, whereas A, C *is* significant.

In such a way we obtain a structure formula at a level of resolution Δ that describes only significant structure. Repeating this procedure for different levels of Δ, we again obtain a finite number of structure formulas that describe the structure at *all* levels of resolution simultaneously. In this case a given detail may not be visible at either coarse apertures (because of spatial resolution) or very fine apertures (because of increased photon noise). The method quarantees the extraction of all significant detail from a scintigram and as such is clearly superior to all methods that use a fixed aperture, no matter how optimally chosen.

On the Optimization of Scintillation Camera Collimators with Parallel Holes

M.J. Yzuel

Departamento de Física Fundamental, Cátedra de Optica
Universidad de Zaragoza, Spain

A. Hernández

Departamento de Radiología y Medicina Física, Sección de Física
Universidad de Zaragoza, Spain

1. Introduction

In the optimization of a collimator of a scintigraphic system there are two main features to take into account: the spatial resolution and the sensitivity. In general the collimator design has been studied independently from the other components of the system and it has been tried either to improve one of these characteristics, generally the sensitivity for a given resolution, or to find a compromise between them [1,2,3]. SIMMONS et al. [4] proposed an optimization method based on the sentivity and on the MTF of the total system. It was applied to objects with the shape of infinite cylinders, placed at a given distance from the collimator. ROLLO [5] introduced an index to compare the performance of scintigraphic imaging systems. This index combines the contrast efficiency function with the plane sensitivity.

In this paper we study an optimization method based on the performance index introduced by ROLLO [5]. Taking into account the scattering medium, the collimator, and the detector, an overall index has been evaluated. During this procedure the collimator parameters vary. An optimum collimator is studied for each particular case (lesion size and object-collimator distance). Finally an optimum collimator for the ensemble of all the situations is obtained.

2. Performance Index

The performance index Φ introduced by ROLLO [5] is the product of the contrast-efficiency function and the square root of the plane sensitivity. The former is the ratio of the image contrast (c_i) to the object contrast (c_o) for a given lesion and the sensitivity measures the response to a plane source. Let us say

$$\Phi = (c_i/c_o)\ \sqrt{S_p} = \left[\int_0^\infty \nu L(\nu)\ \mathrm{MTF}(\nu) d\nu \Big/ \int_0^\infty \nu L(\nu) d\nu \right] \sqrt{S_p} \qquad (1)$$

where ν is the spatial frequency, $L(\nu)$ is the Fourier transform of the emission of the lesion and $\mathrm{MTF}(\nu)$ is the overall transfer function of the scintigraphic system.

Although we will deal with spherical lesions in this paper because they represent the general case, the definition of the performance index (1) is applicable, as observed in [5], to other lesions having any geometry which can be described mathematically.

3. Scattering Medium

In practice, a scattering medium or tissue fills the space between the source and the collimator. If the system is linear, the medium can be studied as a component by introducing its own MTF_M by the expression

$$MTF_M(\nu) = MTF(\nu) / (MTF_C(\nu) \, . \, MTF_D(\nu)) \tag{2}$$

where $MTF(\nu)$, $MTF_C(\nu)$, and $MTF_D(\nu)$ are the overall transfer function, the collimator transfer function and the detector transfer function respectively.

We have obtained experimentally the MTF as the Fourier transform of the image of a linear source. It has been done for the scintigraphic system with a medium (equivalent to the human tissue) placed between the source and the collimator and for the same system and the same source-collimator distance, without the medium. The quotient of both functions will give the transfer function of the medium. The measurements have been repeated for different thicknesses of the medium, different source-collimator distances and different collimators. It is observed that a $MTF_M(\nu)$ can be associated to each thickness of the medium and it can be obtained from the $MTF_M(\nu)$ corresponding to a thickness of 1 cm.

The influence of the medium on the system sensitivity has also been studied. The variation depends on the medium thickness and it can be introduced to the sensitivity as a factor.

4. Optimization Method

The performance index evaluated by (1) is taken as the figure of merit in the optimization procedure. The optimization program is based on the gradient method [6]. The overall $MTF(\nu)$ is obtained in each step by

$$MTF(\nu) = MTF_M(\nu) \; MTF_C(\nu) \; MTF_D(\nu) . \tag{3}$$

$MTF_M(\nu)$ is obtained according to the description of the last paragraph and it is introduced in the optimization program as input data.

The collimator $MTF_C(\nu)$ is computed from the collimator data that vary along the procedure. It is evaluated in the geometric approximation according to the expression given by METZ et al. [7], that, for circular holes of radius R, is

$$MTF_C(\nu) = [J_1(2\pi\alpha R\nu)/\pi\alpha R\nu]^2 \tag{4}$$

where J_1 is the first order Bessel function and $\alpha = 1 + (Z+B)/t$, with Z being the distance from the source to the collimator entrance surface, B the distance from the collimator exit surface to the detector and t the collimator thickness.

The geometric approximation is applied along the optimization procedure because the penetration is very small for low energies (less than 200KeV), and the computation, considering the septal penetration, would need much longer computing time.

We have taken as $MTF_D(\nu)$ in (3) the detector MTF of an OHIO Nuclear Sigma 400 gamma camera.

The system senstivity is computed in the optimization procedure by the geometric efficiency given by ANGER [8] modified by the correction factor due to the medium.

5. Results

Although the method is general, we have applied it to collimators with circular holes in a hexagonal distribution. In the optimization procedure the hole radius and the collimator thickness vary and we keep fixed the sum of the radius and the septal thickness. For the results given in this paper this value is 0.1125 cm. That corresponds to one of the collimators of the mentioned gamma camera. Along the optimization procedure the septal thickness can decrease but the constraint given by ANGER [8] that the minimum way through the collimator material (excluding hole edge effects) corresponds to an attenuation superior to 95% is maintained.

The lesions are considered spherical with diameters between 1 and 3 cm. The object-collimator distances are taken from 1 to 10 cm and the space between the object and the collimator is supposed to be filled with medium.

5.1 Collimator Optimization for Particular Conditions

Table 1 shows the results obtained for the collimator parameters when the performance index is maximized for each lesion size and distance. It is observed that for a given distance the hole radius and the collimator thickness decrease if the lesion size increases. Also for a given lesion size the hole radius and the collimator thickness increase if the distance increases. Besides, for a given distance the performance index is higher for bigger lesions and for a given lesion size the performance index is better for smaller distances.

Table 1 Optimum collimators parameters

Lesion-collimator distance [cm]	Lesion radius [cm]	Collimator thickness [cm]	Hole radius [cm]	Normalized performance index
5	1.5	0.88	0.069	1
5	1.0	0.97	0.072	0.79
5	0.5	1.43	0.082	0.37
10	1.5	1.17	0.077	0.45
10	1.0	1.55	0.084	0.31
10	0.5	2.60	0.094	0.12

5.2 Collimator Optimization for General Conditions

In general, a gamma camera is not provided with many collimators. Then, the optimization technique must lead to an optimum collimator for the ensemble of the usual lesion sizes and distances.

We have studied two figures of merit: the sum of the performance indices for each particular situation and the sum of the inverses of the indices. The method must maximize the figure of merit in the first case and minimize it in the second one.

As an example we give the results for the case of considering three lesion sizes (radius equal to 0.5, 1 and 1,5 cm) being each one able to be placed at 5 or 10 cm from the collimator entrance surface. With the first figure of merit, an optimum collimator of 0.0754 cm of hole radius and 1.10 cm of thickness is obtained. If the second figure of merit is applied the optimum collimator parameters are: hole radius equal to 0.0877 cm and thickness equal to 1.82 cm. Comparison of these collimators with those of Table 1 shows that the parameters of the optimum collimator for the ensemble are in each case more similar to the parameters obtained for the particular conditions which have greater influence on the figure of merit considered.

Nevertheless, if the different terms of the figure of merit are weighted in such a way that the contributions of the particular situations on the overall figure of merit are equal, both methods lead to the same collimator: hole radius 0.082 cm and collimator thickness 1.42 cm.

In general the imposition of less detectable conditions (small lesions and large distances) leads to thicker collimators which means better resolution and less sensitivity. The optimization procedure based on the performance index tries to compensate this fact by increasing the hole radius. Even so, we have verified analytically that the contrast-efficiency of the system increases, along the procedure.

Finally, we have compared, for the different optimum collimators described previously, the MTF in the geometric approximation with the MTF considering the septal penetration. The latter has been obtained from the PSF given by the Fourier analysis method [9]. The effect of the septal penetration is very small for low energy and it justifies the use of the geometric collimator MTF in the optimization technique.

6. References

1. H.O. Anger, J. Nucl. Med., 5, 515 (1964)
2. R.N. Beck, in "Medical radioisotope scanning" (Vienna:IAEA), vol. 1, 35 211 (1964)
3. W.G. Walker, in "Medical radioisotope scintigraphy" (Vienna:IAEA), vol.1 545 (1969)
4. G.H. Simmons, J.M. Christenson, J.G. Kereiakes and G.K. Bahr, Phys. Med. Biol., 20, 771 (1975).
5. F.D. Rollo, J. Nucl. Med., 15, 757 (1974)
6. P.R. Bevington "Data reduction and error analysis for the physical sciences" (New York: McGraw Hill Co. 1969).
7. C.E. Metz, F.B. Atkins, and R.N. Beck, Phys. Med. Biol., 25, 1059 (1980)
8. H.O. Anger, in "Instrumentation in nuclear medicine" ed. G.J. Hine (New York: Academic Press 1967) vol. 1, 485
9. S. Miracle, M.J. Yzuel, and S. Millán, Phys. Med. Biol. 24, 372 (1979)

Computer Analysis of Compton Scattering Images: Acquisition, Restoration and Application in Patients with Pulmonary Oedema

L. Azzarelli, M. Chimenti, F. Denoth, and F. Fabbrini
CNR Istituto di Elaborazione dell'Informazione, Pisa, Italy

M. Pistolesi, M. Miniati, S. Solfanelli, C. Giuntini
CNR Istituto di Fisiologia Clinica e 2° Clinica Medica Università di Pisa, Italy

1. Introduction

The development of methods for non-invasive inspection of the internal organs of a living body is among the most important aims in the biomedical field, and many different solutions have already been suggested.

In Pisa, at the Istituto di Elaborazione dell'Informazione, in cooperation with the Istituto di Fisiologia Clinica, a tomographic method that allows the visualization of frontal and sagittal planes of a chest was developed, in order to detect regional lung density changes due to pathological processes [1]. Sectional visualization of the chest is obtained by employing a collimated linear source of gamma photons and a gamma camera - as imaging device - to detect the Compton scattering at 90° to the primary beam. Imaging 90° scattered rays provide a view of the anatomical cross-section corresponding to the path of the primary rays in chest tissues. Chest tomographic views with 500 K counts are obtained in 1 or 2 min with the spatial resolution of the gamma camera and with tissue density discrimination of 12% on average, as shown by phantom studies. The maximum estimated radiation dose to the patient is about 0.09 rads for each view. This radiation load is comparable to that of standard chest x-ray and considerably less than that of conventional or computerized x-ray tomography.

This technique was applied to 15 patients with pulmonary oedema, a clinical condition in which lung density increases by accumulation of extravascular lung water. In cardiogenic (hydrostatic) pulmonary oedema, Compton scatter tomography shows a predominantly central perihilar distribution of the increased lung density, whereas in injury (increased permeability) pulmonary oedema, a patchy more peripheral and less gravity dependent distribution is observed.

Compared to chest x-ray, the most used clinical method to assess pulmonary oedema, Compton scatter tomography is more accurate in defining the above abnormalities and more sensitive in detecting lung density with time. In conclusion, Compton scatter tomography may be considered an adjunctive diagnostical tool for detecting and monitoring pulmonary oedema.

2. Image Production

Chest images are obtained irradiating a biological target by means of a collimated Ir^{192} linear gamma source and detecting the fluorescence of 90° scattered photons. As the intensity of diffracted beams depends linearly on the density of scattering tissues, a map related to the density distribution of the irradiated section is obtained.

Original views are affected by several artefacts and systematic and random errors:

- incident beam attenuation;
- 90° Compton scattered beam attenuation;
- divergence and non-uniformity of the radiation source;
- detector spatial response;
- room and detector noise;
- multiple Compton scattering.

In addition, another source of error is the photographic process. In fact, at this stage of the work, images formed on the gamma camera display are photographed, digitized by means of a flying spot device [2], into 512x512 matrices at 20 ℓ/mm resolution,and then analyzed.

The processing procedures performed on original images reduce the listed errors. Simple approximation algorithms have been used for two main reasons: the lack of precision in measured data (which is particularly due to the use of a gamma camera), and the necessity of using fast and simple procedures which may be implemented into small-sized, microprocessor based systems.

3. Image Correction

Before correcting the incident and scattered beam attenuation and the image degradation due to multiple Compton scattering, we process the images in order to reduce the above listed errors.

Noise due to the source-detector set and the photographic process is reduced by means of appropriate digital filtering.

Distorsions due to the non-uniformity of the incident radiation and that of the detector spatial response are corrected by normalizing the picture matrix to be analyzed. In order to do that, matrices are used, experimentally obtained by comparison of the spatial distribution of the incident beam and of the gamma camera response.

A model of the multiple Compton scattering was developed in order to derive algorithms which can be applied to clinical data. It was shown [3] [4] that the multiple scattering contamination is a function of depth in the irradiated target. The effect due to this phenomenon in the image production is a reduction of the spatial and photometric resolution, which itself is a function of the distance covered by the incident beam in the subject. Assuming the multiple Compton scattering contamination as a spatial modulation degrading the image contrast in a linear manner, a restoration is performed (Fig. 1 and Fig. 2).

An iterative algorithm was developed [5] to correct the incident beam attenuation. The algorithm applied to each element of every row of the picture matrix, keeps into account the attenuation of every preceding element. Because such an algorithm needs too many computation steps, and consequently a long processing time, a linear regression procedure is necessary which takes into account that the method used is based on optical density values on a logarithmic scale. Such an operation allows to normalize values corresponding to the chest walls. In fact, chest walls originally show the same density, and thus they should have the same value after correction. The entry values are then modified linearly along the primary beam direction according to the given conditions.

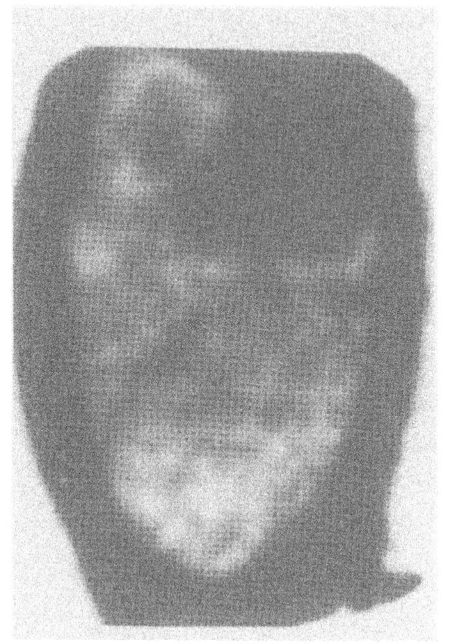

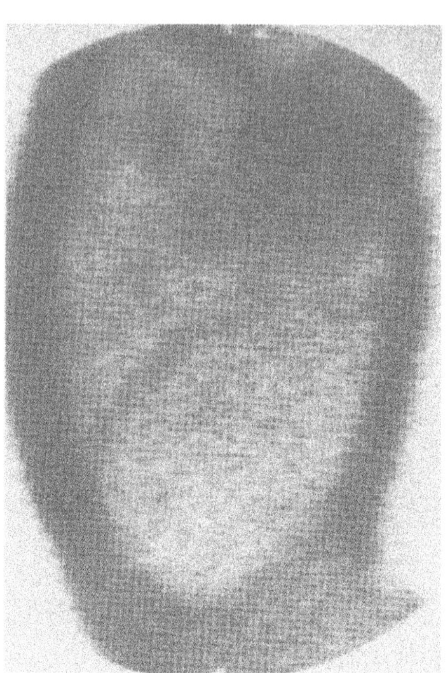

Fig.1 Digitized original image of chest sagittal cross-section

Fig.2 Processed image

In a similar manner the correction for the scattered beam attenuation can be performed. Because, in this case, there is no information about actual densities of first and last sections, we take as reference the matrix corresponding to the section closest to the gamma camera, neglecting the attenuation due to the initial tissue layer. The matrix - values of each following section are modified by the values of the preceding one, and so on (Fig. 3 and Fig. 4).

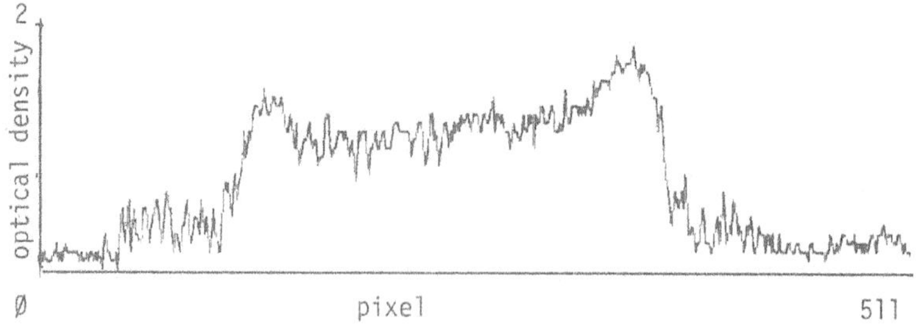

Fig. 3 Plot of row # 256 of Fig. 1

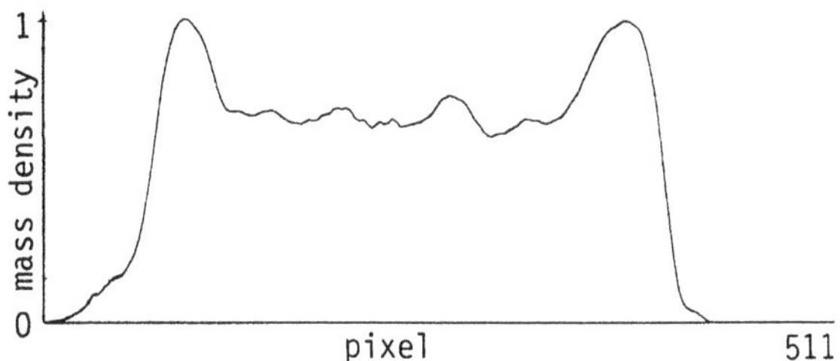

Fig.4 Plot of row # 256 of Fig. 2

4. Conclusion

The processing algorithms used are partially interactive and heuristic. They are the first step in setting up a system which automatically restores gamma camera images. In spite of the simplifications introduced, experience has shown that the obtained results are a useful tool for the physician's diagnosis.

Reference

1 L. Azzarelli, M. Chimenti, F. Denoth, F. Fabbrini, C. Giuntini, M. Pistolesi, M. Miniati: Tomographic Imaging of Regional Lung Density by 90° Scattering. Computers in Critical Care and Pulmonary Medicine, Lund, June 1980

2 L. Azzarelli, M. Chimenti: A Digital System for Image Processing. Mem. S.A.It. (1979) 305-321

3 J.J. Battista, L.W. Santon, M.J. Bronskill: Compton Scatter Imaging of Transverse Sections: Corrections for Multiple Scatter and Attenuations. Phys. Med. Biol., 2(1977) 229-244

4 A. Del Guerra, R. Bellazzini, G. Tonelli, R. Venturi, W.R. Nelson: Application of the Method of Montecarlo to Compton Scattering Radiography in Homogeneous Media. A.I.F.B. Congress, March 16-18, 1981, Firenze

5 L. Azzarelli, M. Chimenti, F. Denoth, F. Fabbrini: Chest Tomography by Compton Scattered Radiation: a Heuristic Enhancement and Restoration. Proceedings of the International Conference on Image Analysis and Processing, October 22-24, 1980 Pavia, pp.141-147

3-D Time Gated Transaxial Tomography in Nuclear Medicine

J. Brunol, N. de Beaucoudrey
Institut d'Optique, Université de Paris XI
F-91405 Orsay, France

1. Introduction

Classical or static transaxial tomography has now become a powerful tool in nuclear medicine [1-3]. However, the necessary limitation in the radioactive dose the patient is exposed to, leads to low statistical quality of the reconstructed slices. Thus, dynamic imaging appears to be very difficult due to the reduction of activity in each individual frame. We present a new four-dimensional process which works simultaneously in the space and time domain and enables time-resolved axial tomography, particularly, in cardiac gated dynamic studies.

2. Emission Computed Axial Tomography: ECAT

ECAT is based on collecting parallel projections of emitted gamma rays, by rotating an Anger camera around the patient. These projections are collected at several heights simultaneously, each slice is reconstructed independently.

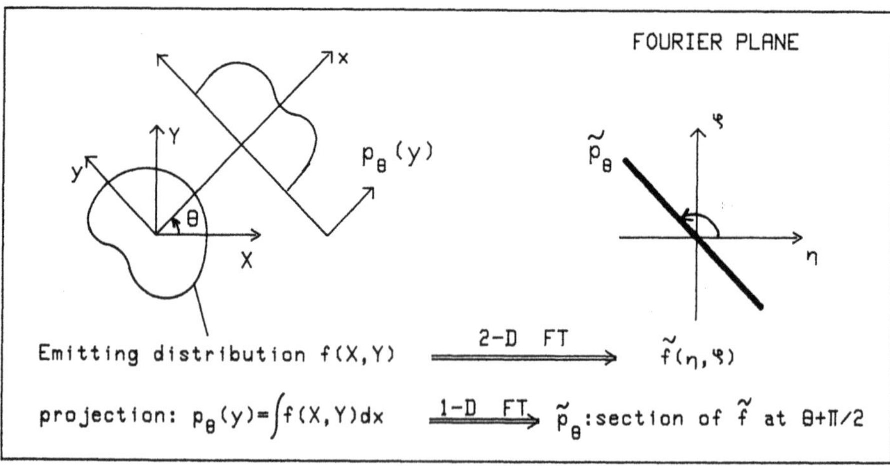

Fig.1 Central Slice Theorem.

Let $f(X,Y)$ be the activity in the slice we want to reconstruct. Neglecting the attenuation of the gamma radiation by the body, the projection at angle θ is given by:

$$p_\theta(y) = \int_\Delta f(X,Y)dx .$$

According to the Central Slice Theorem, the one-dimensional Fourier transform of this one-dimensional projection, $\tilde{p}_\theta(\nu)$, is a section of the 2-dimensional Fourier transform of the object, $\tilde{f}(\nu \cos\theta, \nu \sin\theta)$ (see Fig.1). A complete set of projections in 2π provides all the information in the Fourier space ; by interpolation and inverse Fourier transform, the object $f(X,Y)$ can be reconstructed : this method is called Fourier synthesis [4].

Many other techniques exist for the reconstruction of the slice ; the one we use is filtering of projections followed by back-projection and summation at the different angles.

We do not develop here the correction of self-absorption in order to concentrate our attention on the problem of statistical fluctuations.

3. Statistical Fluctuations in ECAT

Statistical fluctuations are very important in emission imagery because of limitations of patient dose. The main feature is that photon emission has a Poisson distribution. An approximate value of the averaged signal-to-noise ratio in ECAT is given by [6],

$$SNR = \frac{N_o^{1/2}}{n^{3/4}}$$

where n is the number of pixels in the slice and N_o the total number of collected photons. But this equation is valid only for a homogeneous object. In fact, in each point, the noise and therefore the SNR depend on all surrounding points.

Thus, the "noise image", i.e. the variance in each point, can be calculated [6]. (It can be shown that the noise image is nothing but the image obtained by back-projection without filtering).

Figure 2 represents the reconstruction of a slice of lung (Fig. 2-a) and the associated noise image (Fig. 2-b). We can notice that areas without signals are not exempted from noise, because they are affected by nearby areas of high activity.

Therefore, the statistics are good only where the signal is large. We define the local SNR to be the ratio of the signal in each point to the square-root of the variance at the same point. It indicates the significance of the result for each region of the image, particularly for areas of low activity.

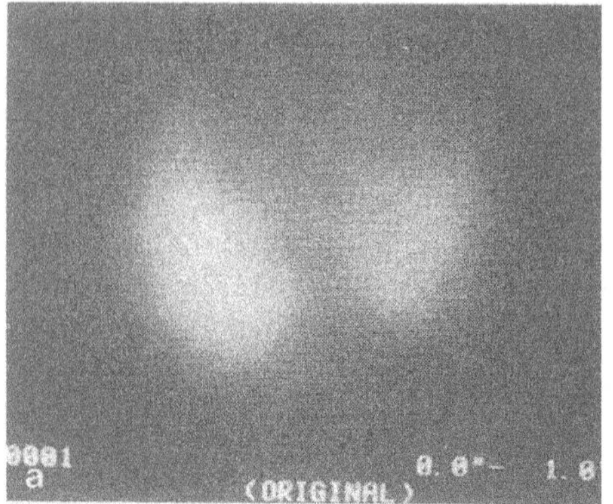

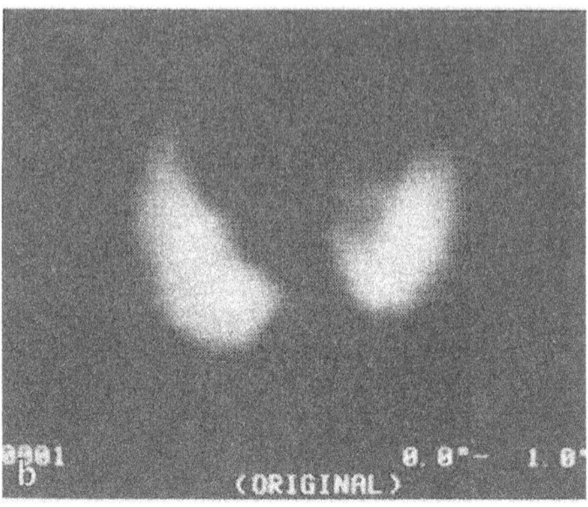

Fig.2 Slice of lung (a) and the associated noise image (b).

4. Dynamic Cardiac Imaging

ECAT, as it is used at present, is suitable for static imagery only, because of the time needed for a whole rotation around the patient. Therefore, it cannot be used without blurring for moving parts of the body, unless signal acquisition can be synchronized to the object motion. Such synchronization is possible in cardiac investigations: the electrocardiogram allows dividing the cardiac cycle into several intervals and storing projections at different moments of the cycle in different memories, using direct gating of the projection acquisition by the ECG.

Thus, several slices at different moments of the cardiac cycle can be reconstructed. However, the total number of photons, which are available to reconstruct a single slice at a given time, is reduced compared to a static examination.

As previously discussed, for the most active zones such as the myocardium, the SNR is sufficient in order to detect the temporal variations, whereas for zones of low activity, such as the borders and the background, the image quality is insufficient for a diagnosis due to the low SNR.

Thus, a four-dimensional processing is performed according to the SNR level: the worse the SNR, the stronger the temporal filtering performed.

The following mathematics are used: if $f(x,y,z,t)$ denotes the activity at point $M(x,y,z)$ and at time t, filtering is performed by multiplying the one-dimensional temporal Fourier transform $\tilde{f}(x,y,z,\alpha)$ by the transfer function

$$g = \frac{\tilde{f}^2}{\tilde{f}^2 + \varepsilon} \quad , \text{ where} \quad \varepsilon = \tilde{f}(x,y,z,o).$$

The final result is obtained by inverse temporal Fourier transform.

"In vivo" results obtained with a floating point AP 1208 array processor in the Hospital Cochin in Paris show the importance of this four-dimensional treatment.

References

1 - H.H. BARRETT and S.W. SWINDELL, Proc. Inst. Elect. Electron. Eng. 65, 89 (1977).
2 - F.F. BUDINGER and G.I. GULLBERG, I.E.E.E. Trans. Nucl. Sci. 21, 1 (1974).
3 - R.A. BROOKS and G. DI CHIRO, Radiology , 117, 561 (1975).
4 - D.B. KAY, J.W. KEYES and W. SIMON, J. Nucl. Med., 15, 981 (1974).
5 - N. de BEAUCOUDREY, J. BRUNOL and J. FONROGET, Opt. Commun., 30, 309 (1979).
6 - L. GARNERO, Thèse de 3è cycle (Orsay) (1981).

Image Processing of Cancer Radiographs

K. Matsuoka, Y. Ichioka, and T. Suzuki

Department of Applied Physics, Faculty of Engineering, Osaka University
Yamadaoka 2-1, Suita, Osaka, Japan

1. Introduction

Presently, medical radiographs play an important role in detection of cancers, especially of those with high mortility rates, such as mammary, gastric, and lung cancers. Nevertheless, the important information required for diagnosis, contained in a medical radiograph, is often indistinct and sometimes hardly detectable for human eyes which seems to be one of the obstacles in finding a cancer in its early stage.

The purpose of this paper is to investigate the applications of image processing techniques for diagnosis of mammary and gastric cancer. A synthetic judgement based on comparison of the processed images and the original radiographs enables a physican to diagnose the cancer more exactly with less labour as compared to the traditional diagnosis based on observation of the original radiographs only.

2. Image Processing

In a radiograph of a mammary cancer, there are two kinds of important structures characterizing the disease. One is the calcification image which shows white granular structures described by high frequency components. The other is the shadow of the tumor which is a relatively blurred part characterized by low frequency components. A radiograph of a mammary cancer is generally a low-contrasted and indistinct image in which both structures are buried in the background. In a radiograph of a gastric cancer, the important structures are the concentration of pleats of the stomach wall, the shapes of these pleats and the shape of the pool of contrast media around the diseased part. In an early stage of a gastric cancer the concentration of pleats is weak, and the pool is small. Therefore, the diseased part in a radiograph is extremely low-contrasted and it happens that the diagnostic information is obstructed by the backbone image in some cases. Thus, it seems that image processing such as contrast enhancement, image enhancement, or feature extraction is suitable for improvement of medical radiographs of mammary and gastric cancer.

2.1 Interactive Image Processing System

Medical radiographs of the same portion of a human body, show quite different image appearances and contrast according to the conditions when they are photographed. In order to manipulate these radiographs and to obtain the processed images with similar appearances and contrast independent of the recording conditions, man-machine interactive image processing is suitable.

We have developed a digital image processing system based on the minicomputer INTERDATA 7/32, which has the capability of man-machine interactive processing. Features of this system are the data acquisition device, the display device and the processing system, which is able to execute two-dimensional (2-D) convolution at high speed.

2.2 Image Processing Techniques

We have attempted five kinds of image processing methods:

(1) local intensity level expansion (LILE),
(2) local histogram equalization (LHE) [1],
(3) local average subtraction (LAS),
(4) adaptive binarization (AB) [2], and
(5) smoothing and first derivative (SD) [3].

Image processing of categories (1)-(3) results in contrast improvement and image enhancement, and that of category (4) provides feature extraction from the image.

Image contrast is not necessarily low over an entire image of a mammogram or a medical radiograph of the abdomen. However, contrast in a local

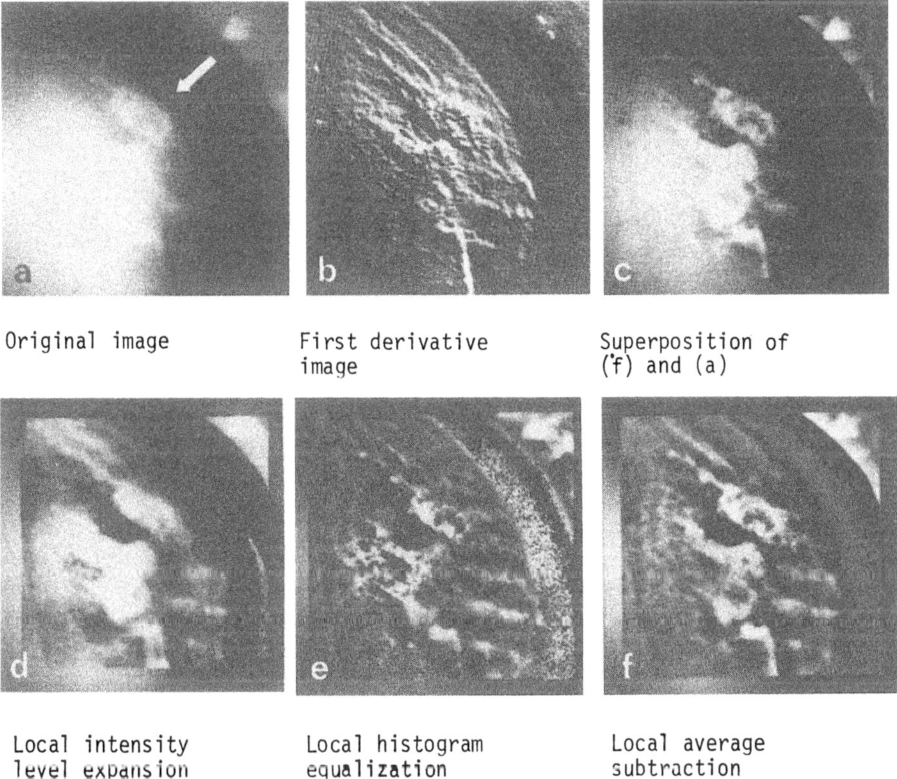

Fig.1 Experimental results of contrast improvement and image enhancement for a radiograph of a mammary cancer (each image has 256x256 pixels)

area containing a diseased part is generally insufficient to extract the desired information. Thus, to improve contrast of such fine structures in a local area, image processing stated in categories (1)-(3) is necessary.

3. Experimental Results

3.1 Mammary cancer

Figure 1 shows the processed results of image enhancement for a mammogram, using the techniques described in the previous section.

Figure 1(a) is the sampled version of an original mammogram, which has 256x256 pixels. Indistinct calcification images and the shadow of tumor can be seen in the area pointed by the arrow. Figures 1(b)-(f) show the images processed by (b) SD, (c) superposition of the image processed by LAS and the original image in (a), (d) LILE, (e) LHE, and (f) LAS, re-

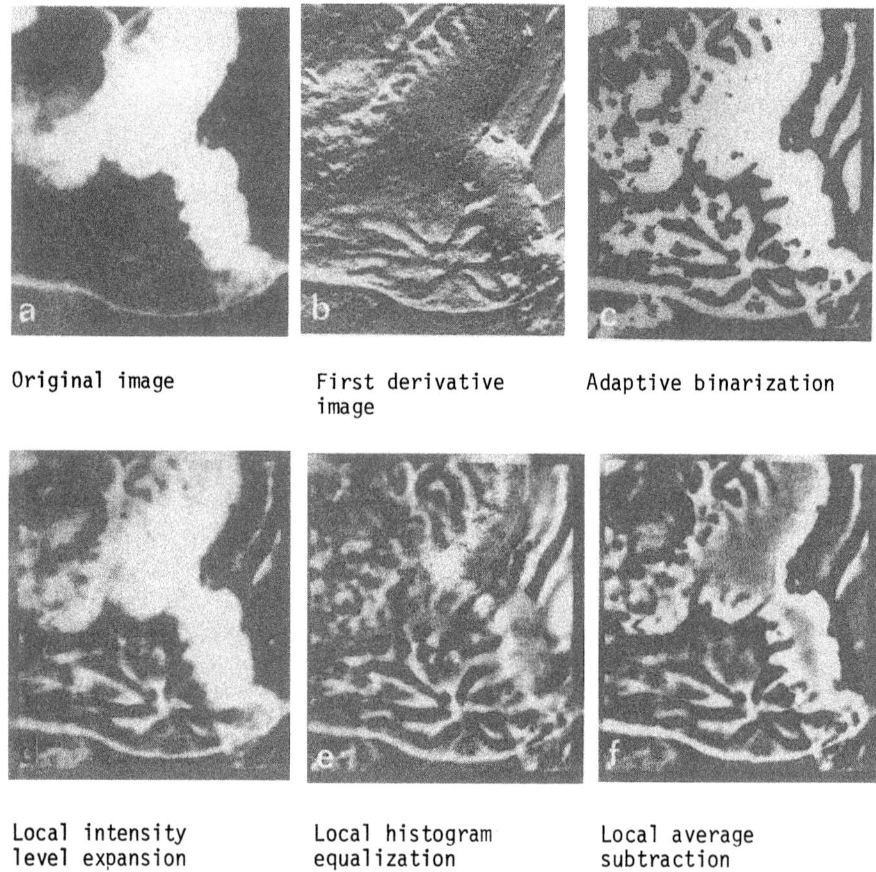

Original image — First derivative image — Adaptive binarization

Local intensity level expansion — Local histogram equalization — Local average subtraction

Fig.2 Experimental results of contrast improvement and feature extraction for a radiograph of gastric cancer (each image has 320x352 pixels)

spectively. From comparison of the original image and five processed results, it can be deduced that the calcification images are well extracted in Fig. 1(b) and (c), and the shape of tumor becomes discriminable in the contrast enhanced images in Fig. 1(d), (e) and (f). It seems that comparison of these processed images and the original one allows to extract characteristics of the calcification image and the shape of the tumor and thus assists diagnosis of mammary cancer.

3.2 Gastric cancer

Figure 2(a) ist the sampled version of a radiograph of a gastric cancer, which has 320-352 pixels. In this picture, the concentration of pleats of the stomach wall can be seen around the area pointed by the arrow. Figures 2(b)-(f) are the images processed by (b) SD, (c) AB, (d) LILE, (e) LHE, and (f) LAS, respectively.

4. Summary

We have attempted image processing of mammograms and radiographs of gastric cancer. Experimental results show to some extent the effectiveness of image processing for medical radiographs.

Acknowledgement

We are grateful to Dr. K. Nakanishi and Dr. T. Suzuki of the Center for Adult Diseases, Osaka, for providing the radiographs and for helpful discussions.

References

1. D. J. Ketcham: Proc. SPIE, Image Processing, 74, 120 (1976)
2. J. Tokumitsu, S. Kawata, Y. Ichioka, T. Suzuki: Appl. Opt. 17, 2655 (1978)
3. S. Kawata, Y. Ichioka, T. Suzuki: Optik 46, 235 (1978)

Assessment of Radiographic Image Quality by a Single Number in Terms of Entropy

H. Fujita

Department of Electrical Engineering, Gifu Technical College
Motosu, Gifu 501-04, Japan

S. Uchida

Department of Electrical Engineering, Faculty of Engineering, Gifu University
Kagamigahara, Gifu 504, Japan

1. Introduction

A new and simple method by which the image quality of radiographs could be evaluated by a single number in terms of entropy was introduced by UCHIDA et al. in 1978, and applied to the problem of evaluating the performance of development processes [1]. This method of evaluation, which is termed the entropy method, has been applied to various problems in radiology [2-7]. In the present paper, the assessment of radiographic image quality or granularity of screen-film systems used in diagnostic radiology by a single number in terms of the entropy is described on the basis of our recent paper [4], and some considerations on the conditional entropy are made.

2. Entropy Method

SHANNON has defined the average amount of information(entropy) by the formula

$$H = -\Sigma\, p_i \log_2 p_i \text{ [bits]}, \qquad (1)$$

where p_i is a discrete probability distribution of an event. If the entropy of the source or the input is $H(x)$ and that of the output is $H(y)$, then there are two conditional entropies of $H_y(x)$ and $H_x(y)$ in the presence of noise, namely, the entropy of the input when the output is known, and vice versa. The joint entropy $H(x,y)$ is defined by

$$H(x,y) = H(x) + H_x(y) = H(y) + H_y(x), \qquad (2)$$

and the transmitted information $T(x;y)$ can be obtained by

$$T(x;y) = H(x) - H_y(x) = H(y) - H_x(y)$$
$$= H(x) + H(y) - H(x,y). \qquad (3)$$

Consider an experiment in which every input has a unique output belonging to one of various output categories. A data matrix of frequency, as shown in Table 1, is employed for calculating the above entropies and the transmitted information. The columns and rows of this table represent various discrete inputs and outputs. The capitals X and Y stand for the number of input and output categories, respectively.The number of times input x_i is presented is symbolized by $n_{i.}$ and frequency, with which the input x_i corresponds to the output y_j, is given by n_{ij}. It is apparent from Table 1 that

$$\sum_j n_{ij} = n_{i.} \tag{4}$$

$$\sum_i n_{ij} = n_{.j} \tag{5}$$

$$\sum_{ij} n_{ij} = \sum_i n_{i.} = \sum_j n_{.j} = n. \tag{6}$$

Now, three information quantities of $H(x)$, $H(y)$ and $H(x,y)$ are calculated by using

$$H(x) = -\sum_i p_{i.}\log_2 p_{i.} \tag{7}$$

$$H(y) = -\sum_j p_{.j}\log_2 p_{.j} \tag{8}$$

$$H(x,y) = -\sum_{ij} p_{ij}\log_2 p_{ij}, \tag{9}$$

Table 1 A data matrix of frequency for Y outputs to X inputs

Output y \ Input x	x_1	$x_2 \cdots x_i \cdots X$	Σ
y_1	n_{11}	$n_{21} \cdots n_{i1} \cdots n_{X1}$	$n_{\cdot 1}$
y_2	n_{12}	$n_{22} \cdots n_{i2} \cdots n_{X2}$	$n_{\cdot 2}$
$\vdots$	$\vdots$	$\vdots$	$\vdots$
y_j	n_{1j}	$n_{2j} \cdots n_{ij} \cdots n_{Xj}$	$n_{\cdot j}$
$\vdots$	$\vdots$	$\vdots$	$\vdots$
Y	n_{1Y}	$n_{2Y} \cdots n_{iY} \cdots n_{XY}$	$n_{\cdot Y}$
Σ	$n_{1\cdot}$	$n_{2\cdot} \cdots n_{i\cdot} \cdots n_{X\cdot}$	n

where $p_{i.}=n_{i.}/n$, $p_{.j}=n_{.j}/n$ and $p_{ij}=n_{ij}/n$. For simplicity, these equations can be rewritten as

$$H(x) = \log_2 n - \frac{1}{n}\sum_i n_{i.}\log_2 n_{i.} \tag{10}$$

$$H(y) = \log_2 n - \frac{1}{n}\sum_j n_{.j}\log_2 n_{.j} \tag{11}$$

$$H(x,y) = \log_2 n - \frac{1}{n}\sum_{ij} n_{ij}\log_2 n_{ij}. \tag{12}$$

Then, the transmitted information $T(x;y)$ is obtained from (3) together with (10), (11) and (12). The relative efficiency of transmission η defined as

$$\eta = [\ T(x;y)/H(x)\ ] \times 100\ [\%], \tag{13}$$

is also obtained.

3. Experiments and Results

The discrete input x_i was considered to be various X-ray exposures obtained by using a 1-mm lucite stepped wedge varying from 0 to 5 mm. The image on the film developed after an X-ray exposure in the screen-film system consists of a graduated scale of photographic densities with different values which correspond to discrete outputs.

Two intensifying screen-film systems which have much the same sensitivity were used. One is HS-A screen-film system of fast speed screen-standard speed medical X-ray film, and the other is LTII-QS system of medium speed screen-fast speed X-ray film. The specified exposure factors were kept at 40 kVp at 130cm for HS-A system and 150cm for LTII-QS system. This difference of distances is due to the slight differences in sensitivity between the two systems. All films exposed were processed by an automatic processor for 3.5 min at 32°C, and their densities were measured by a microdensitometer with a slit area of $8\times720\mu m^2$. The image density of each step was

Table 2 Data matrices of frequency for both systems

HS-A System

Output Y	Input X: 0	1	2	3	4	5	Σ
0.89	1						1
0.88	2						2
0.87	2						2
0.86	7						7
0.85	11						11
0.84	51	2					53
0.83	87	5					92
0.82	122	20					142
0.81	112	33					145
0.80	168	76	5				249
0.79	161	115	15				291
0.78	113	164	26	1			304
0.77	82	167	54	6			309
0.76	46	177	121	9			353
0.75	14	110	150	42	4		320
0.74	17	79	192	70	16	1	375
0.73	2	31	172	104	40	1	350
0.72	2	17	132	139	66	6	362
0.71		4	65	174	105	35	383
0.70			46	186	195	66	493
0.69			14	115	167	105	401
0.68			8	96	210	175	489
0.67				30	87	156	273
0.66				23	66	145	234
0.65				5	33	130	168
0.64					7	103	110
0.63					2	51	53
0.62					2	15	17
0.61						11	11
Σ	1,000	1,000	1,000	1,000	1,000	1,000	6,000
M	0.799	0.770	0.738	0.708	0.691	0.667	
SD	0.025	0.022	0.022	0.023	0.021	0.023	

LTⅡ-QS System

Output Y	Input X: 0	1	2	3	4	5	Σ
0.81		1					1
0.80	1	0					1
0.79	3	7					10
0.78	8	9	1				18
0.77	22	17	2	1			42
0.76	45	31	6	2			84
0.75	61	56	18	4			139
0.74	109	102	32	13	3		259
0.73	108	126	44	25	6	2	311
0.72	155	159	88	44	27	14	487
0.71	136	131	101	62	41	11	482
0.70	135	132	142	84	50	38	581
0.69	115	103	137	123	93	76	647
0.68	57	70	177	179	166	102	751
0.67	24	28	105	134	148	139	578
0.66	15	18	74	138	138	165	548
0.65	6	4	38	78	116	145	387
0.64		5	27	68	101	130	331
0.63		1	5	31	59	77	173
0.62			3	13	34	63	113
0.61				1	11	20	32
0.60					5	14	19
0.59					1	3	4
0.58					1	1	2
Σ	1,000	1,000	1,000	1,000	1,000	1,000	6,000
M	0.717	0.715	0.692	0.678	0.667	0.658	
SD	0.026	0.027	0.026	0.026	0.026	0.025	

M : Mean

SD : Standard Deviation

measured 1000 times by sampling intervals of 30μm. Table 2 shows data matrices of frequency for both systems. The columns and rows represent the thickness and density levels of the lucite steps. By using these frequency matrices, several entropies and the relative efficiency of transmission were calculated and shown in Table 3.

4. Discussion

It is apparent from Table 3 that the ability for transmission of radiographic images in HS-A system is superior to that in LTⅡ-QS system by comparing the values of the transmitted information T(x;y) or the relative efficiency of transmission η in each system. It was shown in our recent paper [4], T(x;y) or η is concerned with both RMS granularity(the standard deviation of density fluctuations) and film contrast(the gradient of H&D curve).

Table 3 Results of entropy calculations

System	H(x)	H(y)	H(x,y)	T(x;y)[bits]	η [%]	$H_x(y)$[bits]
HS-A	2.585	4.330	5.784	1.131	43.8	3.199
LTII-QS	2.585	3.817	5.992	0.410	15.9	3.407

The conditional entropy $H_x(y)$ means the average of entropy y for each value of x, and called ambiguity or noise component. The next equation for $H_x(y)$ in bivariate entropy method could be derived theoretically, when output distributions are Gaussian [8],

$$H_x(y) = \log_2 \sqrt[k]{\sigma_1\sigma_2\cdot\cdot\sigma_i\cdot\cdot\sigma_k} + \log_2 (\sqrt{2\pi e}/\Delta y), \qquad (14)$$

where σ_i, e and Δy are the standard deviation of output distribution for input x_i $(i=1,2,\cdot\cdot,i,\cdot\cdot,k)$, the base of natural logarithms, and the interval value for the output, respectively. This equation means that $H_x(y)$ is concerned with the geometric mean of the standard deviation σ_i (=RMS granularity) of the output density distribution for each input. The greater the value of $H_x(y)$, the greater the degree of granularity, because the conditional entropy $H_x(y)$ is used here as a measure of disorder. From the values of $H_x(y)$ in Table 3, it is found that the granularity in the HS-A system is better than that in the LTII-QS system.

References

1 S.Uchida and D.Y.Tsai,Jpn.J.Appl.Phys.17,2029(1978).
2 S.Uchida and D.Y.Tsai,Jpn.J.Appl.Phys.18,1571(1979).
3 S.Uchida, H.Inatsu and H.Fujita,Jpn.J.Appl.Phys.19,1177(1980).
4 S.Uchida and H.Fujita,Jpn.J.Appl.Phys.19,1403(1980).
5 S.Uchida, S.Katsuragawa and T.Sueyoshi,Jpn.J.Appl.Phys.19,2257(1980).
6 S.Uchida, D.Y.Tsai and H.Abe,Jpn.J.Appl.Phys.19,2247(1980).
7 S.Uchida, A.Ohtsuka and H.Fujita,Jpn.J.Appl.Phys.20,629(1981).
8 H.Fujita and S.Uchida,submitted to Jpn.J.Appl.Phys.

Reciprocity-Law Failure in Medical Screen-Film Systems and Its Effects on Patient Exposure and Image Quality

H. Fujita

Department of Electrical Engineering, Gifu Technical College
Motosu, Gifu 501-04, Japan

S. Uchida

Department of Electrical Engineering, Faculty of Engineering, Gifu University
Kagamigahara, Gifu 504, Japan

1. Introduction

Reciprocity effect or the failure of the reciprocity law is of considerable importance in medical radiography, because most radiographs are made with a pair of intensifying screens in intimate contact with both emulsions of an X-ray film. This failure of the law means that the photographic density depends upon both intensity and exposure time. Recently we have developed an easy and practical method to determine the reciprocity-law failure curve in medical screen-film systems [1-3]. By use of this method, two kinds of reciprocity-law failure curves, expressed as photographic density vs. exposure time for constant exposure and relative exposure vs. exposure time for constant density, can readily be obtained from several time-scale characteristic curves taken experimentally for several focus-film distances(FFDs) [3]. In the present paper, this new method is described with its application to a green-sensitive X-ray film for use in combination with rare-earth screens. The reciprocity effect on the patient exposure will be discussed based on the results of our study.

2. Experimental Procedures and Results

2.1 Change in X-Ray Intensity by FFD-Variation

BUNSEN-ROSCOE law in photography states that the density of the image formed depends on the energy of the exposure, equal to the product of the exposure intensity and exposure time. Films directly exposed to X-rays without intensifying screens obey this law [4]. Thus, X-ray film functions as a good receiver which depends on the energy of the exposure, not on its intensity or time. Figure 1 shows time-scale characteristic curves for FFD values from 0.40 to 5.68 meters. All these curves were obtained without screens. Exposures were made with 0.2mm Cu + 0.5mm Al filtration at the tube and with tube potentials of 80kVp and 50mA by a single-phase full-wave rectified generator(60Hz). The average field size on a film was adjusted to 11.0×13.5cm for each FFD by a multi-leaf-diaphragm. It can easily be confirmed from Fig.1 that the curves all have the same shape. Thus, changing the FFDs from 0.40 to 5.68 meters caused no significant variation of the X-ray beam spectrum. The combinations of FFD and exposure time needed to produce a certain density are obtained from the graph shown in Fig.1. These data are plotted in Fig.2 as a curve of exposure time (t) vs. FFD. This graph shows that the relationship between the logarithm of exposure time and the logarithm of FFD is linear, and the slope of the straight line is about 2.12. The equation in Fig.2 is now obtained. In the equation, E

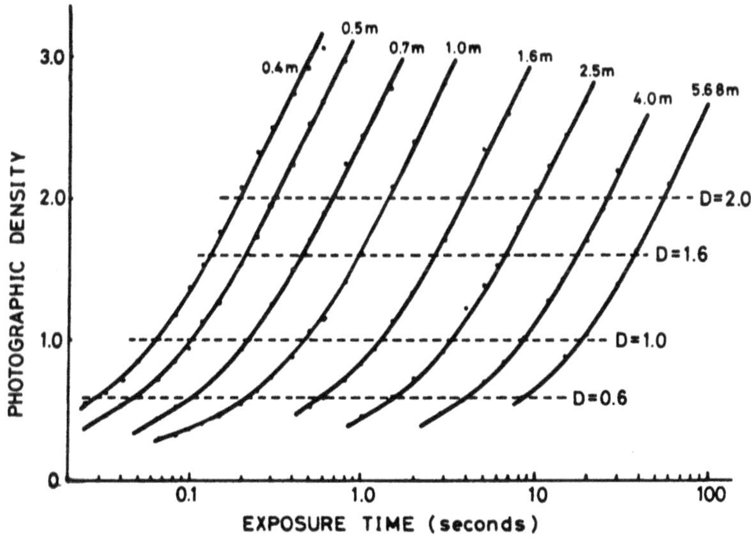

Fig.1 Time-scale characteristic curves obtained without screens

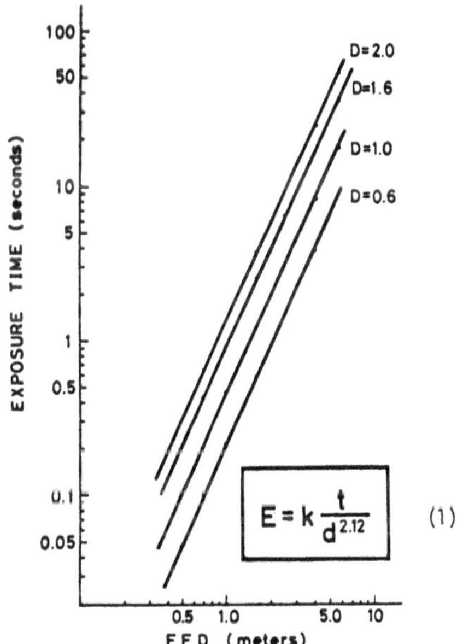

Fig.2 Data obtained from the graph of Fig.1 are plotted as a exposure time vs. focus-film distance(FFD) curve to produce densities of 0.6, 1.0, 1.6 and 2.0

and k are X-ray energy and a constant, respectively. Eq.(1) includes the effects of the focus size and of scattering and absorption by air, and the power value is affected by the exposure conditions, such as the tube voltage and the filtration [1,3].

2.2 Reciprocity-Law Failure Curves

The result shown here is for one brand of medical green-sensitive X-ray film exposed with rare-earth screens [3]. Results for three brands of medical blue-sensitive films exposed with conventional screens [1], X-ray films in reversal processing [2], and two brands of rare-earth screen-film systems [3] are shown in detail in references.

The dashed curves in Fig.3 show time-scale characteristic curves taken experimentally for each of the FFDs, expressed as density vs.t. Experimental conditions were the same as those in subsection 2.1, except for the addition of intensifying screens to the film in the cassette. From these time-scale characteristic curves, reciprocity-law failure curves as illustrated in Fig.3(bold curves) can be obtained from calculations using (1). Each of the curves shows the density produced by a constant exposure, plotted as a function of the exposure time. Figure 4 illustrates reciprocity-law failure curves also obtained from the time-scale characteristic curves shown in Fig.3 by calculations using (1). Each of the curves shows the relative exposure required to produce the indicated density, plotted as a function of the exposure time.

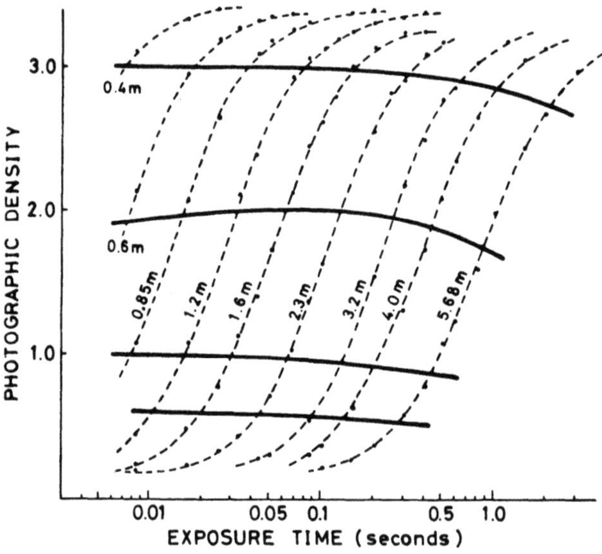

Fig.3 Time-scale characteristic curves obtained in a rare-earth system (dashed curves) and reciprocity failure curves expressed as density vs. exposure time for constant exposure(bold curves)

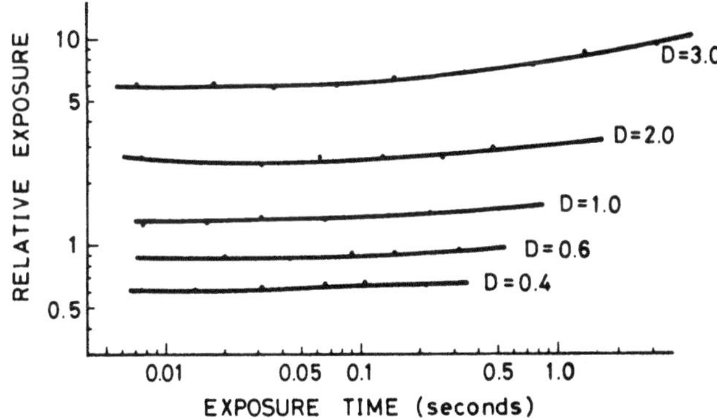

Fig.4 Reciprocity failure curves expressed as relative exposure vs. exposure time for constant density for the same system in Fig.3

3. Discussion

Low-intensity reciprocity failure for exposure times of more than about 0.1 second was observed in Figs.3 and 4, but there was no high-intensity reciprocity failure. If the reciprocity law held, that is, if the photographic effect depended only on the product of time and intensity, these curves would appear as straight line parallel to the horizontal axis. The amount by which the curves depart from horizontal lines is a measure of the magnitude of the reciprocity-law failure. Figure 3 shows the reduction in image density due to this failure effect, and Fig.4 shows the reduction in film speed, which means an increase in patient exposure. Note that radiation exposure is expected to increase by a factor of approximately two when the exposure time is increased from 0.01 to 5 sec, as in Fig.4. Since exposure times of five to ten seconds are sometimes necessary for multi-directional tomography, to which the rare-earth system has recently been introduced, low-intensity reciprocity failure can by no means be ignored. Reciprocity effects must be expected to show variations in different film types, so it is desirable to find the optimum exposure conditions for each system by measuring the reciprocity effects.

Since a significant difference in H&D curve shape due to the reciprocity law failure was noted [3,5], it is necessary to study its effect on image quality. This is the subject for a future study.

References

1 S.Uchida and H.Fujita,Jpn.J.Appl.Phys.18(1979)501.
2 S.Uchida and H.Fujita,Jpn.J.Appl.Phys.18(1979)1641.
3 H.Fujita and S.Uchida,Jpn.J.Appl.Phys.20(1981)227.
4 B.E.Bell,Br.J.Radiol.9(1936)578.
5 J.A.Bencomo and A.G.Haus,SPIE-Apl.Opt.Instr.Med.VII.173(1979)21.

Transmission Imaging with Incoherent Ultrasound

C. Scherg, H. Brettel, U. Roeder, and W. Waidelich

Gesellschaft für Strahlen- und Umweltforschung, Abteilung Angewandte Optik
D-8042 Neuherberg, Fed. Rep. of Germany

Introduction

Ultrasonic imaging methods have found wide acceptance in medical diagnosis. Commercially available systems apply echography by usually measuring the backscattered energy of short ultrasonic pulses.

An alternative and complementary method is to use ultrasound in transmission analog to optical imaging with lenses. Absorbing, reflecting, and scattering losses of the tissue under examination are displayed and are suitably presented in a 2-dimensional view comparable to X - ray images. Transmission sonography however should not be dangerous for biological tissue.

Ultrasonic transmission imaging can be realized in a way similar to optical imaging: an appropriate ultrasonic wavefield insonifies the object and ultrasonic imaging lenses focus a plane of the object upon a detector which transforms the ultrasonic intensity distribution into a visible image. Incoherent insonification makes imaging more reliable since unwanted interferences and speckles are prevented [1]. Like in optics the spatial resolution is determined by the wavelength of ultrasound and the aperture of the lenses.

We describe here a laboratory version of an ultrasonic transmission camera which differs in some important respects from other designs [2,3]: the object is insonified continuously by spatially and temporally incoherent ultrasound, the imaging lenses are roughly corrected for a flat image plane, and the detector consists of a moving linear ultrasonic transducer array with parallel analog processing of the signals.

Our Experimental Laboratory System

The experimental setup shown in Fig. 1 consists of a water tank which contains all ultrasonic components. For insonification we use an improved version of our incoherent ultrasound source described previously [4]: Piezoelectric transducers are driven by power generators at frequencies of about 2 MHz. Temporal coherence is reduced by sweeping the frequency across the

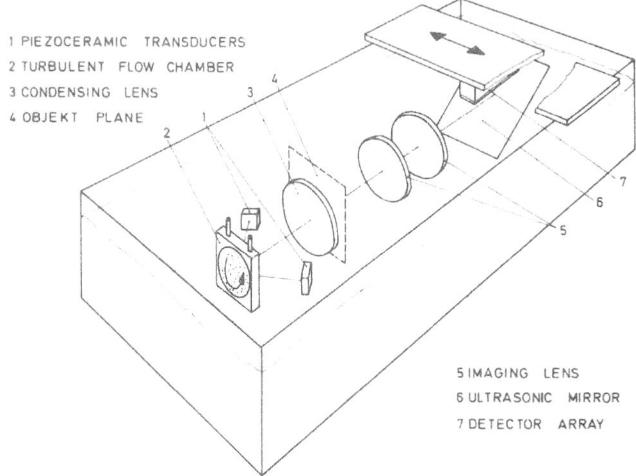

Fig. 1 Arrangement of the ultrasonic components in the water tank.

resonant transducers response. The emitted primary ultrasound waves are directed towards a chamber which contains small glass beads moved randomly by a turbulent water flow. The ultrasonic waves are backscattered at these particles and therefore, spatial coherence is destroyed.

The object under test is insonified through a condensing lens with the spatially and temporally incoherent ultrasound. A pair of polystyrene ultrasonic lenses focus a thin slice from the object onto the detector plane via a mirror. Spherical aberration and curvature of the image field is compensated using specially designed aspherical lens surfaces.

The ultrasonic image is detected by a linear piezoelectric transducer array which is mechanically moved line by line across the whole image plane. The detector array consists of 200 piezoelectric elements which convert the ultrasonic intensity distribution along a line into corresponding electrical signals. These weak signals are preamplified, rectified, and averaged over a period which is long compared to the remaining coherence time. Parallel analog processing is used in order to permit fast scanning across the image even at long times for averaging.

All analog processed image data are multiplexed, A/D - converted, and stored in a video refresh memory under control of a microcomputer. In this way the image is presented flicker free on a standard television monitor and continuously updated during mechanical scanning.

Experimental Results and Conclusion

In first tests of our transmission imaging system we found a spatial resolution of better than 3 mm at a frequency of about 2.1 MHz corresponding to a wavelength of about 0.7 mm in water. The maximum intensity required for imaging human limbs was below 5 mW/cm^2.

Figure 2a shows a transmission image taken from a human hand. Finger bones, finger joints with ligaments, and, and tendons are visible in this image. Due to the small depth of focus, only structures within a thin slice of the object are imaged. In Fig. 2b only the outer part of the leg is focused and some veins become visible whereas bones are slightly smeared. Fig. 3a shows a heel focused on the tendons and Fig. 3b presents a transmission image through the epigastrum focused on the vertebral column. Two ribs crossing the right kidney are visible.

In comparison to coherent transmission images [1] our results with incoherent ultrasound don't suffer from speckles or interference artefacts. Details outside the focus plane disturb the image much less than in coherent imaging. Interpretation of the images for medical diagnosis should therefore be more reliable.

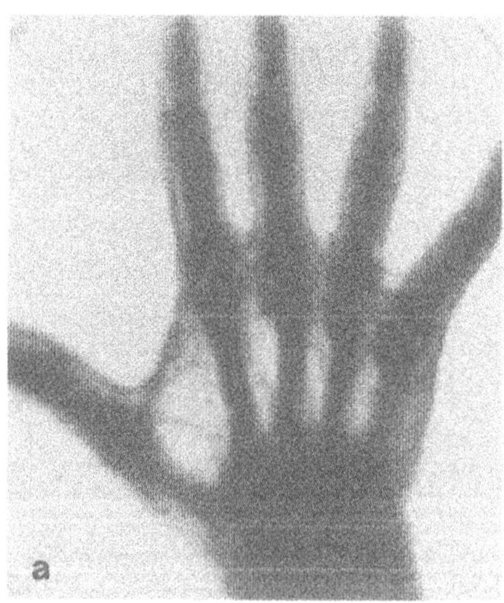

Fig. 2a Hand focussed near the back.

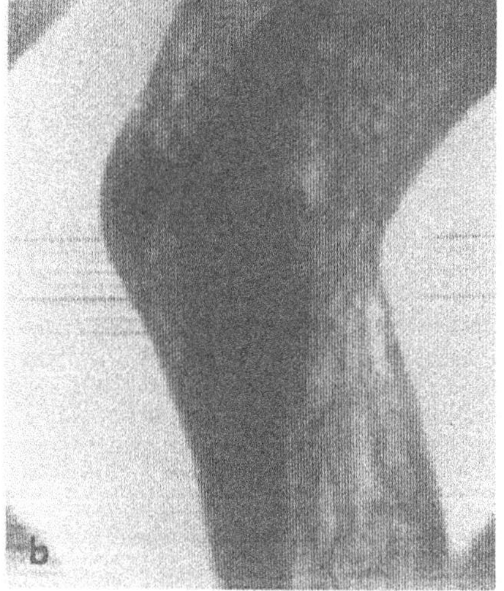

Fig. 2b Knee-joint with veins.

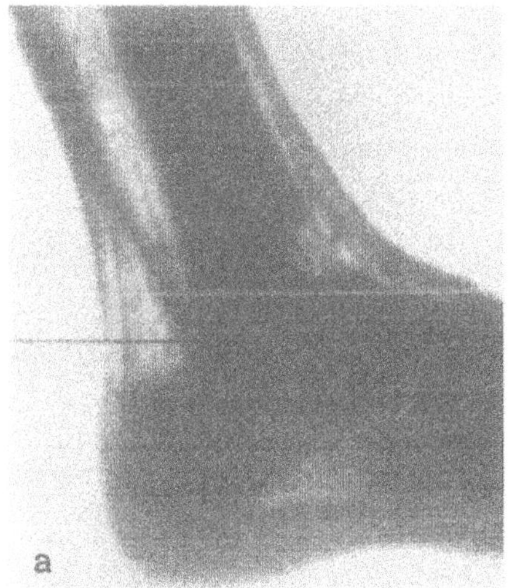

Fig. 3a Heel focused on tendons.

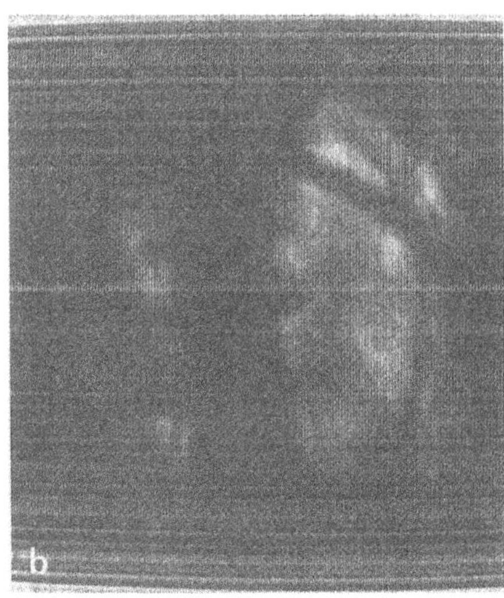

Fig. 3b Epigastrum and vertebral column with rib crossing the right kidney.

References

1 Roeder,U., Scherg,C., Brettel,H., Coherence and noise in ultrasonic transmission imaging. Acoustical Imaging, Vol. 10, Plenum Press, New York (in press).

2 Green,P.S., Schaefer,L.F., Jones,E.D., Suarez,J.R., A new high performance ultrasonic camera. Acoustical Holography, Vol. 5, ed. Green,P.S., Plenum Press, New York (1974), 493-503.

3 Havlice,J.F., Green,P.S., Taenzer,J.C., Mullen,W.F., Spatially and temporally varying insonification for the elimination of spurious detail in acoustic transmission imaging. Acoustical Holography, Vol. 7, ed. Kessler,L.W., Plenum Press, New York (1977), 291-305.

4 Roeder,U., Scherg,C., Scattered ultrasound for incoherent insonification in transmission imaging. Ultrasonics (1980), 273-276.

Part III

Interferometry and Holography

Laser Interferometric Measurement of Basilar Membrane Vibrations in Cats Using a Round Window Approach

S.M. Khanna, and D.G.B. Leonard

Department of Otolaryngology, Columbia University
New York, NY 10032, USA

A comparative method of interferometric measurement was developed to measure basilar membrane vibrations. In this method simultaneous measurements are made on the object to be measured and on a reference mirror vibrating with a known amplitude. The unknown amplitude is then determined by comparing the two measurements.

The basic interferometer arrangement is shown in Fig. 1. A focused laser beam is directed onto a partially silvered reference mirror and a signal mirror. The orientations of these two mirrors are adjusted so that their reflections overlap and interfere at the detector. Vibrations of the signal and reference mirrors produce ac outputs at the detector. The theory for the instrument design is discussed in [1]. The sensitivity and the dynamic range of the method determined with an ear phone is shown in Fig. 2. Photodetector output is shown as a function of the vibration amplitude at 1000 Hz. The lowest level at which vibration can be detected is 3×10^{-12}cm (S/N=0 dB). Precise measurements can be made at amplitudes greater than 3×10^{-11}cm (S/N 20 dB) and up to a maximum amplitude of 10^{-5}cm. Averaging time was 0.1 sec above 10^{-8}cm. The total linear range of measurement is 90 dB. The frequency response of the interferometer is flat from 100 Hz to 40 kHz.

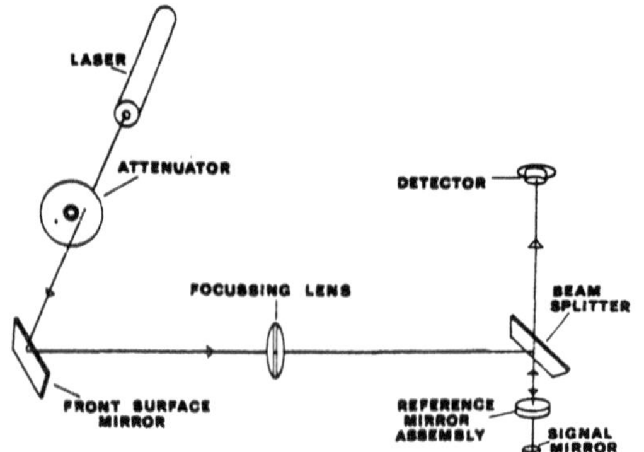

Fig.1

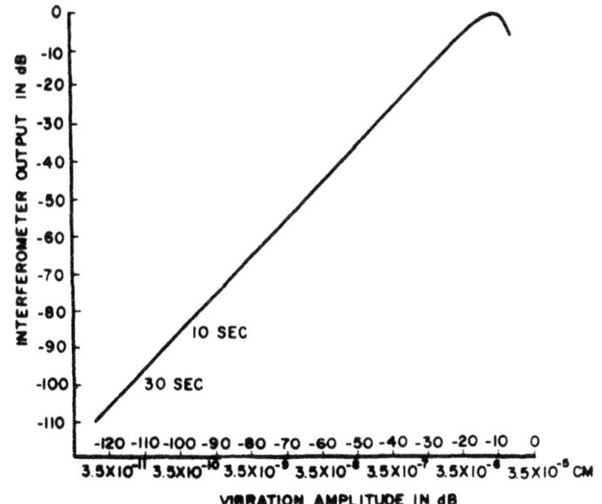

Fig.2

In order to have an experimental preparation with a minimal amount of trauma to the cochlea, a surgical approach to the basilar membrane through the round window was developed [2]. The round window membrane was removed and a microscopic gold crystal weighting less than 10^{-8}gm was dropped onto the basilar membrane without removing the perilymph from the scala tympani. The optically flat gold crystal was used as the signal mirror of the interferometer. Vibrations of the basilar membrane were measured in response to sound applied by the SOKOLICH driver [3].

To assess damage to the cochlea, the round window cochlear microphonic (CM) response was measured several times during the experiment. Also, the cochleas were prepared and examined using an epon-embedded surface preparation method [4]. Many causes of trauma to the cochlea were discovered and these were either eliminated or minimized [5]. At the end of these improvements the CM loss incurred in the preparation was reduced to a few dB. In order to measure vibrations at the lowest possible sound pressure levels (SPL), the sound pressure was adjusted until the S/N ratio at the interferometer output was between 15 and 20 dB. This corresponded to a vibration amplitude of approximately 10^{-9}cm. The lowest SPLs required (20 dB) were in the peak region of response.

Basilar membrane vibration amplitudes calculated for 30 dB SPL are shown for one experiment in Fig. 3 (K & L). The curve shows a shallow maximum in the frequency region between 0.25 and 3 kHz, a plateau region between 5 and 14 kHz and a peak at approximately 23 kHz. In the peak region, the low frequency

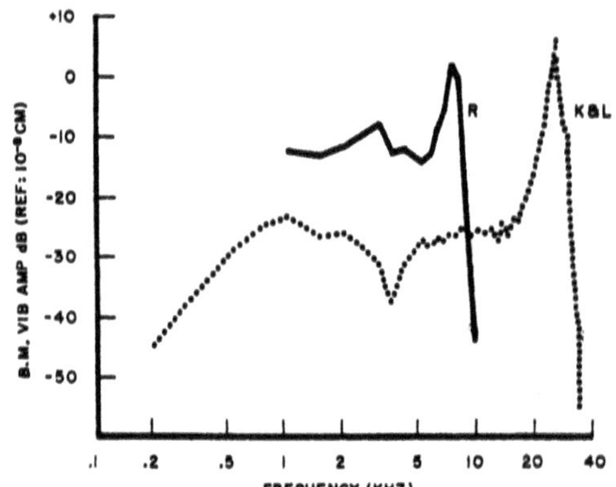

Fig.3

slope is 86 dB/oct, the high frequency slope is 538 dB/oct, the sharpness of resonance (Q_{10dB}) is 5.9 and the height of the peak is 30 dB above the response level at 1 kHz. Peak heights exceeding 25 dB were obtained in at least five experiments. A basilar membrane response measured by RHODE [6] in the squirrel monkey (73-104) is shown for comparison in Fig. 3 (R). The peak measured by RHODE is 15 dB above the response at 1 kHz (Note that both curves show vibration amplitude measured for constant SPL at the tympanic membrane).

A comparison [7] shows that the basilar membrane response begins to look similar to tuning curves of auditory nerve fibers. The response observed in the peak region is still highly variable from one experiment to another. The higher peaks are obtained in cochleas with the least trauma. CODY and JOHNSTONE [8] recording from single auditory nerve fibers, have shown that the peak height is reduced with even small amounts of acoustic trauma. We suspect that the residual trauma is still too large in our experiments to observe the normal response of the basilar membrane.

References

1 Khanna, S.M. A comparison technique of interferometric measurements. (In preparation) (1981)
2 Khanna, S.M., Leonard, D.G.B. Basilar membrane vibrations measured in cat using a round window approach. (In preparation) (1981 A)
3 Sokolich, W.G. Improved acoustic system for auditory research. J. Acoust. Soc. Am. 62, Suppl. 1, p. S21 (abstract) (1977)
4 Liberman, M.C., Beil, D.G. Hair cell condition and auditory nerve response in normal and noise-damaged cochleas. Acta Otolaryng. 88, 161-176 (1979)

5 Khanna, S.M., Leonard, D.G.B. Cochlear damage incurred during preparation for and measurement of basilar membrane vibrations. (In preparation) (1981 B)
6 Rhode, W.S. Some observation on cochlear mechanics. J. Acoust. Soc. Am. 64, 158-176 (1978)
7 Khanna, S.M., Leonard, D.G.B. Basilar membrane tuning in the cat cochlea. (In preparation) (1981 C)
8 Cody, A.R., Johnstone, B.M. Single auditory neuron response during acute acoustic trauma. Hear. Res. 3, 3-16 (1980)
9 Supported by NIH Grants # 5 K04 NS 00292 and # 5 R01 NS 03654

Displacement Measurement of the Basilar Membrane in Guinea Pigs by Means of an Optical-Fiber Interferometer *

F. Albe, J. Schwab, P. Smigielski, and A. Dancer

Institut Franco-Allemand de Recherches de Saint-Louis (ISL)
F-68301 Saint-Louis, France

1. Introduction

For a few years the Franco-German Research Institute (ISL) has been active in investigating lesions and traumatisms generated under the action of strong acoustic noise such as occurs in the environment of aeronautics. In this context the acoustic stimulus transmitted from the external medium to the ear must be investigated. One possibility of carrying out successfully those investigations consists in using optical methods in order to measure the very weak displacements of the various components which form the auditory chain [1].

In a first phase a map was drawn of the deformations of the tympanic membrane generated under the action of transient acoustic impulses. Toward this end holographic interferometry with double exposure was used [2].

In a second phase the displacement of the umbilicus could be measured with a Michelson interferometer [3]. The acoustic stimuli produced in this case were either pure tones or impulse noise.

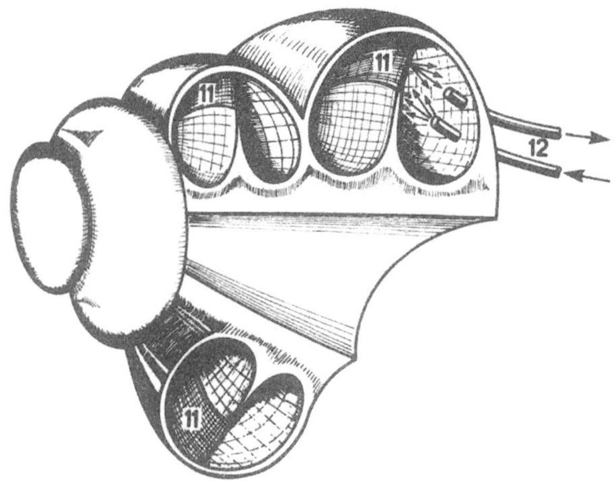

Fig. 1: Partial section of the cochlea of the guinea pig with the optical fibers

* This work has been supported by DRET, Paris

For the study of the displacements of the basilar membrane an optical-fiber interferometer was used [4] [5]. The cochlea of a guinea pig is illustrated in figure 1 with the optical fibers implanted.

2. Schematic Diagram and Principle of the Interferometer

The schematic diagram of the experimental arrangement used is shown in fig.2. The beam of the light emanating from a He-Ne Laser (5 mW, λ=6328 Å) is split into two parts by means of a semi-transparent plate. One part (called object beam) arrives on an optical fiber with a graded index (outer diameter: 100 μm; core diameter: 60 μm) and illuminates the basilar membrane. The back-scattered light which is modulated in response to the displacements of the basilar membrane is then collected by the second fiber and interferes, at the input of the photomultiplier, with the part of light (called reference beam) modulated via the piezoelectric mirror M_r at the frequency f_r. A sound wave generated by a loudspeaker leads to the excitation of the ear at the frequency f_0. We shall use the following designations:

$d_0 = a\sin(\omega_0 t + \Phi_0)$ the basilar membrane displacement,

$d_r = b\sin(\omega_r t + \Phi_r)$ the piezoelectric mirror displacement,

$I_p = I_r + I_0 + 2\varepsilon\sqrt{I_r I_0}X(t)$ the intensity detected by the P.M.,

$I_r =$ the intensity due to the reference beam,

$I_0 =$ the intensity due to the object beam,

$X(t) = \cos(\varphi_0 + \varphi_r + \Omega)$ where

$\varphi_0 = \frac{4\pi}{\lambda} a\sin(\omega_0 t + \Phi_0)$ the optical phase variation due to the displacement of the object,

$\varphi_r = \frac{4\pi}{\lambda} b\sin(\omega_r t + \Phi_r)$ the optical phase variation due to the displacement of the reference mirror,

$\Omega =$ optical phase including a constant expression due to the adjustment of the interferometer, and a variable expression due to the disturbing motion of the guinea pig or to other possible perturbing effects (vibrations, fluctuations in temperature, etc.).

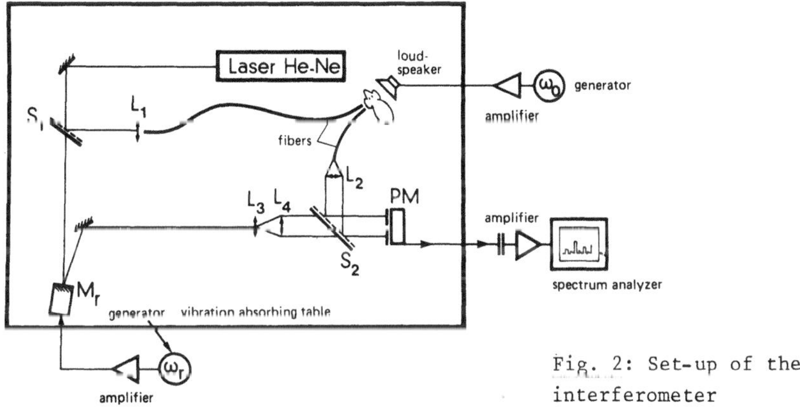

Fig. 2: Set-up of the interferometer

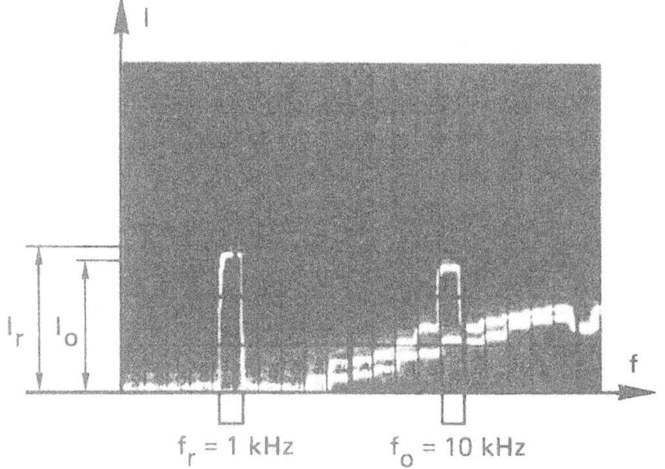

Fig. 3

An example of the spectrum analysis

The first order development in Bessel function yields [6]

$$\begin{aligned} X(t) &= \cos\Omega \; J_0\left(\frac{4\pi}{\lambda}a\right) J_0\left(\frac{4\pi}{\lambda}b\right) \\ &- 2\sin\Omega \; J_1\left(\frac{4\pi}{\lambda}a\right) J_0\left(\frac{4\pi}{\lambda}b\right) \sin(\omega_0 t + \Phi_0) \\ &- 2\sin\Omega \; J_1\left(\frac{4\pi}{\lambda}b\right) J_0\left(\frac{4\pi}{\lambda}a\right) \sin(\omega_r t + \Phi_r) \\ &- 4\cos\Omega \; J_1\left(\frac{4\pi}{\lambda}a\right) J_1\left(\frac{4\pi}{\lambda}b\right) \sin(\omega_0 t + \Phi_0)\cdot\sin(\omega_r t + \Phi_r) \; \ldots \end{aligned}$$

The intensity delivered by the photomultiplier is subjected to a frequency analysis. Thus, two signals are obtained at the frequencies f_0, f_r. Their amplitude ratio M does not depend on Ω, when $\sin\Omega$ differs from 0. Figure 3 shows a spectrum recorded at f_0=10 kHz and f_r=1 kHz. We have

$$M = \frac{I_0}{I_r} = \frac{J_1\left(\frac{4\pi}{\lambda}a\right) J_0\left(\frac{4\pi}{\lambda}b\right)}{J_1\left(\frac{4\pi}{\lambda}b\right) J_0\left(\frac{4\pi}{\lambda}a\right)} \approx \frac{a}{b}$$

provided a and b are sufficiently small (< 100 Å).

This technique allows a continuous calibration [6]. A control system is not required. The spectrum of the signal can be observed on the remanent screen of the oscilloscope during a relatively long time period (several minutes). The maximum values of the intensities I_0 and I_r are read at the frequencies f_0 and f_r which correspond to $|\sin\Omega| = 1$ and to $\cos\Omega = 0$ indicating the correct operation of the interferometer.

3. Experiments

After the anaesthesia of the animal, the bulla is opened. Thus the scala tympani and the round window are visible. With the aid of a microdrill two little holes $\sigma f \approx 0.15$ mm are made in the bone wall of the cochlea at 1.5 mm under the round window. The optical fibers are introduced through these apertures. Each fiber is used, with the aid of an additional laser beam, to

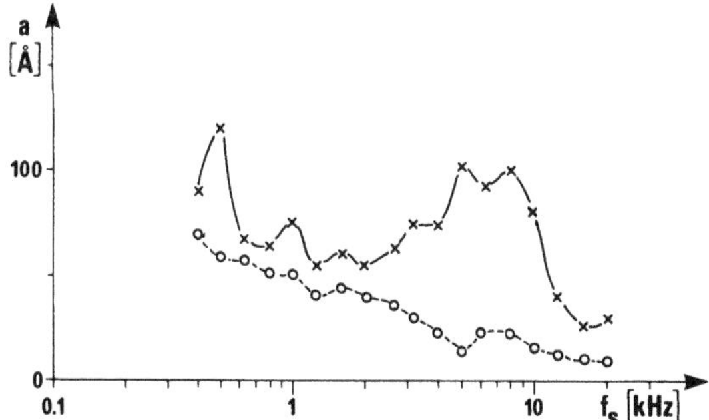

Fig. 4: Diagram of the displacement of the basilar membrane of a living guinea pig as a function of the frequency. Sound level: 94 dB

adjust the position of the other fiber. The correct position is obtained when the output beam collected by the second fiber is suitable for the measurements. Then the fibers are sticked on the edge of the bulla bone.

This tricky operation is made under a binocular microscope. In the first phase of the experiments great difficulties were encountered in determining, in an accurate way, the position of the fiber ends with respect to the basilar membrane. The animal with the fibers is placed into the interferometer.

The loudspeaker placed in the vicinity of the guinea pig delivers the acoustic stimuli. This loudspeaker is calibrated such that over the whole frequency range explored (by third octaves from 500 Hz to approximately 16 kHz) a constant sound level is attained. Thus the diagrams of the displacements of the basilar membrane could be recorded as a function of the excitation frequency (fig. 4) for the following two cases: obstructed ear canal (dotted line) and not obstructed ear canal (full line).

4. Conclusion and Perspectives

The feasibility of this method could be shown. Complementary studies will be necessary, however, to obtain improved knowledge of the transfer function in the inner ear.

References

1. S.M. KHANNA et al. J. Acoust. Soc. Am. 44-6-1968 p. 1555-64
2. A. DANCER et al. J. Acoust. Soc. Am. 58-1-1975 p. 223-28
3. A. DANCER et al. Acustica 46-2-1980 p. 178-187
4. M.A. NOKES et al. Rev. Sci. Instrum. 49.6.1978 p. 722
5. F. ALBE et al. Congrès "Horizon de l'Optique" Pont à Mousson 22 - 25 Avril 1980
6. P.R. DRAGSTEN et al. J. Acoust. Soc. Am. 60.3.1976 p. 665-71

On Easy-to-Handle Holocameras for Endoscopic Applications

P. Greguss

Applied Biophysics Laboratory, Technical University Budapest
H-1111 Budapest, Hungary

1. Introduction

In spite of the fact that in recent years the performance of the various endoscopes - especially after the introduction of fiber optics - has considerably been improved, they still have several drawbacks. One of these is the very limited depth of focus, another that they have a rather restricted field of view, thus, if information from the entire wall of the cavity to be investigated is needed, the endoscope has to be rotated around its optical axis.

It was in the early 70s that the idea of using holography to solve some of these problems emerged. It was theoretically and experimentally analyzed in detail by HADBAVNIK [1], who concluded that the holoendogram should be recorded in the cavity, on a plate fixed in it, which is illuminated by a reference beam too. In contrast to him, GRUNEWALD et al. [2] used a technique where the optical information is lead out via normal endoscope optics for holographic recording. Recently, YONEMURA et al. [3] concluded that fiber bundles instead of single fibers can be used to transmit high power laser light for object illumination to prevent the light path break induced by the high power density of the concentrated laser light.

Comparing the inside recording method with the outside recording method one finds that both have their own advantages and disadvantages. Theoretically the inside recording method allows 3-D reconstruction, while in the outside recording method, where the object light is imaged on the entrance end of the fiber bundle to be transmitted to the exit end, 3-D reconstruction is impossible. In practice, however, in the first case, due to the generally very restricted recording surface area not enough parallax can be achieved for 3-D observation. On the other hand, using the outside method the recording technique becomes far easier (e.g., rewinding the holographic film, the size of which is not really restricted).

However, whichever method is chosen, it is based upon conventional endoscope construction, only a reference wave is added, - i.e., the endoscope has to be rotated around its optical axis if 360° panoramic view is needed, thus none of them is really ready for clinical use. During our endeavors to develop an easy-to-handle holocamera we have come across an optical element having the unique property of converting an incoming collimated light beam into an outgoing diverging annular wavefront. The scope of this report is to describe two holoendoscopes based on this optical element. Before this, however, we have to become acquainted with some of its important properties.

2. Majoros-type Toroidal Lens

Lens systems allowing panoramic view have been known since the end of the last century when MANGIN introduced the first panoramic telescope. This was followed by the astronomic panoramic telescope of HILL [4], and SCHULZ [5], and the panoramic photographic camera of MERLE [6]. A forgotten Hungarian lens designer, MAJOROS, constructed about 30 years ago a lens having only flat and spherical surfaces, out of which three acted as mirrors, and the others as refractive surfaces [7]. As a consequence, these refractive-reflective combinations form from the space around the optical axis of the lens a virtual annular image with a viewing angle the value of which depends upon the curvature ratio of the surfaces and the refractive index of the material of which it is made (Fig. 1). Ray tracing shows that if a collimated beam impinges on the flat surface of the lens, then, the outgoing bundle spreads to fill an annulus of cone. However, in the formation of this wavefront only an annular part of the collimated beam is responsible for the outgoing conical beam, i.e., the core remains unused. This is the reason why we refer to it as the Majoros-type toroidal lens. These remarkable properties inspired us to try to design panoramic holoendoscopes based on this lens.

3. Panoramic EHHC-35

Previously we have developed an easy-to-handle holocamera (EHHC) [8] labeled Holocamera-35, the principle of which was that the laser beam diverged by a microobjective passed directly through

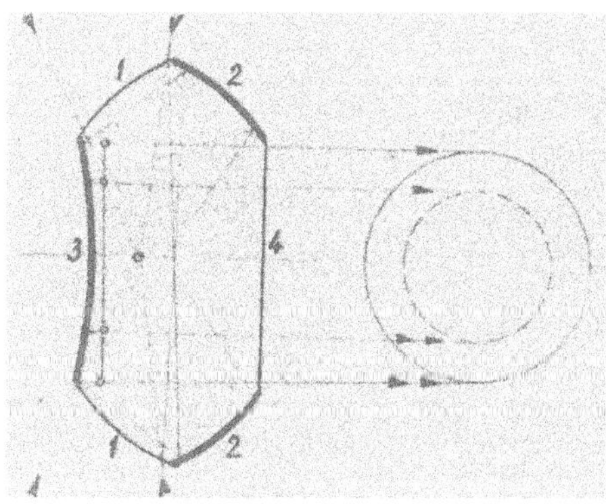

Fig. 1 Ray tracing of a Majoros-type toroidal lens (1 = refractive surface, 2, 3 = mirror surface, 4 = flat surface)

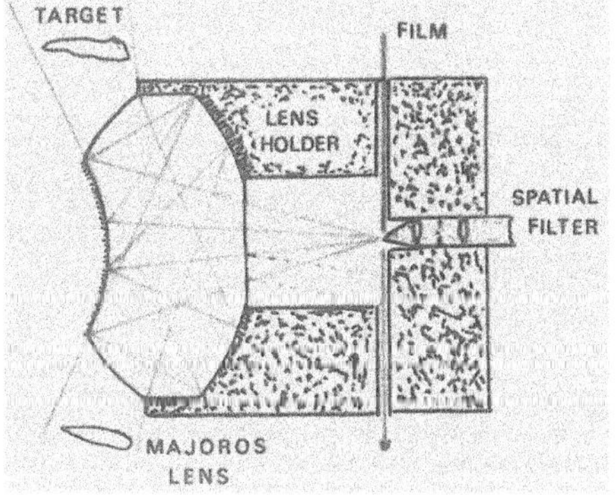

Fig. 2 Combination of an EHHC-35 with a Majoros-type toroidal lens

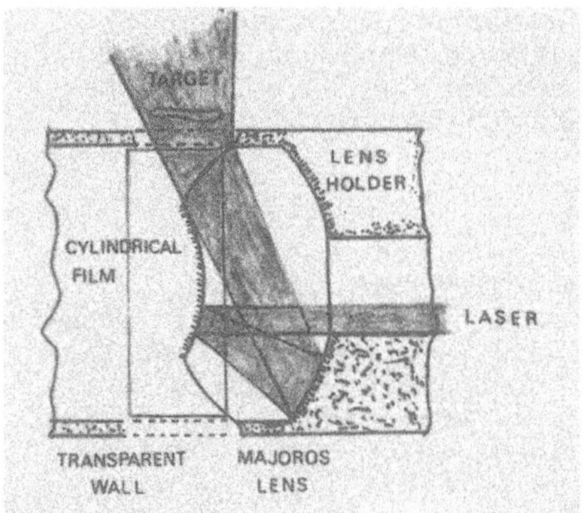

Fig. 3 Conceptual draft of a panoramic holographic endoscope based on the Majoros-type toroidal lens

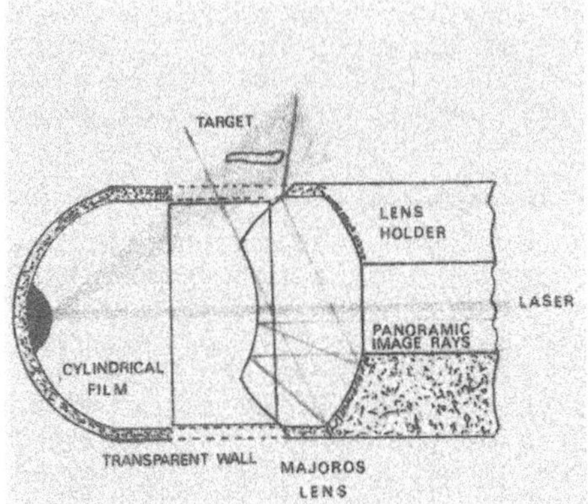

Fig. 4 Easy-to-handle panoramic holographic endoscope (EHPHE)

a hole in the photographic emulsion before it striked the beam splitter which provided the reference beam to record the hologram of the object illuminated by the not reflected (transmitted) part of the diverging beam. Thus, it seemed to us that by replacing the beam splitter with a Majoros-type toroidal lens in such a way that its plane surface - as shown in Fig. 2 - coincides with the plane where the beam splitter was before its removal, a panoramic easy-to-handle holocamera could be constructed. However, the quality of the recorded hologram was in general rather poor, mainly because it was difficult to optimize the reference to subject energy ratio. (In our original EHHC-35 this is achieved by simply changing the beam splitter.)

4. Panoramic Holoendoscope (PHE)

In order to overcome these difficulties and to record a panoramic volume reflection hologram, we placed the holographic film - as shown in Fig. 3 - in the tip of the endoscope, with the emulsion side facing the transparent wall. The laser beam passing through the unused center part of the lens is reflected from a suitably designed mirror at the end of the tip of the endoscope and impinges on this holographic film. The part absorbed by the film serves as a reference beam, while the part that passed through the film illuminates the target in the cavity, thus a cylindrical 360° volume reflection hologram is formed. 360° reflection holograms have already been recorded [9], however, only from objects inside a cylinder, and not from targets around it.

To distinguish this type of 360° reflection hologram from that described in [9] we call it panoramic hologram. As long as there is no difficulty of viewing the reconstructed image of a 360° reflection hologram, since it is located inside the cylinder or cone formed by the developed film hologram, the reconstructed image of a panoramic hologram has to be viewed from the inside of this cylinder, which is difficult to realize. However, with the Majoros-lens we have practically no difficulty to look around in the reconstructed 360° space, provided that the developed film has been placed back in the tip of the endoscope - or in a cylinder with the same inner diameter - and has been illuminated with approximately the same light pattern as that produced by the PHE.

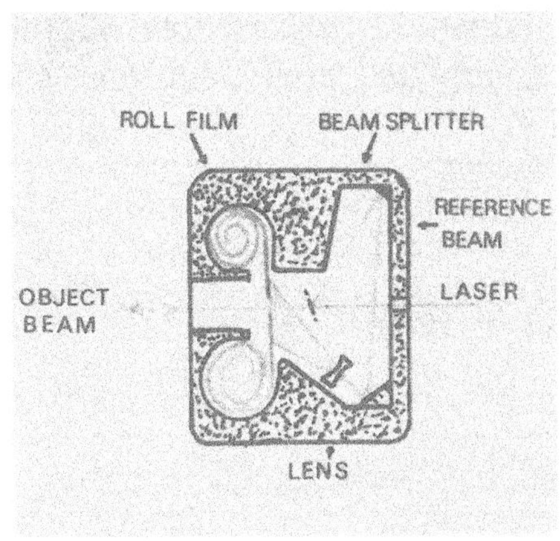

Fig. 5 PHE with sheet film recording

Since the PHE uses cylindrical film, its changing is rather clumsy, in contrary to the panoramic EHHC-35, while film rewind can even be automatic. This limitation can be overcome by suitably combining the two recording techniques, as shown in Fig. 5. The panoramic image of a cavity provided by the Majoros-lens is projected onto the film, while the reference beam reaches it from behind, thus forming a reflection hologram.

5. Conclusions

Various types of holocameras having the common feature of using a specially designed toroidal lens to make 360° transmission or reflection holograms from cavities have been discussed. These panoramic holograms recorded on acetate based roll films have the advantage of giving information from the entire wall of a cavity without the need for rotating the endoscope around its optical axis.

References

1. D. Hadbawnik: Dissertation, Stuttgart 1975
2. K. Grunewald, H. Wochutka: in Holography in Medicine and Biology, ed. by G. von Bally, Springer Series in Optical Sciences, Vol. 18 (Springer, Berlin, Heidelberg, New York 1979)
3. M. Yonemura et al.: Appl. Opt. 20:1664-1667, 1981
4. P. Greguss: Optics News 5:26-27, 1979
5. R. Hill: Brit. Pat. 225.398 (1924)
6. H. Schulz: DRP 620.538 (1935)
7. A. Merle: DRP 672.393 (1939)
8. S. Majoros: Hung. Pat. 143.538 (1954)
9. S.T. Hsue et al.: Am. J. Phys. 44:927-928, 1976.

Holographic Measurement of Deformation in Complete Upper Dentures - Clinical Application

I. Dirtoft

Department of Prosthetic Dentistry, Karolinska Institutet, Box 3207
S-103 64 Stockholm, Sweden, and

Department of Production Engineering, Royal Institute of Technology, Fack,
S-100 44 Stockholm, Sweden

1. Introduction

Prosthodontists are responsible for dentures which are well retained, stabile and show optimal fit even after extensive use in order to prevent injury to the underlying tissues. Various processing techniques result in unequal physical properties and behaviour in different directions due to the anisotropy of the polymer. Dimensional changes cause resorption and are indirectly responsible for a decrease in retention and stability. Numerous investigators have compared the dimensional and other differences in some common base materials [1, 2, 3, 4, 5, 6]. Previously described methods for dimensional measurement are sometimes based on repositioning on the mastercast, [7, 8, 9, 10] which may not be sufficiently accurate [11, 12, 13]. The determination of three-dimensional shape by use of a measuring microscope has also been described [11, 12, 13]. Other methods have also been used for the same purpose [14, 15, 16]. In any case, measuring deformations in polymers is a very intricate problem due to the anisotropy of the material, and knowledge of the total deformation pattern of the object is very limited. Holographic interferometry seems to solve many of these special polymer problems in a satisfying way [11, 16, 17, 18, 19].

The purpose of the present study was to determine if the previously developed holographic method [11, 16], heretofore used in vitro, was suitable for clinical purposes as well.

2. Patient and preoperative treatment

An edentulous female patient aged 62 was chosen for the study. She was wearing a 20 years old loose complete upper denture. A careful clinical examination which included history and status was carried out. She was informed in homecare of the denture and soft tissues. Extra and intraoral photographs were taken, and preprosthetic treatment of denture stomatitis, angular cheilitis, height of the bite and retruded position were made. A recommendation was made to contact a doctor because of the shiny, dry appearence of the tongue, ragades and dental sore mouth.

3. Materials

A new denture was carefully made by use of conventional technique. The palate was painted with gold-bronze and acrylic as described in previous articles.

The steel balls were placed at the papilla insiciva and at the place of the first molars to obtain as stable repositioning as possible. Denture base material was SR-Ivocap, a heat-cured resin for injection moulding. The material was treated according to the manufacturers recommendations at the School of Dental Technicians, Huddinge, Sweden.

4. Methods

For purposes of comparison and conclusion, three holographic techniques were used, i.e. real-time, double-exposure and sandwich holography. Both conventional and white-light holograms were made at the same time with these different techniques. The techniques are described in previous articles [11, 16].

4.1 Real-time holography

This technique was used in order to control the repositioning of the denture and plate, as well as the stability of the equipment. The appearence of interference fringes after deformation indicates that the change is within the range of measurement.

4.2 Double-exposure technique

These holograms were made to store information on deformation at every measuring occasion.

4.3 Sandwich holography

Sandwich holograms were made to allow the possibility of eliminating unwanted rigid body motion during the observation phase by an optically performed coordinate transformation. A stepwise study could thus be carried out in different combinations within a ten times larger range of measurement than when conventional holography was used.

4.4 White-light holography

The advantages with this method are that the observation angle can be considerably increased compared to conventional holography. The image of the object is naturally enlarged due to the short distance between the plate and the object. In addition, it is possible to evaluate and observe the hologram without the use of a laser.

5. Experimental setups

Two different setups are used, one conventional and one of Lippman type. The conventional setup is shown in Fig. 1.
The laser used was a He-Ne 50 mW. The three mirrors behind the denture were tilted to obtain an observation angle of 45^{0}. The object beam illuminates the denture from the top mirror. The following conditions must be fulfilled: 1. The object must be observed from three different directions at the same time. 2. All equipment must be extremely stable. 3. The stand and the denture must be equipped to allow an accurate repositioning. The white-light setup is shown in Fig. 2.

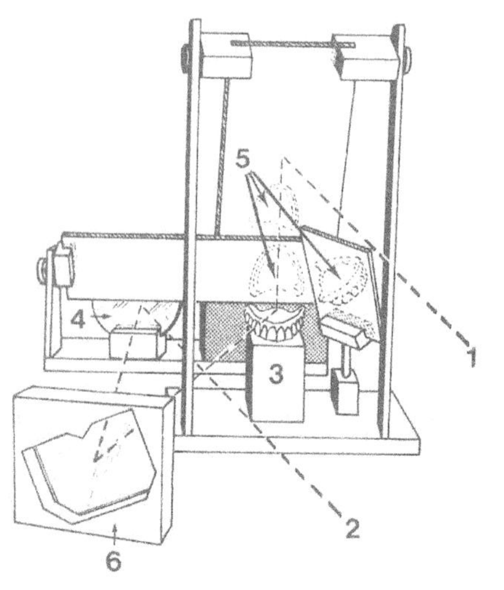

Fig. 1 1. Object beam 2. Reference beam 3. Denture 4. Reference mirror 5. Mirror images 6. Plate holder

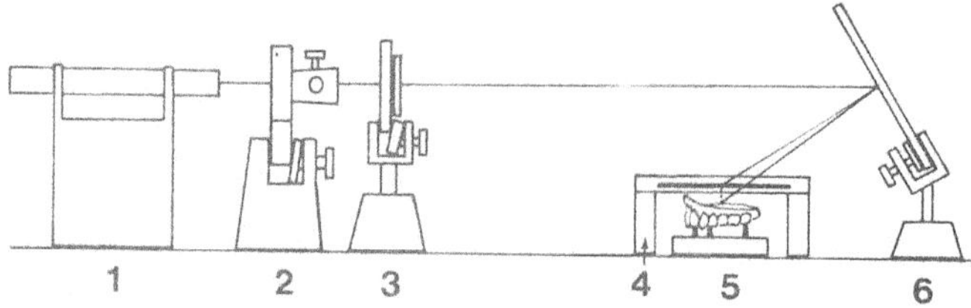

Fig.2 1. Laser He-Ne 7 mW 2. Spatial filter 3. Beam shutter 4. Plate holder 5. Denture on the stand 6. Mirror

In this case the laserbeam interfers in the plate with the beam reflected from the object. Measurements were made on 46 separate occasions over a period of 37 days and resulted in 466 holograms.

6. Results

The previously developed holographic method for measuring deformation in complete upper dentures has been tested here in a real clinical situation for 37 days with very promising results. The patient was wearing a denture day and night except for cleaning it and had no problem with the steel-balls. The gold-bronze remained a well defined reflection surface during the whole experiment. This was determined at every measurement occasion by real-time holography. At the same time the stability of the equipment and the repositioning of the denture were checked by observing the painted metal surface of the stand. An accurate repositioning without interference fringes is shown in Fig. 3. This figure shows the three views of the denture in the conventional setup after 31 days. Interference fringes may be clearly seen all over the object in all views in the hologram.
The double-exposure holograms were repositioned as well. The accuracy of this was checked in the image of a part of the plate holder, where interference fringes could be counted. The Lippman type holographic method used for measurements was very well suited for this type of experiment. 44 double-exposure holograms were made and 12 had an extremely good quality (see Fig. 4), although the photographs do not show as many details as the holograms.

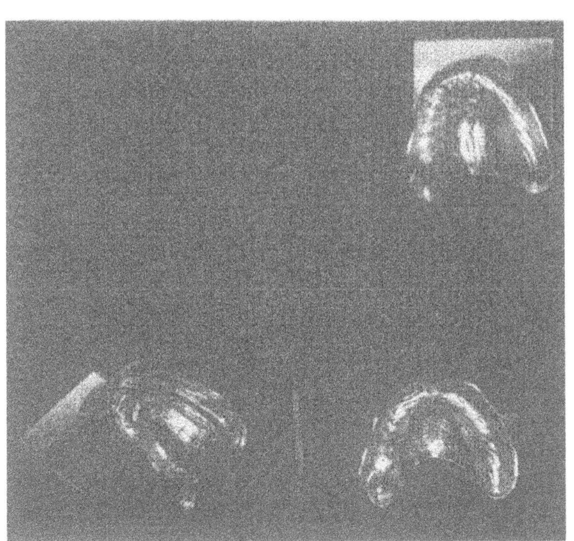

Fig. 3

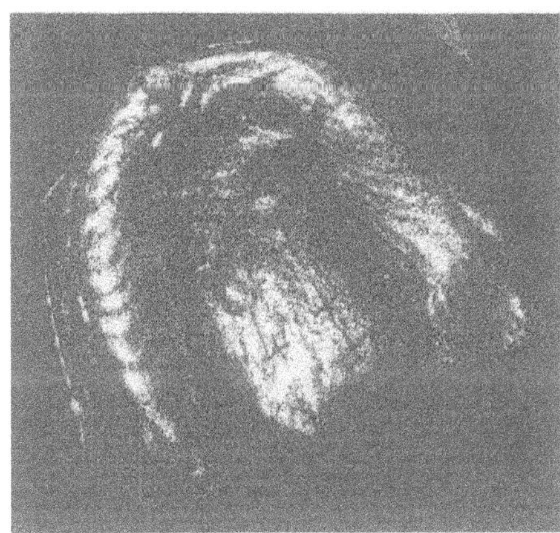

Fig. 4

7. Discussion

The increased demands on dentists from patients for well-fitting dentures made from thoroughly studied dental materials have made it necessary to develop a technique for measuring changes in three-dimensional form. The problems are the recurrent changes of pH, temperature and chemical properties of the saliva, combined with swelling and intermittent loading of the complexed formed denture. The use of holography for clinical measurements of deformation in complete upper dentures is a new application for this sensitive method. The advantages are that the whole deformation pattern of the anisotropic resin can be determined and followed three dimensionally, realistically and nondestructively at every point of the object [18, 19, 11, 16]. The stability of the equipment and the design for the repositioning is accurate for the purpose. The white-light holograms are very promising for measuring purposes because of the size of the image and the extremely good observation angle. Comparisons with some other base materials will be made on the same patient in a clinical follow up now in progress to verify and expand the use of this method. Evaluation of the holograms is in progress in collaboration with the Technical University, Budapest, Hungary.

8. References

1. J.E. Ruyter, S.A. Svendsen: J. Prosthet. Dent. 43, 1, 95 (1980).
2. F. Hardy: J. Prosthet. Dent. 39, 4, 375 (1978).
3. P.O. Glantz, G.D. Stafford: Swed. Dent. J. 66, 129 (1973).
4. C.M. Becker, D.E. Smith, J.I. Nicholls: J. Prosthet. Dent. 37, 3, 330 (1977).
5. C. Bessing, B. Nilsson, M. Bergman: Swed. Dent. J. 3, 221 (1979).
6. I.E. Ruyter, S. Espevik: Acta Odontol. Scand. 38, 169 (1980).
7. J.B. Woelfel, G.C. Paffenbarger, W.T. Sweeney: J.A.D.A. 65, 495 (1962).
8. J.B. Woelfel, G.C. Paffenbarger: J.A.D.A. 59, 250 (1959).
9. J.B. Woelfel, G.C. Paffenbarger, W.T. Sweeney: J.A.D.A. 62, 643 (1961).
10. P.M. Soni, J.M. Powers, R.G. Craig: J. Oral Rehabil. 6, 35 (1979).
11. I. Dirtoft, N. Abramson, U. Sandström: SPIE 211, 106 (1979).
12. C.M. Becker, D.E. Smith, J.I. Nicholls: J. Prosthet. D. 37, 4, 450 (1977).
13. A.J. de Gee, E.C. ten Harkel, C.L. Davidson: J. Prosthet. D. 42, 2, 149 (1979).
14. L. Hollender, P. Lockowandt: Odont. Revy 17, 55 (1966).
15. P.O. Glantz, G.D. Stafford: Swed. Dent. J. 66, 137 (1973).
16. I. Dirtoft, N. Abramson: in print.
17. S. Nakadate, N. Magome, T. Honda, J. Tsujiuchi: Opt. Eng. 20, 2, 246 (1981).
18. F. Albugues, C. le Floch: ICAO Bulletin may 1981.
19. A.E. Ennos, K. Thomas: Plastics and Rubber Processing, 165, Dec 1978.

Holographic Investigation of Tooth Deformations

T. Matsumoto, and T. Fujita
Osaka Prefectural Technical College, Department of Mechanical Engineering
26-12, Saiwaimachi, Neyagawa, Osaka, Japan, 572

R. Nagata, and K. Iwata
University of Osaka Prefecture, Department of Mechanical Engineering
College of Engineering, Mozu, Sakai, Osaka, Japan,591

T. Sugimura, and Y. Kakudo
Osaka Dental University, Department of Oral Physiology, Kyobashi
Higashiku, Osaka, Japan, 540

1. Introduction

A bad approximal contact may cause food impactions between teeth by masticatory forces. This may give rise to caries or other paradental diseases. Many studies on approximal contact relation have been presented based on clinical experience [1] and considered from a morphological standpoint [2] up to now. However, there have been only few studies on approximal contact relation investigated from a functional viewpoint. Therefore, investigations of the deformations of teeth subjected to occlusial or masticatory forces and the mechanical effect to their adjacent teeth are very important. Deformations have been measured, until now, with mechanical instruments [3], [4] such as dial gauges and strain gauges. Recently, double-exposure holographic interferometry has been found useful to measure three-dimensional deformations of teeth [5], [6]. This method, however, is restricted to the measurement of deformations within a certain time interval. Real-time holographic interferometry, on the other hand, allows measuring such deformations continuously.

Thus, real-time holography has been applied to measure movements of teeth subjected to occlusial or masticatory forces. The purpose of these measurements was to clarify the effect of contact between approximal teeth. Dry human mandibles were used as specimen and concentrated vertical load was applied to the second premolar.

2. Experiments

In the dry human mandible, used as specimen, each tooth was contacted closely with its approximal teeth. The alveolar crest did not show any resorption. After extracting the teeth silicone impression material was injected in the place of the periodontal membrane. Then, each tooth was replaced. The buccal side of the mandible was removed using a dental trimmer so that the deformation of the dental root could be observed. The mandibular base was fixed to a board made from a super hard plaster by means of alpha cyanoacrylate adhesive (Aron Alpha), so that the occlusal plane became parallel to the surface of the plaster board as shown in Fig. 1.

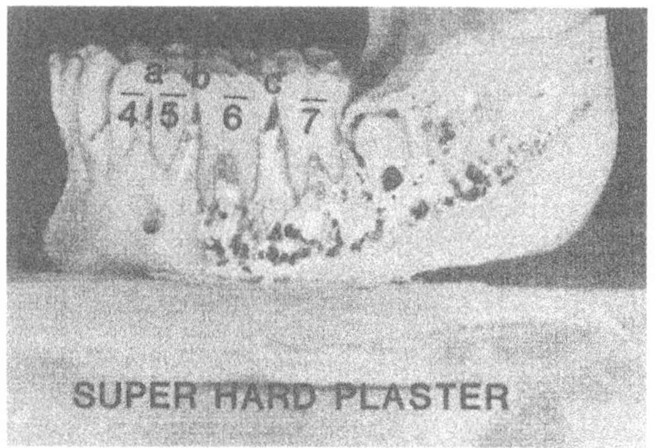

Fig. 1 Dry human mandible after removing its buccal side; (4): lower first premolar, (5): lower second premolar (6): lower first molar, (7): lower second molar. Symbols a, b and c show contact points between the teeth.

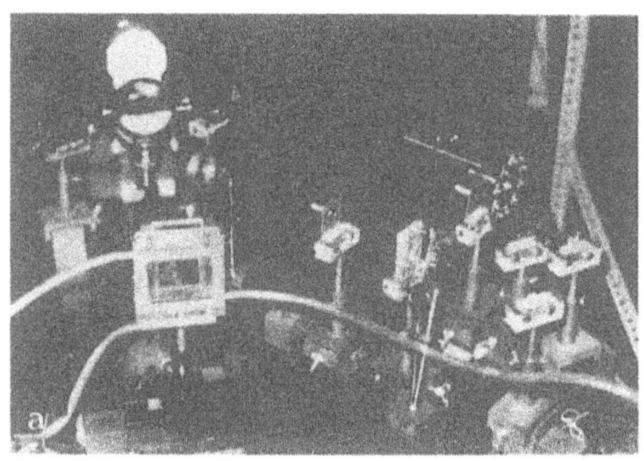

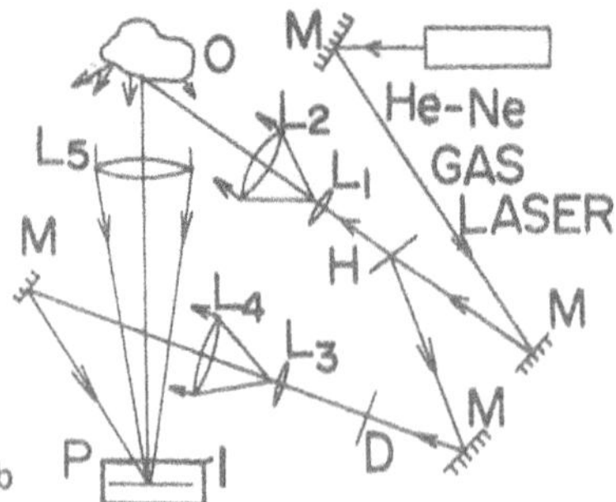

Fig. 2 Optical system (a) experimental apparatus, (b) schematic diagram of the optical system. M: plane mirrors; L_1, L_3: microscope objectives; L_2, L_4: collimating lenses; D: density filter; P: photographic plate; O: specimen; H: half mirror; L_5: Fourier transform lens; I: immersion cell.

Figure 1 shows the specimen to be examined. Vertical concentrated load acts on tooth (5). Symbols a and b denote mesial and distal contact points of tooth (5), respectively, and c characterizes the distal contact point between teeth (6) and (7).

Figure 2 outlines the optical system for real-time holographic interferometry.

The photographic plate P is kept in an immersion cell for in situ development [7]. Density filter D is used to control the intensity ratio of object and reference waves.

The loading device consists of an aluminium rod and a magnet stand having slide ball bearings in which the rod slides vertically. The rod is about 80 mm in height, 10 mm in diameter, and its weight is 46 g. Known weights are placed on the flat stage (50 mm in diameter) fixed to the upper surface of the rod. The lower end of the rod has conical shape so that the vertical force is concentrated on the central groove of the second premolar (5). Firstly, a hologram is recorded in the initial unloaded state. Then, weights are increased from 46 g to 226 g in steps of 5 g or 10 g. Teeth movements for various loads are observed by real-time holography. Thus, displacements at each point on the specimen can be measured quantitatively [8]. The measuring accuracy of displacement is 0.34 μm in this experiment.

3. Results

Figure 3 shows interference patterns which respesent the movement of teeth for various loads. In Fig. 3(a), we can see a similar number of fringes on the second premolar (5), the first premolar (4) and the first molar (6).

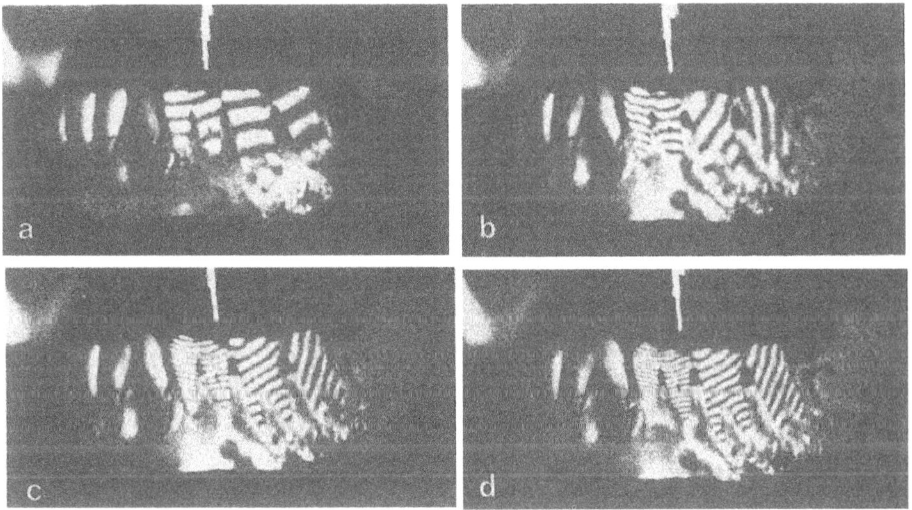

Fig. 3 Real-time holographic interferograms obtained with various loads; (a) 61 g, (b) 176 g, (c) 196 g and (d) 216 g.

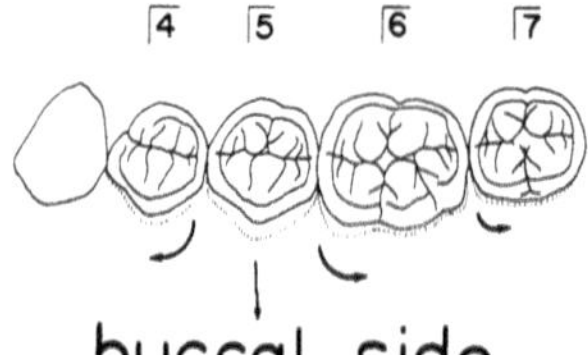

Fig. 4 Schematic diagram of the movement of the investigated teeth when the second premolar is subjected to vertical load

This means that the three teeth tilt slightly with the same gradient. Figure 3(b) shows a remarkable movement of approximal teeth caused by concentrated loading of the second premolar (5), and a little larger tilt of the mesial side towards the buccal side than that of the distal side of the first molar (6) can be observed. Furthermore, a large displacement of the first and second premolar can be found in Fig. 3(c) and 3(d).

In Fig. 3(d), medial and distal sides of tooth (5) have the same buccal inclination. From the interferogram in Fig. 3(d) movements of the teeth can be deduced as outlined in Fig. 4. In this figure dotted contours express teeth tilt towards the buccal side and the arrows mark teeth rotation or inclination directions.

Relative displacement of the second premolar (5) can be calculated by using the interferograms in Fig. 3. The result is shown in Fig. 5, where the abscissa illustrates the distance from the root apex of this tooth.

This figure shows that only tilt but not translation of the tooth occurs, because the impression material plays the role of a buffer.

Next we removed contact points of the second premolar (5). Figure 6 shows a real-time holographic interferogram of that specimen, where the load acting on the second premolar (5) was 56 g.

Comparing Fig. 6 and Fig. 3(a), it can be noticed that the interference fringe density on tooth (5) is higher in Fig. 6 than that in Fig. 3(a). Therefore, it can be deduced that occlusal force is distributed via contact points to approximal teeth, thus reducing the force acting on the bone.

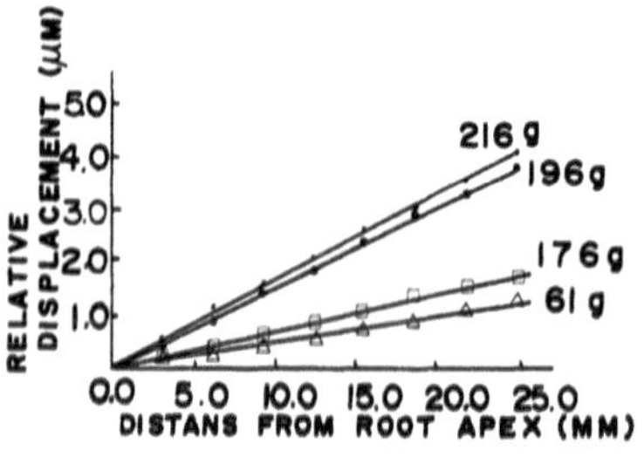

Fig. 5 Relative displacement of the second premolar (5) for different loads

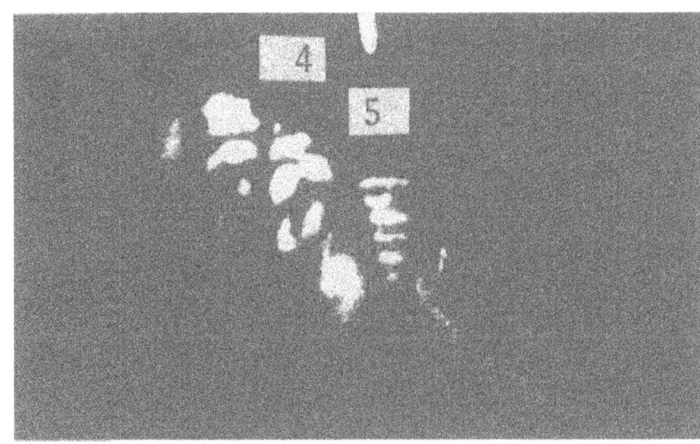

Fig. 6 Real-time holographic interferogram obtained for 56 g load after removing contact points of the second premolar (5)

4. Conclusion

Real-time holographic interferometry turned out to be useful in dental investigations. Using this holographic method for quantitative measurements of the movements of the teeth in a dry human mandible, we found that close contact between approximal teeth reduces stress concentration in the teeth caused by occlusal or masticatory forces.

5. References

1 T. Hirshfeld: J. A. D. A. 17, 1504 (1930).
2 S. Murata: J. Prosthet. Dent. 4, 673 (1954).
3 D. C. A. Picton: Arch. Oral Biol. 8, 109 (1963).
4 H. R. Mühlemann: Oral Surg. Oral Med. Oral Path. 4, 1220 (1951).
5 P. R. Wedendal and H. I. Bjelkhagen: Appl. Opt. 13, 2481 (1974).
6 T. Matsumoto, T. Fujita, R. Nagata, T. Sugimura, and Y. Kakudo: In *Holography in Medicine and Biology*, ed. by G. von Bally, Springer Series in Optical Sciences, Vol. 18 (Springer, Berlin, Heidelberg, New York 1979) pp. 170
7 P. Hariharan and B.S. Ramprasad: J. Phys. E. 6, 699 (1973).
8 T. Matsumoto, R. Nagata, and K. Iwata: Appl. Opt. 12, 961 (1973).

Otoscopic Investigations by Holographic Interferometry: A Fiber Endoscopic Approach Using a Pulsed Ruby Laser System

G. von Bally

University of Münster, Ear-Nose-Throat-Clinic
Medical Acoustics and Biophysics Laboratory, Kardinal-von-Galen-Ring 10
D-4400 Münster, Fed. Rep. of Germany

1. Introduction

Otoscopy, as the name implies, deals with the visual inspection of the ear, especially the middle ear. Our research group is involved in this field, since we are interested in a holographic analysis of the vibration patterns of the eardrum in man. Our interest in such an analysis is caused by its cabability to investigate the sound transfer function of the middle ear, as well as to follow up healing processes after tympanoplastic surgery, and to provide the possibility of a differential diagnosis of disorders in the middle ear mechanics without the necessity to open the tympanic cavity surgically. The latter can be accomplished by comparison of "regular" and "pathological" patterns of tympanic membrane vibrations as could be proved by investigations in man in vitro [1] as well as in vivo [2].

In using holographic techniques for vibration analysis of human tympanic membranes in living man quite a number of problems arise from the anatomical and physiological properties of the human peripheral hearing organ. Some of the main obstacles are:

- the difficult optical access through the narrow and somewhat curved outer ear canal;
- unsymmetrical and non-sinusoidal vibrations of biological membranes;
- technical difficulties like unwanted light reflexes and speckles caused by the necessarily small apertured otoscopic instruments; and
- the unfavourable optical, especially spectral properties of tissue - here that of human eardrums - for holographic recording purposes.

Ways to overcome the mentioned difficulties are not only important in otoscopic investigations, but also of general interest for applications of holography in the biomedical field, as can be gathered from this and the following articles in this book [3,4,5,6], demonstrating possible solutions worked out in our group. This article here is confined to the problem listed first: the difficult optical access to the tympanic membrane.

2. Experiments

The first approach to record double-exposure interferograms of tympanic membrane vibrations in living man by means of a Q-switched ruby laser, which is necessary for in vivo applications of conventional holographic techniques, was the development of a special closed acoustic system for phase determination of the eliciting sound pressure oscillation at the location of the tympanic membrane. The optical arrangement of this apparatus was a rigid one, and a usual speculum acted as an adapter to the outer ear canal.

The rigidity of the set-up turned out to be inconvenient for the patient as well as for the investigating physician. Thus, the next step was the development of a flexible system with a fiber bundle as light guide for the illuminating object beam. Although holographic recordings with good image quality could be achieved, proving the applicability of standard endoscopic fiber bundles for object illumination using a Q-switched ruby laser [2], this arrangement still showed shortcomings for holographic otoscopy:

- Using a speculum results in a system, which provides only straight-line observation with a small angular field. Thus, due to the great variation in curvature of the outer ear canal only selected patients could be investigated.
- The reference beam guiding system was still rigid since mirrors were used in the reference beam path.

Thus, a holographic arrangement with an endoscope as imaging device and a monomode fiber in the reference beam path was developed (Fig. 1), although in the literature (e.g. in [7]) the limited power transmission capability of monomode fibers is described as the limiting factor for the use of Q-switched ruby laser systems in the holographic recording process. Basic features of the fiber-holo-endoscopic set-up are:

A beam splitter (BS) separates the light, emanating from a Q-switched ruby laser (RL) (for recording) or a He-Ne laser (HL) (for adjustment), respectively, into an object beam, that is guided into the illuminating light guide fiber bundle (IFB) of the endoscope (E) for illumination of the object (O), and a reference beam, which is launched via a specially coated lens (L_1) into the monomode fiber (MMF) (type = Schott NW 12/2; length = 30 cm; core: diameter = 2.5 µm. n = 1.6484; cladding: diameter = 154 µm, n = 1,6274; theor. NA = 0.262). The light emerging from the monomode fiber is collimated by a lens (L_2) before propagating onto a holographic plate (H) (Holotest 10E75). A photodiode (PD) picks up the laser pulses for control purposes. For in situ reconstruction a Krypton laser (KL) is used. The reconstruction beam passes a lens (L_3), which focusses this beam onto the entrance face of the monomode fiber via a mirror (TM) that can be appropriately tilted and positioned in the pulsed ruby laser reference beam path.

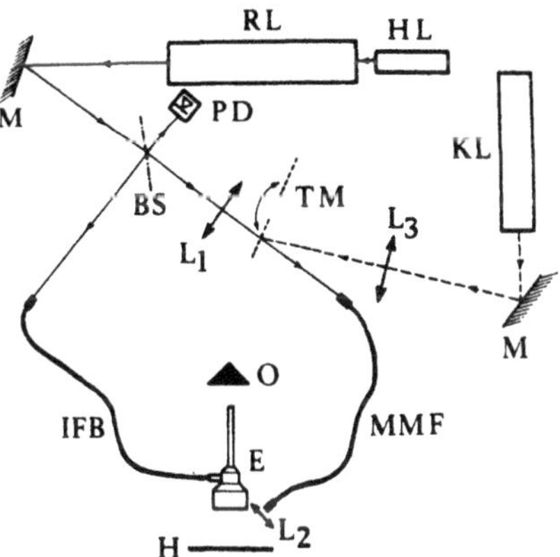

Fig. 1: Experimental set-up for fiber-holoendoscopic recording using a Q-switched ruby laser.

The energy density in the plane of the entrance face of the monomode fiber has to be kept within an upper threshold, that does not have to be exceeded in order to avoid cracks of that fiber, and a lower one, which provides minimum light energy at the exit face of the monomode fiber necessary for exposure of the holographic plate. This energy density range turned out to be rather small. The appropriate energy density on the entrance face of the fiber was selected by variation of the distance between that surface and the launching lens (L_1) (f = 200 mm). An optimal distance was found at 300 mm. Using a thermopile a double-pulse energy of the total ruby laser

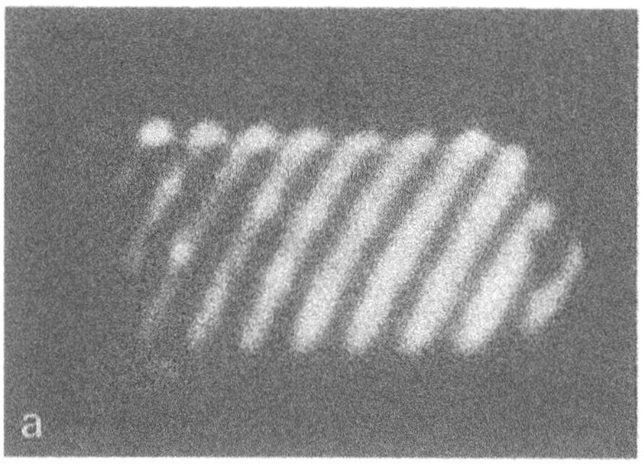

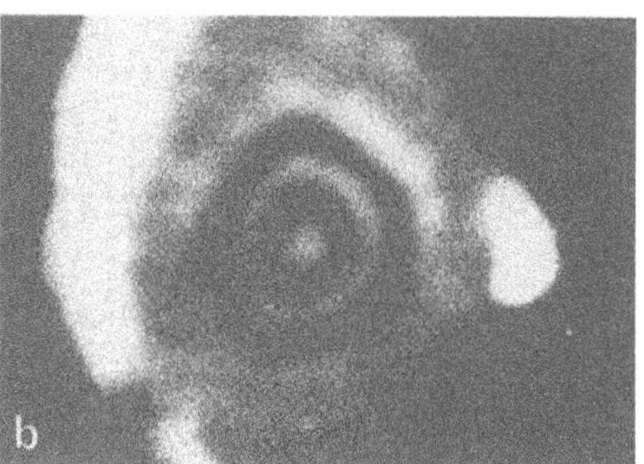

Fig. 2: Double-exposure holograms recorded with the experimental arrangement outlined in Fig. 1, but no endoscope used for imaging; vibration pattern a) of a bone conductor, b) of an artifical tympanic membrane in a plane scale model of the human outer ear canal recorded from the "middle ear" side.

beam (ø = 1.65 mm) of 4 mJ could be measured in the entrance face plane. From simple geometrical considerations an energy density in that plane of 750 mJ/cm² can be estimated. The light energy in the exit face plane of the monomode fiber was below the measuring range of the energey meter used, i.e. <0.2 mJ.

First tests without the endoscope, but with object illumination via the illuminating light guide (multimode) fiber bundle and the monomode fiber in the reference beam path, resulted in double-exposure holograms with good image quality and fringe contrast, as shown in Fig. 2, demonstrating the vibration patterns of a bone conductor (Fig. 2a) and those of an artificial tympanic membrane in a plane scale model of the human outer ear canal recorded from the "middle ear" side (Fig. 2b).

In a next step a so-called otoscope, which is a small endoscope, specially developed to provide an easy optical access to the tympanic membrane

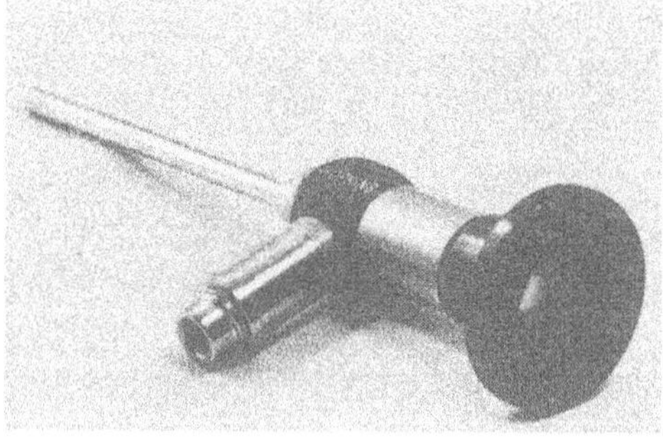

a)

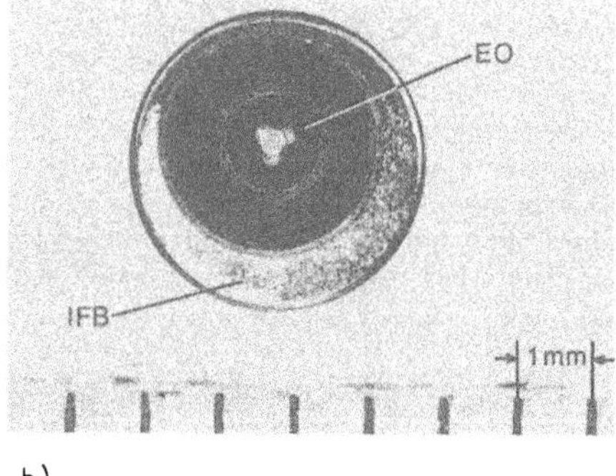

b)

Fig. 3: Special otoscope for holographic recording purposes:
a) instrument with adapter for the illuminating light guide fiber bundle;
b) tip of the instrument (EO: endoscopic objective, IFB: illuminating light guide fiber bundle).

through the curved and narrow outer ear canal for otoscopic inspection, was used as imaging device. Figure 3a shows this otoscope, the ocular of which is specially modified for holographic recording purposes (enlarged exit pupil and image distance compared to the standard version), so that a holographic plate can be inserted and the reference beam be introduced between the observer's eye and the otoscope ocular. The length of this instrument is 10 cm. Image transmission is provided by a Hopkins optic system. A closer look at the tip of the otoscope is presented in Fig. 3b, demonstrating that the endoscopic optic has a limiting aperture of less than half a millimeter. Thus, a considerable speckle noise can be expected in holograms recorded with this instrument.

Figure 4 shows a double-exposure hologram of the vibrating artifical tympanic membrane in the mentioned plane scale model of the human outer ear canal, recorded through this otoscope by means of a pulsed ruby laser and the reference beam guided by the monomode fiber. As expected speckle noise is increased compared to the holographic interferograms shown in Fig. 2, which are taken without the limiting aperture of the otoscope, but the fringe contrast appears still satisfying.

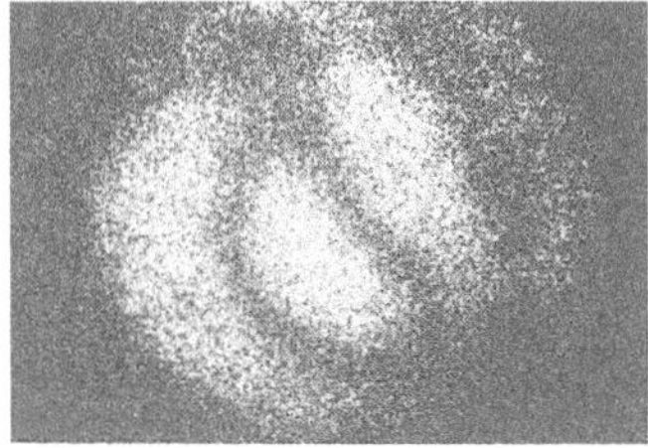

Fig. 4: Double-exposure hologram of the artifical tympanic membrane in the plane scale model of the human outer ear canal (cp. Fig. 2b), recorded with the set-up outlined in Fig. 1, through the special otoscope shown in Fig. 3

Further reduction of speckle noise may be achieved - besides by the use of rainbow holography [4] - by application of graded index fibers in order to increase the effective aperture compared to that of standard otoscopes without changing the outer diameter of the instrument. This can be attained by combining the function of an illuminating light guide and an image guide system using the lens-like imaging properties of a graded index fiber. Preleminary tests with special samples of such fibers show promising results, encouraging the development of a special holo-otoscopic instrument.

3. Conclusions

It could be proved - to the author's knowledge for the first time - that double-exposure holograms can be recorded using a monomode fiber as flexible reference beam guide in combination with a Q-switched ruby laser as coherent light source. Thus, a holographic set-up using a light guide (multimode) fiber bundle for object illumination, a special otoscope as imaging device, and a monomode fiber for reference beam guiding could be developed. First model experiments demonstrated the feasibility of the arrangement for holographic recording of human tympanic membrane vibrations. This optical set-up can be regarded as a first step in the development of a flexible and thus easy-to-handle holo-endoscopic camera for application on living objects.

The author would like to thank Karl Storz GmbH, Tuttlingen, FRG, and Schott Glaswerke, Mainz, FRG, for placing at our disposal the special otoscope and samples of the monomode fiber, respectively.

REFERENCES

1 VON BALLY, G.: in: E. Marom and A.A. Friesem (eds.): Applications of holography and optical data processing, Pergamon Press (1977), 593-602.
2 VON BALLY, G.: in: G. von Bally (ed.): Holography in Medicine and Biology, Springer Series in Optical Sciences, Vol. 18 (Springer, Berlin, Heidelberg, New York 1979) pp. 198-205
3 SIEGER, C. and R. RÖHLER: in this issue.
4 MARTIN, C. and C. SIEGER: in this issue.
5 SIEGER, C.: in this issue.
6 HEEKE, G.: in this issue.
7 HADBAWNIK, D.: Optik 45, (1976) 21-38.

This work was supported by grants of the Deutsche Forschungsgemeinschaft (SFB 88/B3).

Vibration Analysis of Tympanic Membranes: Measurement of Vibration Waveforms Using Sampling Series of Double-Pulse Interferograms

C. Sieger, and R. Röhler

Institut für medizinische Optik der Universität München, Theresienstr. 37
D-8000 München 2, Fed. Rep. of Germany

1. Introduction

With all known holographic recording techniques, including those using temporally modulated reference beams, either time averaged amplitudes or object deformations between two exposures were measured, but not the waveform itself.

For the analysis of vibrating systems it is generally necessary to measure their waveform, too, i.e. to carry out a temporal Fourier analysis of their motion. Preliminary investigations show that especially from alterations of single Fourier components of a complex waveform specific changes in these systems can be detected much more sensitively and accurately than from the overall amplitude.

Especially pathological alterations in the sound transmitting system of the human auditory organ will have a specific influence on the vibration behaviour of the tympanic membrane. Methods for measuring this alteration in amplitude and form could make possible a non invasive differential diagnosis of the cause for a conduction deafness.

Until now three holographic methods were reported, suited to measure vibration waveforms only in a few special cases. All of them were only applicable to interferometrically stable objects. One technique uses the well-known real time interferometry together with cinematography to record the continously changing interference fringes [1]. The second method uses a rotating disk with a number of radial slits, placed in front of the holographic plate [2,3]. If the rotation speed is adjusted to the object vibration, a corresponding number of object positions were recorded along the one holographic plate. A second exposure of the same plate with the object at rest yields a trace of very narrow and grainy interferograms, similar to those, obtained by a normal double-exposure technique. The third method uses holograms with frequency shifted reference beams [4,5,6,7]. This has some additional disadvantages [6,7], especially that it is suited only for waveform measurements of objects, vibrating with very small amplitudes.

We present a very simple technique for the measurement of the temporal Fourier coefficients of any periodic movement of arbitrary 3-dimensional objects. Among other things the method has the great advantage that it allows, in combination with a pulse laser, in vivo measurements of interferometrically unstable objects.

2. Theory of Waveform Measurements Using Double-Pulse Holograms

An oscillating object can be recorded, as known, by light pulses, released in 2 arbitrary positions M_1 and M_2 of its vibration cycle. In reconstruction the 2 holographically stored object waves produce an interference fringe system. This is essentially a function of the distance between both positions M_1 and M_2, independent of the executed vibration waveform. From the intensity variation I of the fringe system the instantaneous amplitudes or deflections of each object point r can be determined according to

$$I(\vec{r}) = I\ (\vec{r})\ \cos^2 \frac{\{[2\pi/\lambda][\cos\phi_1 + \cos\phi_2][M_1(\vec{r}) - M_2(\vec{r})]\}}{2}\ . \quad (1)$$

I describes the intensity of a point r on the reconstructed image, I_B is the image intensity of the same point if the object is stationary. M_1 and M_2 describe the two fixed positions, occupied by the membrane during both exposures. λ denotes the laser wavelength, ϕ_1 and ϕ_2 are the angles between the vibration vector of the membrane and the directions of illumination and observation, respectively.

A complete series of such double-pulse interferograms can be recorded, using a pulse laser or a Pockels cell together with a cw-laser. The first pulses $M_{1,n}$ must always be performed at the same, but arbitrary point of the vibration cycle. The corresponding second pulses $M_{2,n}$ have to run through the entire vibration cycle, released at equidistant time intervals (see Fig.1). From these interferograms the instantaneous amplitudes, i.e. the distances of each object position $M_{2,n}$ relativ to the ground position M_1 can be determined easily.

The problem to determine the true object oscillation M(t) from n such values ΔM_n was solved in information theory by the sampling theorem [8,9]. The set of instantaneous amplitudes ΔM_n

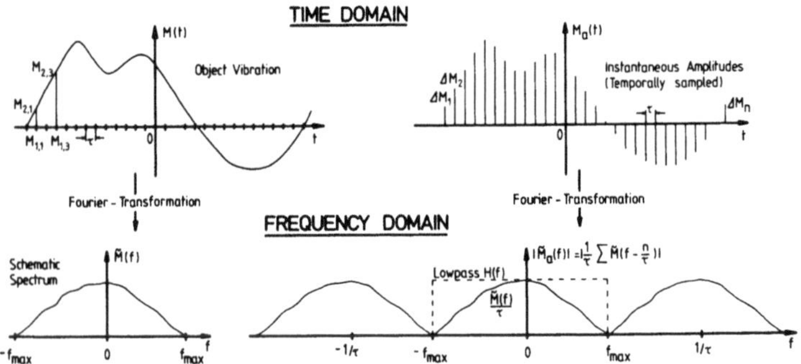

Fig.1 Theory of waveform measurement by a sampling series

forms the sampling signal $M_a(t)$. The frequency spectrum of this signal can be obtained formally by Fourier transformation. Thus, passing from the time to the frequency domain it is seen that the sampling signal M_a has a periodic spectrum with frequency $1/\tau$.

If the following 2 weak requirements are fullfilled:

1) The function M(f), i.e. the Fourier transform of the object motion function M(t), has to be frequency limited;
2) The exposure frequency of the interferograms has to be at least twice as large as the highest frequency component occuring in the object spectrum;

then the sampling theorem states: Each periodic motion function, containing no higher frequency than the limiting frequency f_{max}, is already defined definitely, if one knows its values in the sampling points, whose distance from each other is determined by $1/f_{max}$.

3. Conclusions for Practical Application of a Holographic Sampling Series

For practical application the following items can be derived:

1) The absolute positions of the illuminating pulses (=sampling points) have no influence on the result.
2) For the measurement of n temporal Fourier coefficients of an oscillation one needs 2n interferograms.
3) Sampling an oscillation with more interferograms than required is allowed.
4) If a vibration is sampled by less interferograms than required, then, in most cases, only the corresponding high frequency components of the vibration are not recorded. In such cases the basic form of the vibration waveform remains unaffected.
5) The amplitude of the investigated oscillation has no influence on the number of sampling interferograms required.

Using enough double-exposure interferograms, each real and therefore frequency limited object vibration can be measured. A very simple evaluation scheme results, based upon the sampling theorem, if a sampling series of 12 interferograms is recorded. Fig.2 shows 7 interferograms of such a series, recorded on thermoplastic film. The instantaneous amplitudes ΔM_n, measured by means of these interferograms, were inserted in the evaluation scheme of (Fig.3). This yields at once 6 Fourier coefficients of the mechanical waveform, executed by the object. For the measurement of vibrations containing higher frequencies, a quite analogue scheme exists, using 24 sampling values and yielding 12 Fourier coefficients.

As demonstrated by our group earlier [10], in vivo interferograms of the human tympanic membrane can be achieved, using sufficiently short double-pulse illumination. By recording a sampling series of such in vivo interferograms the waveforms of tympanic membrane vibrations can be measured, too. In such a series the

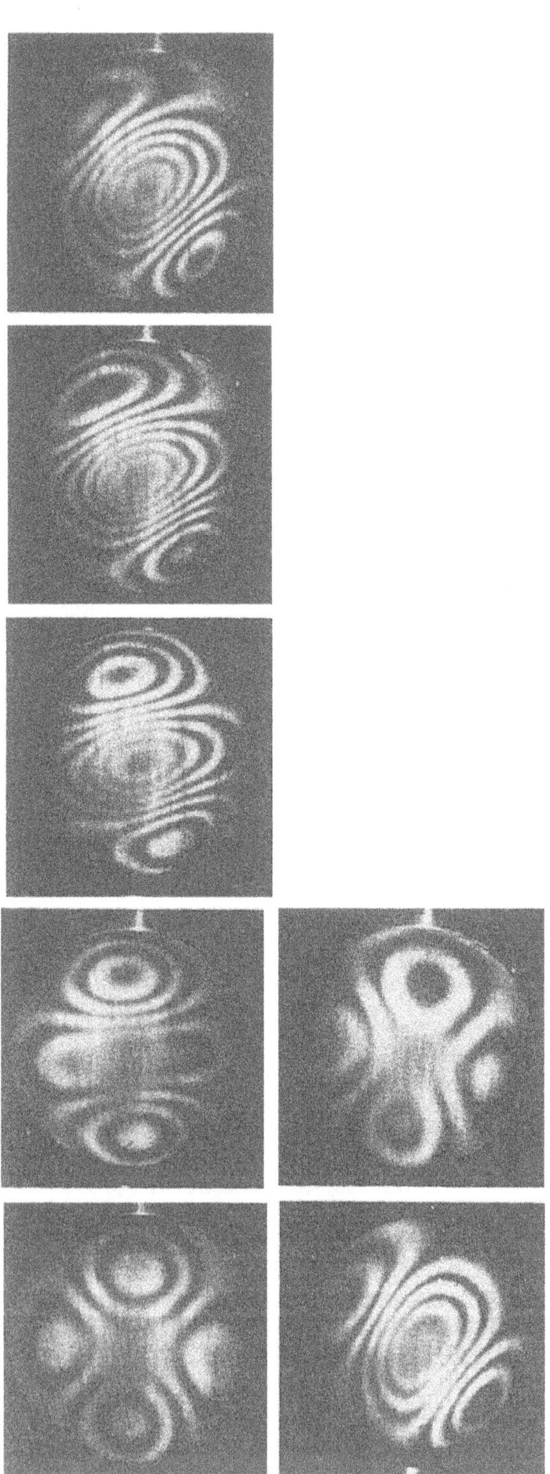

Fig.2 The first 7 holograms out of a total sampling series of 12 double-exposure interferograms, recorded over the entire vibration period at 12 equal time intervals

Calculation Scheme

$M - M_n = \Delta M_n \qquad n = 12$

	ΔM_0	ΔM_1	ΔM_2	ΔM_3	ΔM_4	ΔM_5	ΔM_6
		ΔM_{11}	ΔM_{10}	ΔM_9	ΔM_8	ΔM_7	
Sums	S_0	S_1	S_2	S_3	S_4	S_5	S_6
Differences		d_1	d_2	d_3	d_4	d_5	

	S_0	S_1	S_2	S_3	d_1	d_2	d_3
	S_6	S_5	S_4		d_5	d_4	
Sums	σ_0	σ_1	σ_2	σ_3	ε_1	ε_2	ε_3
Differences	δ_0	δ_1	δ_2		γ_1	γ_2	

	Cosinus								Sinus					
1	σ_0 σ_2	σ_1 σ_3	δ_0		σ_0	$-\sigma_3$	δ_0	δ_2	ε_3				ε_1	ε_3
0,866				δ_1						ε_2	γ_1	γ_2		
0,5			δ_2		$-\sigma_2$	σ_1			ε_1					
Sums	I	II	I	II	I	II	I	II	I	II	I	II	I	II
I+II	$6A_0$		$6A_1$		$6A_2$				$6B_1$		$6B_2$			
I−II	$12A_6$		$6A_5$		$6A_4$		$6A_3$		$6B_5$		$6B_4$		$6B_3$	

$$\Phi(x,t) = \frac{A_0}{2} + \sum_{m=1}^{6} A_m \cos m\omega t + \sum_{m=1}^{5} B_m \sin m\omega t$$

Fig.3 Evaluation scheme for the Fourier analysis of waveforms

time interval between both light pulses amounts to maximally half of the vibration period. This means, that for high frequency vibrations rigid body motions should have a neglegible influence on the quality of the interferograms. If fast rigid body motions should occur, however, then their influence on the interference fringe pattern can be removed completely by using double-pulse sandwich holograms [11].

4. Experiments

By means of a sampling series of stroboscopic double-exposure interferograms the vibration waveform of a loudspeaker, driven by a known electric signal, was investigated (Fig.4). The full line shows the electric signal. ΔM_n are the measured amplitudes. Different vibration directions relative to the position at rest have to be marked by different signs of the corresponding values. The choice of the positive direction has no influence on the values of the calculated Fourier coefficients. The dotted line represents the vibration waveform measured. In a second

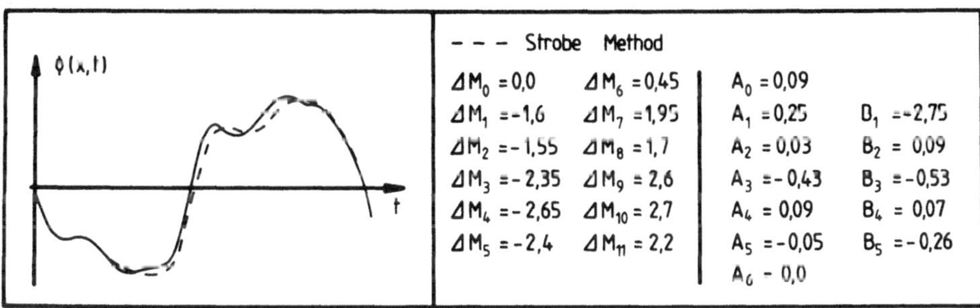

Fig.4 Holographic Fourier analysis of a given waveform

experiment an unknown waveform was analysed. The result was compared with that of another method, which was reported earlier [6], [7]. The agreement in this case was quite well, too.

5. Accuracy and Further Improvement of the Method

The accuracy of measuring the waveform is essentially limited by uncertainties in the interpolation between two successive interference fringes, which amounts to ca.1/6 of a fringe. That means, that amplitudes can be measured with an accuracy of about 30 nm, independent of the absolute value. Therefore the method proposed yields the better results the higher the amplitudes. Thus waveforms of high amplitude vibrations can be measured very accurately, provided, of course, that enough interferograms were recorded.

For the analysis of low amplitude vibrations, as e.g. vibrations of the tympanic membrane under acoustical stimulation at physiological sound pressure levels, the interpolation accuracy between interference fringes has to be increased. This improvement can be achieved by means of the dual-frequency or heterodyne interferometry [12]. This technique can also be applied to in vivo measurements and reduces the interpolation error at least by the factor 15. Together with the heterodyne technique holographic sampling of tympanic membrane vibrations can become a sensitive tool for detecting deviations from a normal or expected vibration mode.

6. References

1 Pierattini, G.: Opt. Commun. 5, 41 (1972)
2 Fryer, P.A.: Appl. Opt. 9, 1216 (1970)
3 Fryer, P.A.: Appl. Opt. 11, 1642 (1972)
4 Dallas, W.J. and Lohmann, A.W.: Opt. Commun. 13, 134 (1975)
5 Cutter, D.: Thesis of the Friedrich-Alexander-Univ., Erlangen (1976)
6 Sieger, C. and Röhler, R.: In *Holography in Medicine and Biology*, ed. by G. von Bally, Springer Series in Optical Sciences, Vol. 18 (Springer, Berlin, Heidelberg, New York 1979) pp. 247
7 Sieger, C.: Thesis of the Ludwigs-Maximilians-Univ., München (1980)
8 Whittaker, E.T.: Proc. Royal Soc. Edinburgh 35, 181 (1915)
9 Shannon, C.E.: Bell Syst. Techn. J. 27, 379 and 623 (1948)
10 Von Bally, G.: In *Holography in Medicine and Biology*, ed. by G. von Bally, Springer Series in Optical Sciences, Vol. 18 (Springer, Berlin, Heidelberg, New York 1979) pp. 198
11 Abramson, N. and Bjelkhagen, H.: Appl. Opt. 17, 187 (1978)
12 Dändliker, R., Marom, E. and Mottier, F.M.: J. Opt. Soc. Am. 66, 23 (1976)

This work was supported by grants of the Deutsche Forschungsgemeinschaft (SFB 88/B 3)

Application of Rainbow Holography for Speckle Reduction in Tympanic Membrane Interferometry

C. Martin, and C. Sieger

Institut für medizinische Optik der Universität München, Theresienstr. 37
D-8000 München 2, Fed. Rep. of Germany

1. Introduction

When examining tympanic membrane oscillations with methods of holographic interferometry the well known phenomenon of speckle patterns is encountered, as is the case in most fields of coherent optics [1]. Although the average size of speckles is about the size of the Airy discs [2], the image quality is affected by speckle granulation [3]. Principally the problem generated by speckles may be removed by choosing large apertures. But there are situations where a small aperture is unavoidable. Such limited apertures exist, e.g. in endoscopic holography. In such situations an improvement of image quality can only be achieved by the application of speckle reduction procedures during hologram reconstruction.

Some speckle reduction procedures are already well known. They use partially coherent light or change of the modulation transfer function. Every speckle reduction procedure works by superposition of a number of independent or partially correlated speckle patterns. E.g., speckle reduction is achieved by superposition of holograms taken at different wavelengths.

The aim of this work is to show how rainbow holography, a white light holographic technique, may be used for speckle reduction. In this technique a continous spectrum of wavelengths achieves the speckle reduction. Normally the rainbow technique has the disadvantage, that the aperture has to be reduced by a narrow slit diaphragm. However, this is no additional disadvantage in cases, where the system aperture is already small enough, that it may be used instead of the slit diaphragm. An example for such situations might be the ear speculum in tympanic membrane interferometry.

2. Rainbow Holography

Rainbow holography is an elegant simple technique for producing white light holograms. The name orginates from the fact, that during reconstruction the white light is selected into a continuous spectrum and the viewer can see the reconstructed image in rainbow colors.

There are two techniques of constructing rainbow holograms, a one-step [4,5] and a two-step process [6,7]. We consider only

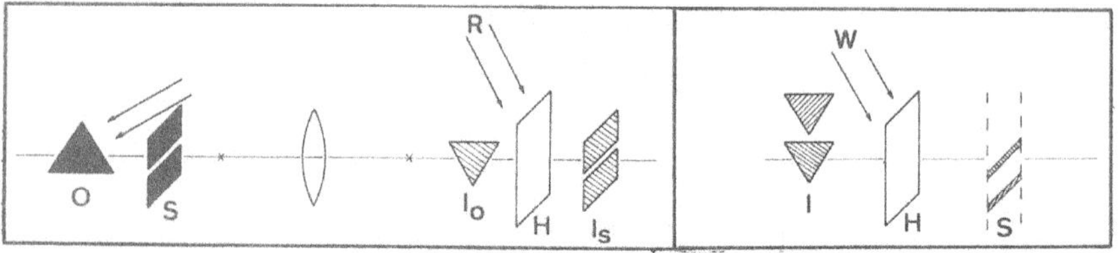

Fig.1 Construction and reconstruction of a rainbow hologram

the one-step process investigated by YU and CHEN. Fig.1 shows this technique.

With suitable projection of the object and a slit the same situation is achieved as in the two-step rainbow technique. The lens produces real images I_o, I_s of the object O and the slit S during recording of the rainbow hologram (Fig.1, left). When the hologram is illuminated with a white light source W, each wavelength forms an image of the slit at different vertical positions S (Fig.1, right). An observer would see the entire image I in only one color when he positioned his eyes at any one slit image. The wavelength selection can be achieved only with the help of a small slit. In rainbow holography the geometry of an optical spectroscope is imitated (see Fig.2).

3. Color Blur of Rainbow Holograms

Although the image produced by a rainbow hologram is rather sharp, some image blur is generated by spectral-wavelength spread. The

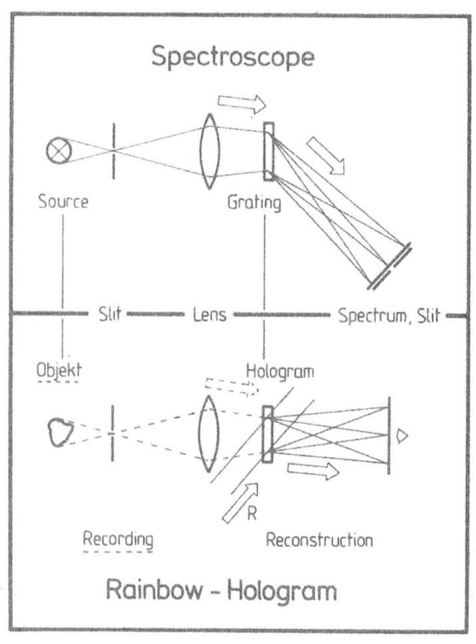

Fig.2 All elements of a spectroscope are present also in the rainbow holographic process. The difference is, that in rainbow holography these elements are not present simultaneously, and that not the object beam, but the reference beam is diffracted.

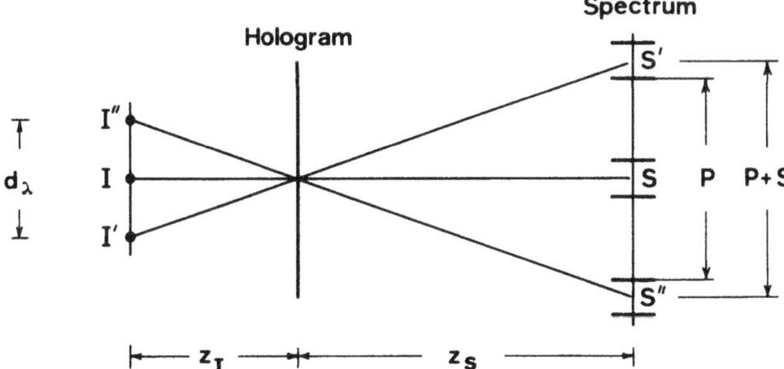

Fig.3 Color blur of rainbow hologram

color blur may be estimated by a simple geometrical consideration (see Fig.3) without complicated analysis [8,9,10].

Consider any image point I at the distance z_I from the hologram plane and the corresponding slit image S at the distance z_S. During white light reconstruction not only one slit image S of one color, but a continuous set of slit images and corresponding image points is generated. Since image points on the hologram plane are fixpoints, any change of the wavelength is equal to a small rotation of the beams around their fixpoints on the hologram. Which colors are seen from the viewer depends on the viewers pupil diameter P. The two slit images S' and S" are just no more seen through the pupil. They have the mean distance P plus S. From two triangles we get a formula for the color blur d_λ:

$$d_\lambda = z_I \left(\frac{P+S}{z_S}\right).$$

The blur is proportional to the distance z_I. Therefore, sharp rainbow holograms of deep objects can only be recorded with narrow apertures S and P.

The color blur is not only a disadvantage of rainbow holography. There is an important positive aspect, because the color blur reduces the speckle pattern. By superposition of speckle patterns of different wavelengths the contrast will be reduced remarkably.

4. Optimal Image Quality

For optimal image quality the color blur should only be as large as it is necessary for good speckle reduction. Any additional color blur would deteriorate the image. For a good speckle reduction the blur must be only as large as the speckle size, i.e. the blur due to diffraction d_D should satisfy the following relation:

$$d_D = 2\lambda(z_I + z_S)/S.$$

With the requirement for optimal imaging, i.e. color blur equal to diffraction blur, a formula of optimal image distance from the hologram plane is calculated under the assumption P equals S, i.e. the viewers pupil is equal to the slit image size:

$$Z \cong \lambda / \alpha^2 .$$

Here λ is the average reconstruction wavelength and α denotes the aperture. Z is not only the optimal image distance but also the depth of good sharpness and good image quality. The formula shows that an optimal rainbow image is obtained only for small apertures at finite distance from the hologram plane. For example, if the aperture is only as large as 1:100, the optimal image distance is only one centimeter for blue light observation. For larger apertures the best imaging is at a distance near to zero. In other words, the rainbow holograms have to be recorded as image plane holograms.

5. Application of Rainbow Holograms

After this short discussion of the properties of rainbow holograms we give an example, how image quality may be improved by rainbow technique.

The following two requirements have to be met: the object depth and the inherent aperture should be small. Both requirements are met in tympanic membrane interferometry, which is chosen as the example to be discussed. The small aperture of the ear speculum may be used instead of the slit stop in normal rainbow holography. The object depth is not large, because for small oscillations the interference fringes are located near the object surface. Fig.4 shows schematically the arrangement for the construction of a one-step rainbow hologram of the tympanic membrane.

A real improvement of the image quality may be achieved by this application of rainbow holography, because no decrease of aperture is necessary. In these applications the rainbow holo-

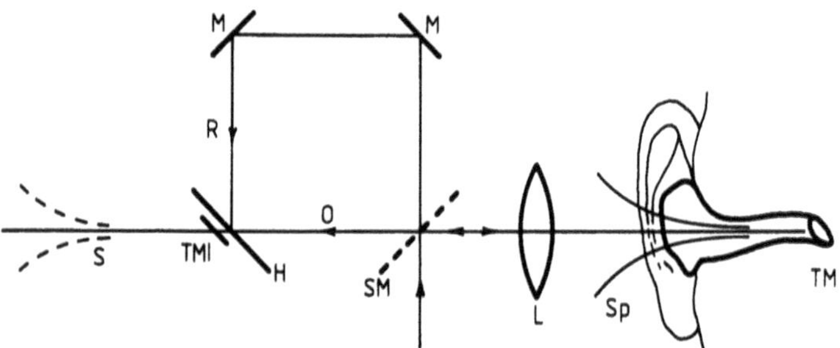

Fig.4 Application of rainbow holography in tympanic membrane interferometry

TM tympanic membrane, TMI image of TM,
Sp speculum, S image of Sp,
L lens, M and SM mirrors,
H hologram

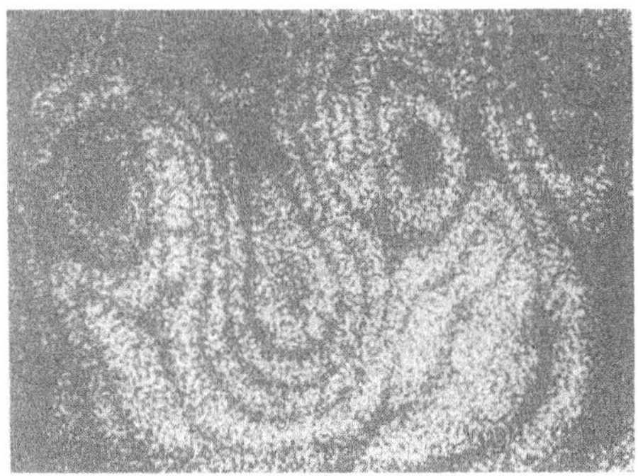
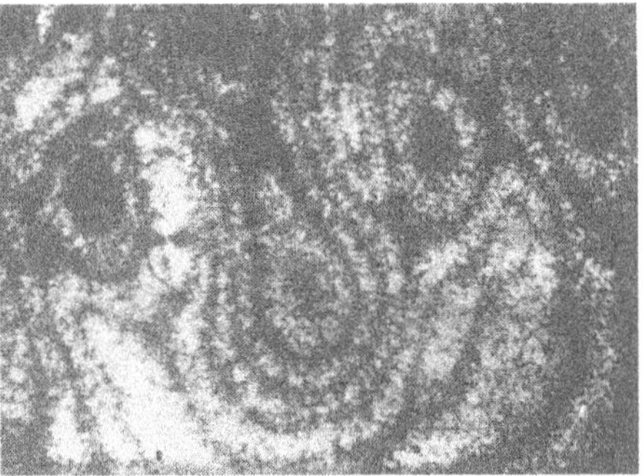

Fig.5 Speckle reduction by rainbow holography

graphy is equal to a quasi image plane holography, where an optimal image is achieved by a small blur.

In model experiments we recorded rainbow holograms of an oscillating membrane by use of circular apertures. In Fig.5 the left side image shows the coherent reconstruction with large speckle noise. For the right side image the rainbow hologram was illuminated with white light. The speckle pattern is reduced and the detectability of the interference fringes is improved. But certainly, the contrast of the interference fringes is somewhat lower as with coherent illumination.

In conclusion we can say, rainbow holography is suitable for objects of small depth and small inherent apertures. The advantages are the reconstruction with white light sources, the resulting speckle reduction and the easy processing. An additional advantage is the comfortable viewing, because the small aperture is smeared into a broad spectrum.

6. References

1. G. von Bally (ed.): *Holography in Medicine and Biology*, Springer Series in Optical Sciences, Vol. 18 (Springer, Berlin, Heidelberg, New York 1979) pp. 183-222
2. J.C. Dainty (ed.): *Laser Speckle and Related Phenomena*, Topics in Applied Physics, Vol. 9 (Springer, Berlin, Heidelberg, New York 1975)
3. M. Young, B. Faulkner, J. Cole: J. Opt. Soc. Am.60,137(1970)
4. F.T.S. Yu, H. Chen: Opt. Commun. 25, 173-175 (1978)
5. F.T.S. Yu, A. Tai, H. Chen: Appl. Opt. 18, 212-218 (1979)
6. S.A. Benton: J. Opt. Soc. Am. 59, 1545A (1969)
7. E.N. Leith: Sci. Am. 235, 80-95 (1976)
8. H. Chen: Appl. Opt. 17, 3290-3293 (1978)
9. P.N. Tamura: SPIE Proc. 126, 59 (1977)
10. J.C. Wyant: Opt. Lett. 1, 130-132 (1977)

Suppression of Disturbing Light Reflexes in Holography, Applied to Practical Recording Problems in Medicine and Technology

C. Sieger

Institut für medizinische Optik der Universität München, Theresienstr. 37
D-8000 München 2, Fed. Rep. of Germany

1. Introduction

In many applications of holographic interferometry it is unavoidable that reflecting media are positioned between object and hologram (see Figure 1). These may be lenses (e.g. in an otoscope or a microscope) or semireflecting mirrors, as often used for a direct observation parallel to the illuminating beam. Sometimes during investigation an object has to be placed in a specific test chamber (e.g. a pressure chamber or a liquid cell containing physiological solutions).

Depending on the coating quality of the lenses and glasses used, more or less of the light illuminating the object is reflected directly onto the recording media. Since usually this light is coherent to the reference beam it will be recorded, too, and is reconstructed as disturbing light reflexes veiling the object of interest.

These very disturbing and normally unavoidable reflexes can be completely removed: Depending on the problem the object under investigation has to be recorded either by the holographic subtraction technique or by a phase modulated reference beam (Fig.2).

2. Avoiding Reflexes by Means of Holographic Subtraction

As was shown first by COLLINS [1] two complex amplitudes may be subtracted from each other by shifting the phase of either the

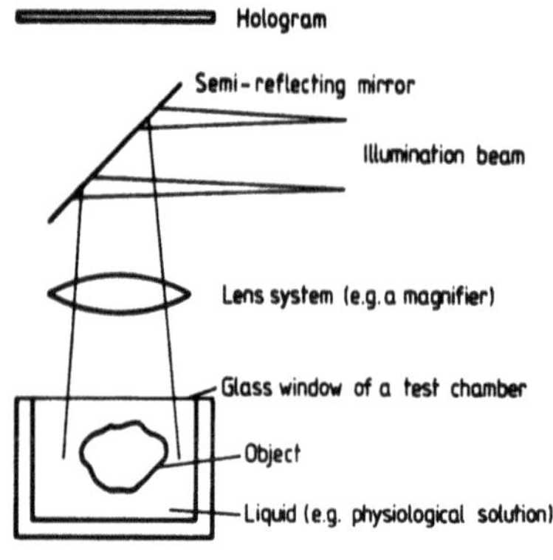

Fig.1 Some components of a holographic set-up possibly creating unwanted reflexes

Suppression of Disturbing Light Reflexes		
in Holograms of	Usual Recording Technique	Technique for Reflex Suppression
Static Objects	Single Exposure	Holographic Subtraction and Moving the Object
Deformed Objects	Double Exposure	Holographic Subtraction
Vibrating Objects	Single (Time Average) Exposure	Modulation of the Reference Beam or Holographic Subtraction

Fig.2 Methods to avoid unwanted light reflexes in the reconstruction of holograms

reference or the object beam by π between both exposures of a double-pulse record. If both recorded amplitudes of the reflected waves are identical, i.e., if an object point has not changed between both recordings, then this point will not be reconstructed and will appear dark in reconstruction. Therefore all light reflections originating from resting parts of the holographic set-up will no longer be reconstructed. Fig.3a shows an object recorded by double-exposure technique through a muddy glass plate. Because of the unwanted scattered and reflected light the interference fringe pattern of the interferogram can hardly be evaluated. Using the holographic subtraction, all reflexes arising from stationary parts of the object scene are suppressed (Fig.3b).

In the same way this subtraction technique can be performed, of course, together with double-pulse or stroboscopic hologram records of a vibrating membrane [2].

Also in recording static objects,e.g. in the application of holography in ophthalmology, the holographic subtraction technique can be used in many cases to avoid reflexes. Again, instead of a single exposure, the static object has to be recorded by a double-pulse subtraction exposure. Between both illuminating

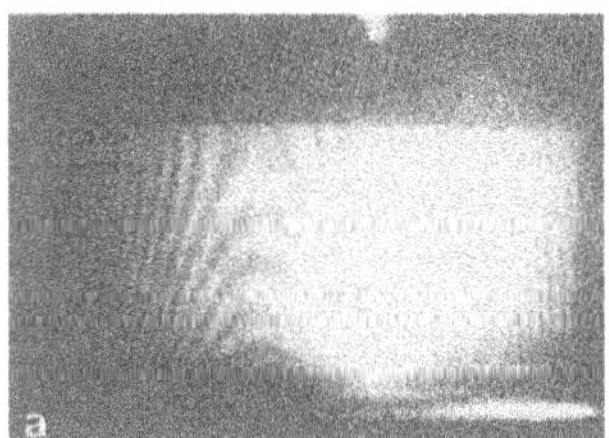

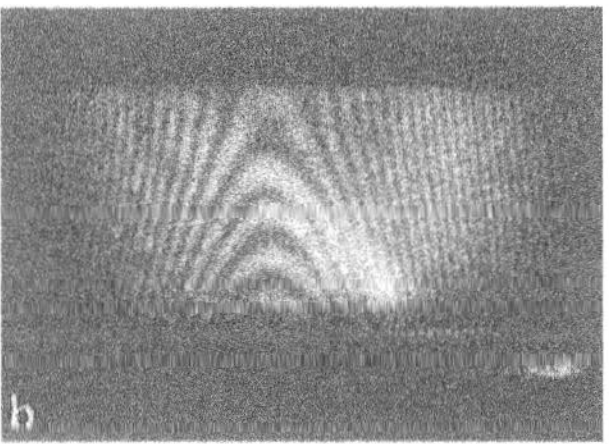

Fig.3a Light reflexes in a ordinary interferogram, disturbing the evaluation

Fig.3b Suppression of these reflexes by means of the holographic subtraction technique

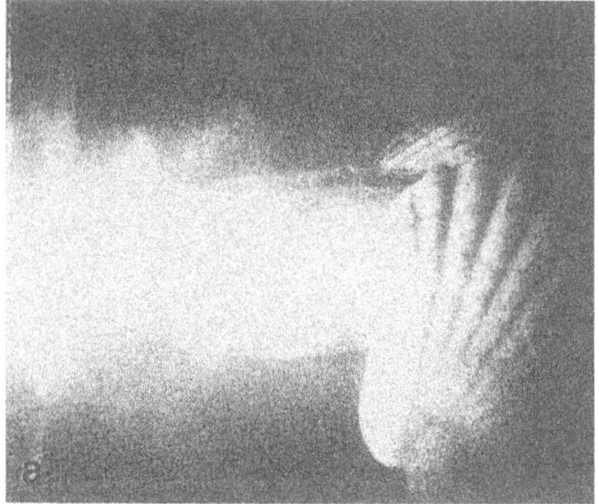

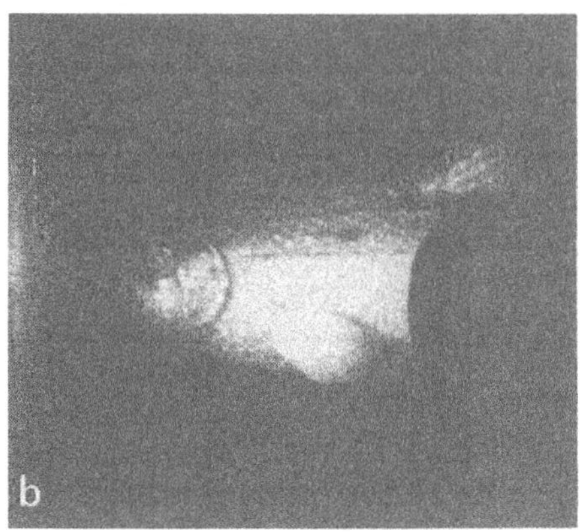

Fig.4a Ordinary hologram of an artificial fish

Fig.4b Same object, recorded under the same conditions as described in the text

pulses, the object which should be recorded free of reflexes has to be displaced by about a quarter of a wavelength. For the recording of interferometrically unstable objects, e.g. the eye or the tympanic membrane, it may be sufficient to choose the time interval between both illuminating pulses suitabl long, so that the object will move approximately by this amount during this interval.

In Fig.4a an artifical fish in an aquarium is recorded. Because of the intense reflection of the illuminating beam at the air-glass boundary, it can hardly be recognized in an ordinary hologram. In the equivalent double-exposure subtraction hologram all static details of the object scene, as the shell in the foreground, but also all disturbing reflexes appear dark (Fig.4b). The head of the fish shows a dark interference fringe, generated by the displacement of the object between both records.

The subtraction method can also be applied to time average interferograms of vibrating objects [3]. For half of the total exposure time, the object is recorded in a stationary state, for the other half in a vibrating state. Between the two records a phase shift of π is introduced to the reference beam again. The contrast of the resulting interference fringes, however, is considerably worse than in a corresponding time average interferogram. This can be a serious drawback if interferograms of biological objects with poor and nonuniformly reflecting surfaces are to be evaluated.

3. Suppression of Light Reflexes by Utilizing the Temporal Filtering Properties of Holograms

In order to examine vibrating objects, mainly time average interferograms are used for following reasons:

1) There is no need need for expensive equipment pulse lasers or Pockels cells with the appropriate control and driving systems.
2) From the interference fringes of these interferograms the maximum vibration amplitude can be determined directly.
3) In time average interferograms oscillation nodes are reconstructed with maximum intensity.

It was pointed out by GOODMAN [4] and ALEKSOFF [5], that a time average hologram acts as a temporal band pass filter for

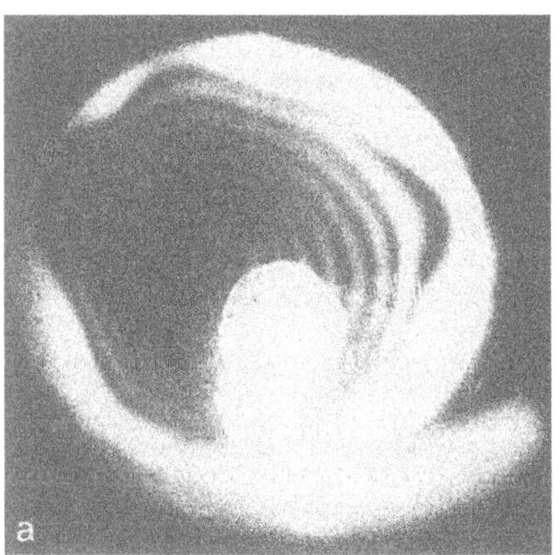

Fig.5a

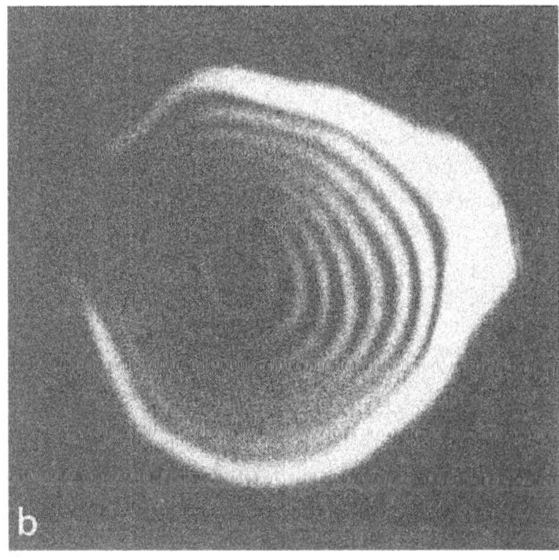

Fig.5b

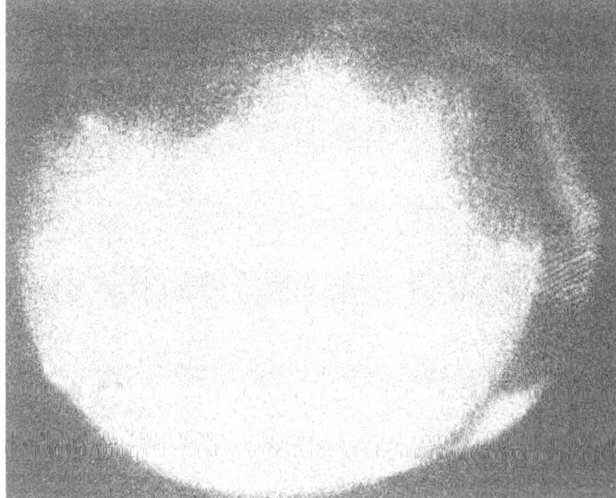

Fig.6a

Fig.6b

Fig.5a + Fig.6a Reflexes of different intensities in an ordinary time average interferogram of a vibrating membrane

Fig.5b + Fig.6b Corresponding interferograms, recorded with a frequency-shifted reference beam

the laser light. Its initial frequency is shifted according to Doppler effect by the object vibration or any other movement. The interferences of that portions of the object spectrum, which have not exactly the same frequency as the reference wave, cannot contribute to the hologram forming process, if the exposure time is sufficiently long. If the frequency of the reference beam is shifted, all non-frequency-shifted portions of the object spectrum, especially light reflexes caused by stationary parts of the object scene, are not reconstructed any more.

To demonstrate this effect, a magnifying lens was adjusted so, that parts of the illuminating wave were scattered with different intensities directly on the hologram plate (Fig.5a and Fig.6a). In the corresponding interferograms with frequency-shifted reference beam, the light reflexes can no longer contribute to the holographic record and therefore do not appear any longer in the reconstruction (Fig.5b and Fig.6b).

4. Conclusions

As shown, different methods exist for nearly all holographic recording problems (see Fig.2), to avoid completely reflexes in the hologram reconstruction. The particular advantage of these methods is that neither the recording geometry has to be changed nor any optical component has to be replaced. Additionally the interference fringe pattern, arising in the methods used, can be evaluated in just the same manner as in ordinary interferograms.

5. References

1 Collins, L.F.: Appl. Opt. 7, 203 (1968)
2 Sieger, C. and Röhler, R.: In Holography in Medicine and Biology, ed. by von Bally, Springer Series in Optical Sciences, Vol. 18 (Springer, Berlin, Heidelberg, New York 1979) p. 247
3 Hariharan, P.: Appl. Opt. 12, 143 (1973)
4 Goodman, J.W.: Appl. Opt. 6, 857 (1967)
5 Aleksoff, C.C.: Appl. Opt. 10, 1329 (1971)

This work was supported by the Deutsche Forschungsgemeinschaft (SFB 88/B 3)

Holographic Vibration Analysis of the Frontal Part of the Human Neck During Singing

R. Pawluczyk, and Z. Kraska
Central Optical Laboratory, Kamionkowska 18, 03-805 Warsaw, Poland
Z. Pawlowski
Fryderyk Chopin, Academy of Music, Okólnik 2, 00-368 Warsaw, Poland

1. Introduction

The knowledge of the particular role in sound formation of the different parts of the human vocal organ and of other parts of the body are of interest for medical applications as well as for prediction of the vocal timbre of students studying voice. Investigations on the human vocal organ are complicated by the fact, that there is no direct optical access with the naked eye to this organ and that its functioning is influenced by any contact with an optical instrument. Assuming that the air vibrations in particular resonance cavities as well as the vibrations of bones and tissues will be transferred to the nearest skin areas, it can be expected that investigations of vibrations of the skin surface may give some information on internal parts of the body, here especially on the vocal organ. Thus, the problem of a non-contactive analysis of the function of the human vocal organ could be partially reduced to investigations of the skin vibrations of the frontal part of the neck. The double-pulse holographic interference method was used for such investigations, because of many advantages described in [1,2].

2. Experiments

A double-pulsed Q-switched ruby laser was used as the source of low-energetic, coherent light pulses. It generates two laser pulses with a pulse duration of about 30 ns each, and a pulse separation adjustable in the range of 0.15 to 0.25 ms. After preliminary experiments the pulse separation interval was set to 0.2 ms. Selecting the combination of the ruby laser oscillator in one-, two-, or three-pass mode and the ruby laser amplifier in

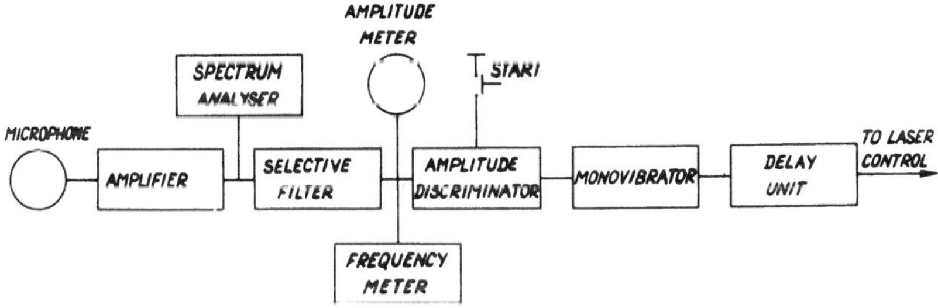

Fig. 1 Block scheme of the synchronization and voice analysis parts of the electronic system.

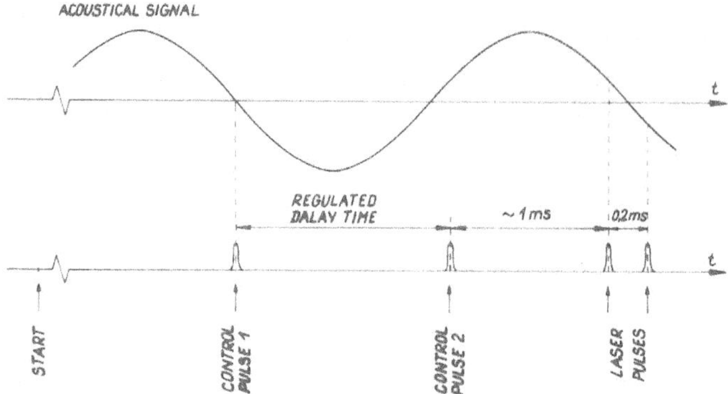

Fig. 2 The sequence of control signals and laser pulses in relation to the fundamental of the voice.

one-pass mode allows regulating the output energy of both pulses in a wide range (10 mJ 1.0 J), in order to optimize the exposure for holographic recordings. An electronic system (Fig. 1) provides the possibility of releasing the first laser pulse at every preselected phase of the fundamental of the sound (Fig. 2). The optical arrangement was basically a conventional one for holographic recordings but with a diffuse object illumination through a ground glass.

Some holographic interferograms recorded with the described set-up are presented in Fig. 3 and Fig. 4, demonstrating the vibrations of the frontal part of the neck of a female and male person, resp., when singing vowels at different fundamentals. All vowels were sung in Polish pronunciation.

3. Conclusions

A preliminary analysis of the obtained results shows that, using holographic interferometry, reproducable interference patterns characteristic for every particular person and a basic frequency of the sound can be obtained. It seems that this method may be useful for investigations of the sound formation mechanism of the human vocal organ. In the future this technique may

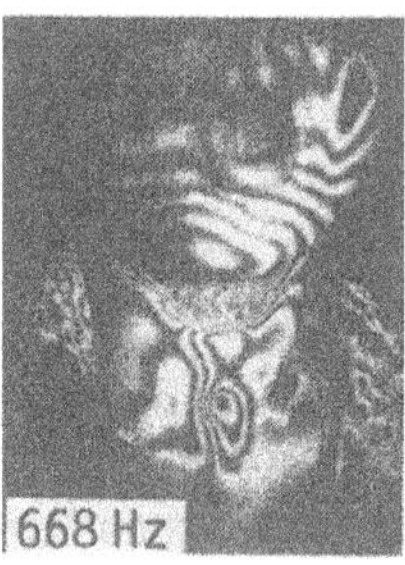

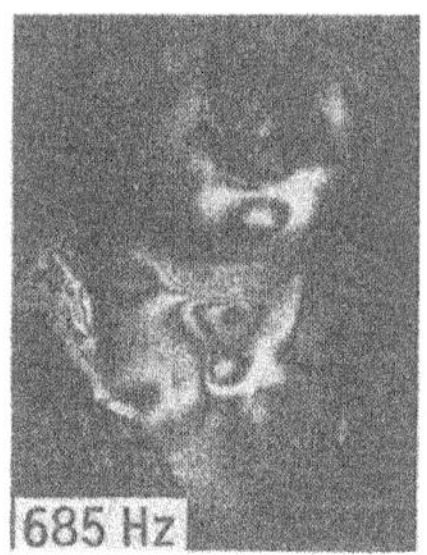

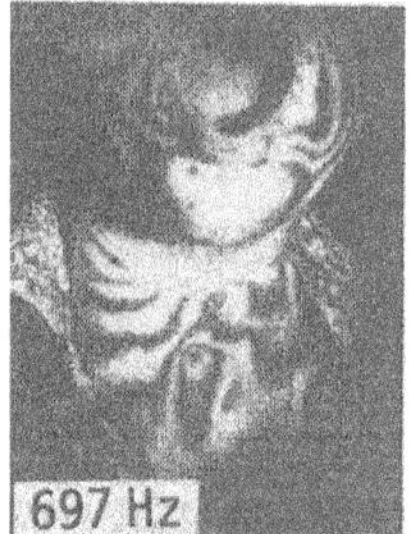

Fig. 3 Holographic interferograms of the vibrations of the frontal part of the neck of a female during singing the vowel "a" at different fundamental frequencies.

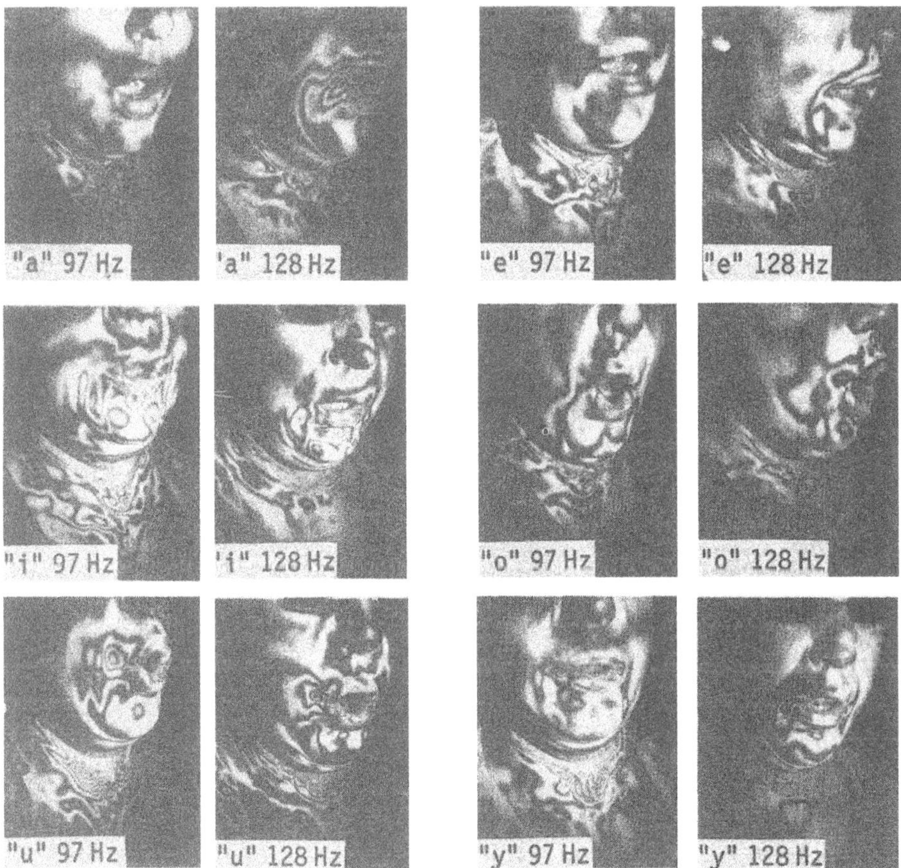

Fig. 4 Holographic interferograms of the vibrations of the frontal part of the neck of a male singing different vowels at two fundamental frequencies 97 Hz and 128 Hz, respectively.

become a tool in teaching correct articulation, especially in vocal training of deaf persons.

The authors wish to express their sincere thanks to all colleagues from the Central Optical Laboratory who have contributed to this work, as well as to the students of the Vocal Department, Fryderyk Chopin Academy of Music in Warsaw for their interested participation in the investigations.

References

1 Von BALLY, G.: Holographic analysis of tympanic membrane vibrations in human temporal bone preparations using a double-pulsed ruby laser system," in: E. Marom, A.A. Friesem, E. Wiener (eds.): Applications of Holography and Optical Data Processing, Pergamon Press (1977) pp. 593.

2 Von BALLY, G. (ed.): Holography in Medicine and Biology, Springer Series in Optical Sciences, Vol. 18 (Springer, Berlin, Heidelberg, New York 1979).

Holographic Measurement of Rabbit-Eyeball Deformations Caused by Intraocular Pressure Change

T. Matsuda, S. Saishin, and S. Nakao
Nara Medical University, Department of Ophthalmology, Kashihara
Nara, Japan, 634
T. Matsumoto
Osaka Prefectural Technical College, Department of Mechanical Engineering
26-12, Saiwaimachi, Neyagawa, Osaka, Japan, 572
K. Iwata, and R. Nagata
University of Osaka Prefecture, Department of Mechanical Engineering
College of Engineering, Mozu, Sakai, Osaka, Japan, 591

1. Introduction

Intraocular pressure change influences the morphology and function of the ocular system [1]. Therefore, the measurement of deformations of eyeballs caused by changes of the intraocular pressure (I.O.P.) is important in opthalmological investigations of glaucoma, intraocular surgery, and ametropia [2-5]. In our study holographic interferometry was applied for deformation measurements on eyeballs of white rabbits.

2. Experiments and Results

In this preliminary study, diurnal variation of I.O.P. was measured in 6 normal rabbits (weight 2500-3000 g) in vivo during 24-48 hours by using an applanation tonometer under instillation anesthesia.

Then the rabbits were anesthetized intravenously with Nembutal and their eyeballs were enucleated. For holographic investigations the eyeball under study was fixed at its limbus in a hole drilled in a metal plate as shown in Fig. 1. A teflon tube, which was connected to a reservoir filled with physiologic saline solution, was inserted into the anterior chamber.

Changing the height difference between the eyeball and the reservoir, applanation tonometry was repeatedly performed and a calibration curve was determined.

The optical system for recording double-exposure holographic interferograms is shown in Fig. 2. A laser beam is divided into two waves by a beam splitter (BS).The transmitted (object-) beam becomes a parallel wave by means of a collimating lens (Ls).

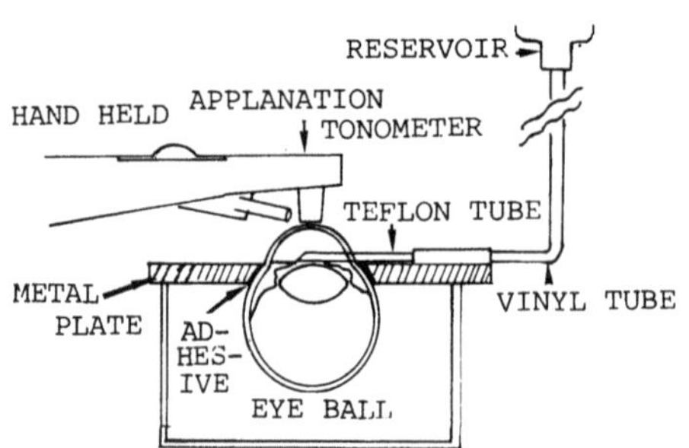

Fig. 1 Experimental set-up for comparison of applanation tonometric and manometric measurements

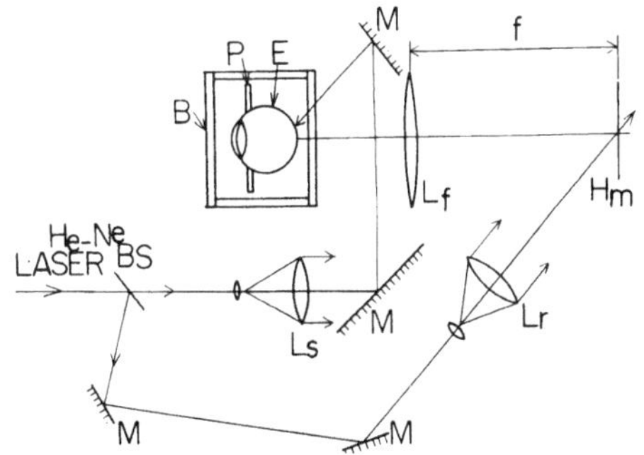

Fig. 2 Optical system for holographic recording

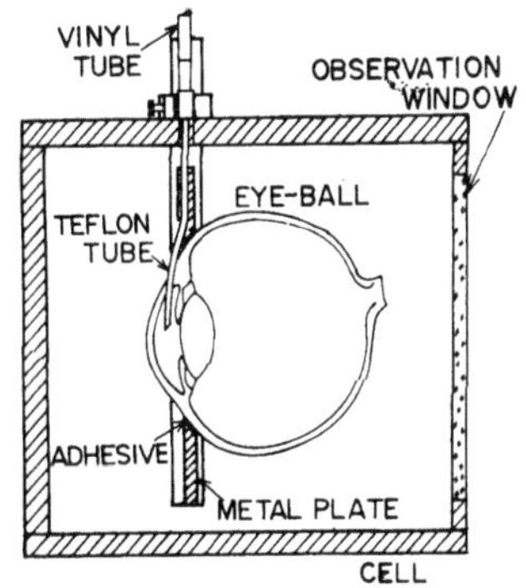

Fig. 3 Test chamber

This wave is reflected by mirrors (M) and illuminates the eyeball from the rear. The wave front reflected from the object passes a (Fourier transform) lens (Lf) (focal length (f)) and is collected on a photographic plate (Hm) located in the focal plane of the lens (LF). The beam reflected from the beam splitter, i.e. the reference beam, is collimated by the lens (Lr) and illuminates obliquely the photographic plate (Hm). After a first exposure the I.O.P. is changed and a second exposure is carried out. In the reconstruction process, the reference beam illuminates the hologram and the reconstructed object is recorded by a camera. Figure 3 shows the test chamber, which is used to measure the scleral deformation by double-exposure holographic interferometry.

Since the used eyeballs vary in size, several plates with different hole diameters were prepared. The eyeball under investigation was cemented at the limbus in the tapered hole of the metal plate with alpha cyanoacrylate adhesive (Aron Alpha). A teflon tube of 1.1 mm in outer diameter, stiffened by a stainless steel trochar, was inserted into the anterior chamber at the limbus of the cornea through a metal guide tube of 1.5 mm in inner diameter which had been adjusted to the center of the hole in the metal plate. After withdrawing the trochar, the teflon tube was also cemented to the metal guide tube with the adhesive in order to avoid mechanical deformation of the eyeball resulting from movements of the teflon tube. The metal plate with the eyeball and the teflon tube was immersed in a large chamber fitted with a flat glass observation window, in which physiological saline solution was filled.

A first holographic exposure was performed at the same I.O.P. as that just after the enucleation. A second exposure was done after a 10 mm upwards shift of the reservoir (equivalent to a pressure increment of 0.7 mmHg). This procedure was repeated about 20 times lifting the reservoir stepwise with increments of 10 mm until the I.O.P. was elevated to about 40 mmHg, resulting in about 20 double-exposure hologram recordings. Thus, the deformation of the sclera produced by every pressure increment of 0.7 mmHg could be estimated from the interference fringe patterns.

Figure 4 shows an example of diurnal variations of the I.O.P. with time. The difference between the data from both eyeballs of this individual rabbit are small. The I.O.P. variations show an upper limit of 22 mmHg and a lower limit of 15 mmHg.

In this experiment, a Perkins tonometer [6], which is designed for human eyes, was used to measure the I.O.P. of rabbits. Therefore, it was necessary to calibrate the tonometer. The result is shown in Fig. 5. The I.O.P. in Fig. 4 was transferred to manometric readings by using the calibration curve in Fig. 5.

The interference fringe patterns obtained are shown in Fig. 6. In the stage of low I.O.P. (Fig. 6(a)) the interference fringes are closely arranged. This shows a steep displacement gradient, paticularly in the area of the posterior pole and indicates a great deformation in this area. The sclera has an easy and wide distensibility. With increasing I.O.P. the number of interference fringes decreases indicating a reduction in sclera distensibility (Fig. 6(b)-(d)). A local distension in the area of the posterior pole could not be found.

The fringe pattern in Fig. 7 indicates the deformation of the sclera for a pressure increment of 0.7 mmHg starting from an I.O.P. of 15 mmHg. In Fig. 8, the displacement in the observing direction along line A-A' in Fig. 7 has been calculated from the fringe pattern [7].

On the basis of these experiments further measurements of scleral deformations due to pressure changes in the vitreous humor are planned.

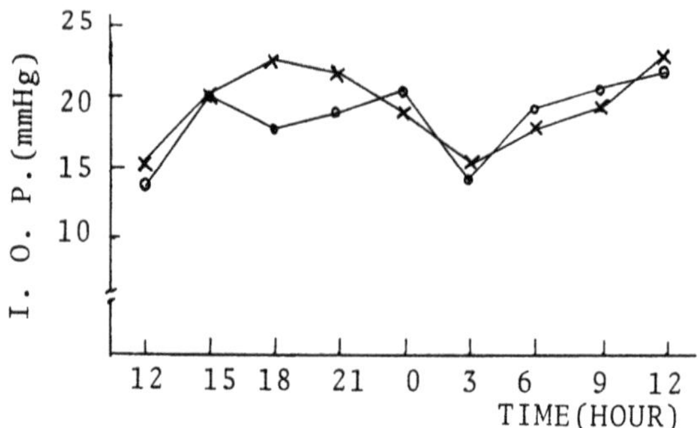

Fig. 4 Diurnal variation of intraocular pressure;
o: right eyeball
x: left eyeball

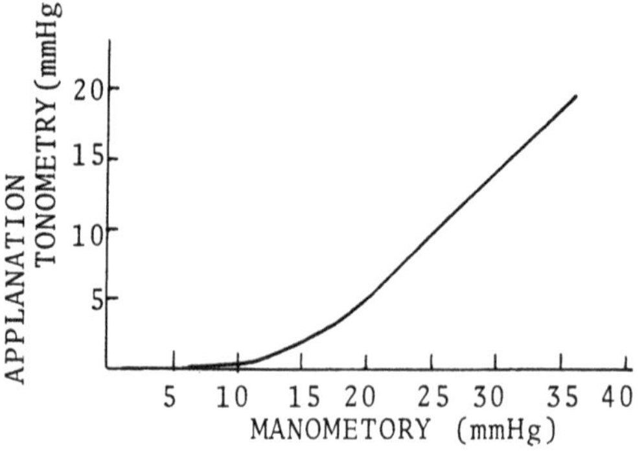

Fig. 5 Calibration curve between applanation tonometry and manometry

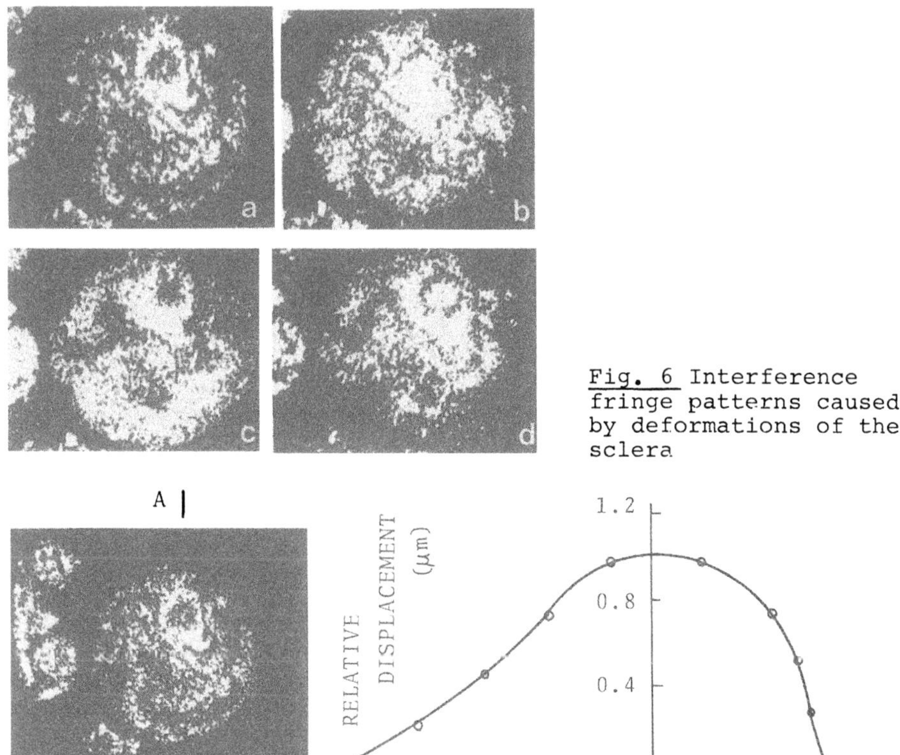

Fig. 6 Interference fringe patterns caused by deformations of the sclera

Fig. 7 Interference fringe pattern for pressure increase of 0.7 mmHg from 15 mmHg

Fig. 8 Displacement of sclera surface along A-A' in Fig. 7

3. References

1 R. A. Moses, *Adler's physiology of the eye clinical application* ed. R. A. Moses(Mosby Co., St. Louis, 1975)179.

2 P. Greguss, Opt. Laser Technol. 7, 253(1975).

3 P. Greguss, Opt. Laser Technol. 8, 153(1976).

4 T. Matsumoto, R. Nagata, M. Saishin, T. Matsuda and S. Nakao, Appl. Opt. 17, 3538(1978).

5 H. Ohzu, *Holography in Medicine and Biology*, ed. by G. von Bally, Springer Series in Optical Sciences, Vol. 18 (Springer, Berlin, Heidelberg, New York 1979) p. 143.

6 Clement Clarke International Ltd., *Operating Instructions of Perkins Hand-Held Applanation Tonometer*.

7 T. Matsumoto, K. Iwata and R. Nagata, Appl. Opt. 12, 1668(1973).

Holographic Investigations of the Human Pelvis

D. Vukičević, S. Vukičević, I. Vinter, and K. Sanković

Institute of Physics of the University, Bijenička 46, and Department of Anatomy
University of Zagreb, Salata 11, Zagreb, Yugoslavia

1. The Pelvic Ring

From Hippokrates to EVANS and LISSNER /1/ scientists have been interested in the function of the human pelvic ring. Some of them just believed in the existence of them; particular functional movements others tried to measure or calculate.

Mechanical characteristics of the pelvic bones are essentially unknown. Until EVANS and LISSNER /1/ showed that the stress coat cracks were not similar to BENNINGHOFF's /2/ split lines no thorough evaluation of the stress distribution in the pelvis has been made. The theoretical concept of the arch like construction of the pelvis /3/ and PAUWELS' /4/ criticism of the former interpretation were supported in part through EVANS' /1,5/ experimental studies on pelvic deformations.

Experiments with unrealistic models of the pelvis made of acrylics /6/ showed that a very substantial part of the actual load is transferred through the corticalis /7/. The trabeculae of the skeleton orient themselves in the direction of the principal stresses and represent stress trajectories /8/. Analysing the trabecular structure of the hip bone HOLM /9/ evaluated the stress distribution. This structure consists of a systematic pattern of trabeculae that corresponds well with the expected pattern of the theoretical stress trajectories and is much more complicated than in any other major bone. HOLM /10/ tried to confirm his assumptions by photoelasticimetry, loading transparent two-dimensional models.

The question of relative movement of the sacrum and the hip bone in adults has been the subject of many discussions over the years. Morphologic findings have suggested the mode of sacral movements /11,12/. Single beam radiography did not provide any proof or final conclusions /13,14,15/. Adapting Thompson's stereo radiography, a method judged satisfactory for the radiographic determination of pelvic movements to about ±0.3 mm /15/, it became evident that rotational, flexural, and torsional movements of individual bones exist. However, it was not possible to correlate them and to define an dynamic model in the light of previous theoretical calculations and experimental observations.

So, experimental techniques failed to provide sufficiently accurate results and to build an appropriate model of the pelvis. Analytical methods supplied some theoretical understanding, but the models were oversimplified and unrealistic. Numerical techniques have not provided any significant result /16,17/, up to now.

If we want to understand the functional meaning of this complicated system and to determine its loading characteristics, just as to understand the mechanism of a total hip prosthesis it is important to know the functional

Fig. 1 The optical arrangement and the loading device

behaviour and the stresses in the pelvis. Optical methods could help. Why holography, as CAULFIELD /18/ asked ?

2. Specimen Preparation and Loading Conditions

Embalmed and fresh specimens of human pelves with preserved lumbar spines, hip joints, and all the ligaments were used. Previous e.m.g. studies have proved the abdominal muscles, the m. quadratus lumborum, and m. gluteus maximus to be inactive in a symmetrical standing position /19,20/. A muscle showing no potential can be neglected in a realistic model. Specimens were tested under static vertical loading by pulsed laser interferometry /21/. The loading device was designed specially for introducing force to the specimens through the lumbar spine (Fig. 1).

With that loading device we could apply coarse preload and fine load at will. In order to eliminate great movements of the object and to obtain only deformations of particular parts, different amounts of preloading were used ranging from 15 to 200 kp /22,23/. The amount of load was accurately measured by a specially designed force transducer consisting of a hydraulic pressure reducer (10:1) and a bridge pressure sender. The base of the reducer was shaped according to a cast of the 1st lumbar vertebrae. Reading of 1mV on a five digit DVM was equal to 5 kp. The measuring range covered 0 to 43mV, corresponding to 0 to 215 kp. The force transducer signal was observed on a DVM and recorded by a strip chart recorder. In this way we could permanently follow the loading history and behaviour the specimen, as well as registrate the two laser pulses detected by a photodiode which are necessary to record a double-exposure hologram. The arrangement of the optical set-up allowed simultaneous recording of the inner and outer side of the entire specimen by placing two large mirrors behind the ala ilia.

3. Holographic Measurements

At first, the specimens were tested by double-exposure interferometry in order to detect the mobility of different pelvic parts. As indicated in Fig. 2, even greatest preloadings could not eliminate large movements of the specimens between two exposures. Deformation patterns of the lumbar

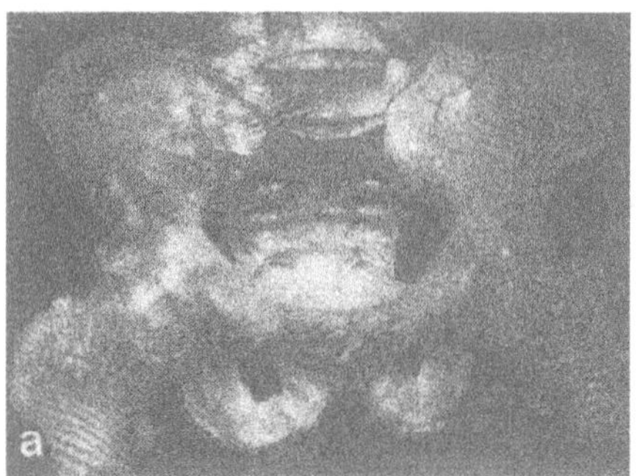
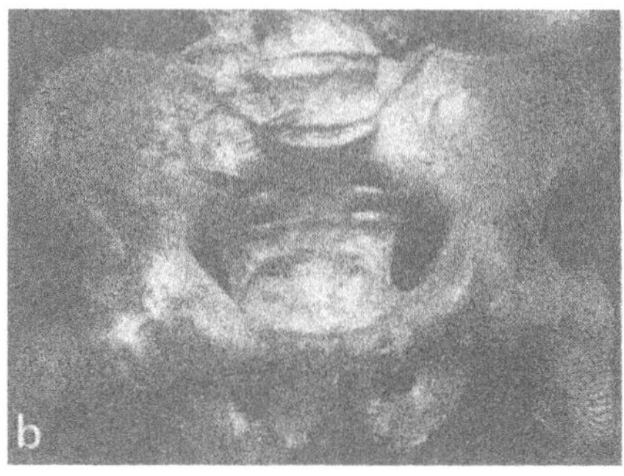

Fig. 2 Deformation patterns of the femurs and lumbar vertebrae obtained with preloadings of 70 kp (a) and 135 kp (b). Fringes on the pelvis are missing.

vertebrae and both femurs were obtained with preloadings greater than 20 kp, while fringes on the pelvic ring were missing. Sandwich holography showed similar results.

In order to define the directions and magnitude of these gross movements we placed around the ala ilia and the sacrum several movement indicator watches. Tracing the movements resulting from different preloads it became evident that the ala ilia showed a lateral and backward displacement. The translation of the sacrum was in the range of several milimeters even under maximal preloading. In order to eliminate the great backward movements we fixed the hip joints by impressing Palacos (bone cement) into the articular cavities. In such conditions, under great preloading (130 to 165 kp) and small loading between two exposures (2 to 5 kp), we could detect rigid body motion of the pelvis without any deformation (Fig. 3).

The movement indicators showed now that the ala ilia and the sacrum were displaced in the range of 10 µm to 2 mm. As the backward movements of the specimen were still too large we supported both ischial tuberosities, thus reducing these movements. Then we could apply the double-exposure method for small loads between two exposures, independently of the preloading state, and sandwich holography as a fringe control method became fully applicable.

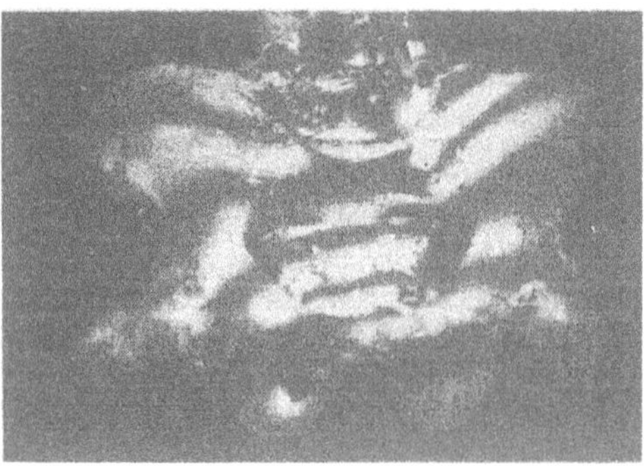

Fig. 3 Rigid body motion of the pelvic ring after fixation of the hip joints

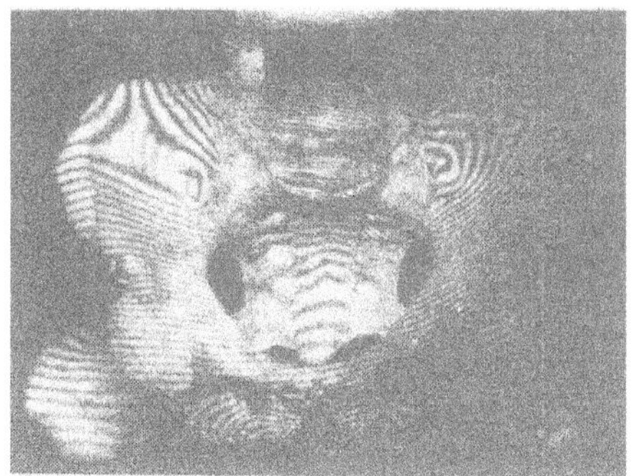

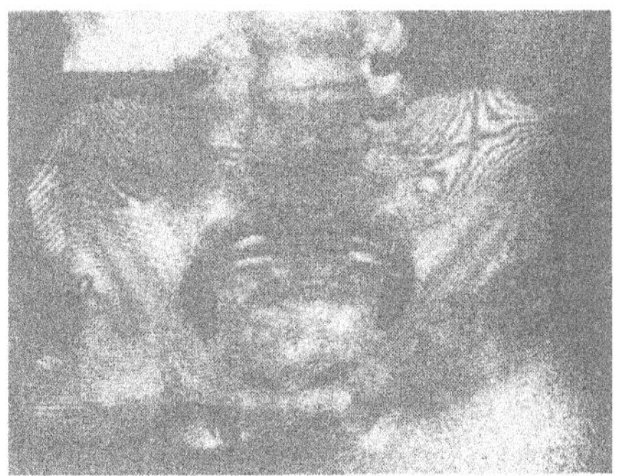

Fig. 4 The deformation pattern of the entire pelvic ring under gradually increased (left to right picture) loading

4. Preliminary Observations

Deformations of the femurs and lumbar vertebrae could be determined without any changes of the intact specimen. Preliminary analysis of the sandwich holograms showed, indicated by different fringe densities, that the sacrum moved relatively to the ala ilia. Its direction and translation were not determined till now (Fig. 4). Ala ilia underwent a wavelike motion or undulation. The deformation pattern showed the central part of the ala ilia to be displaced outwardly (Fig. 4). The bone is so thin there that no structure can be found. This area is interpreted as a region with low stress and, thus, only little bone is needed to withstand it /9/.

To confirm the reliability of any conclusion derived from holographic interferometry of such an object correlation with complementary strain-gauge analysis is a must. The actual deformation pattern will be constructed through the iterative application of a computer model interferogram procedure /24/.

5. Conclusions

We hope that based on the preliminary experiments presented here further investigations will contribute to the understanding of the internal stress distribution in the entire pelvic ring. Such studies may form a basis for a comparative analysis of the function of various types of hip joint prostheses. Even considering the problem of severe rigid body motion of the object, coherent optics seems to be the only solution for such investigations - therefore holography.

References

1 EVANS, F.G., H.R. LISSNER: Anat. Rec. 121, 141 (1955)
2 BENNINGHOFF, A.: Anat. Anz. 60, 189 (1925)
3 BRAUS, H.: Anatomie des Menschen, Vol. I (Julius Springer, Berlin 1929)
4 PAUWELS, F.: Z. Anat. 44, 167 (1948)
5 EVANS, F.G., H.R. LISSNER: J. Bone Jt. Surg. 41-A, 278 (1959)

6 HUGGLER, A.H., A. SCHRIEBER, C. DIETSCHI, H. JACOB: Z. Orthop. 112, 44 (1974)
7 DIETSCHI, C.: Zur Problematik des künstlichen Hüftgelenks, Thesis, Zürich University (1978)
8 KUMMER, B.: Bauprinzipien des Säugerskeletts (Thieme, Stuttgart 1959)
9 HOLM, N.J.: Acta. Orthop. Scand. 51, 421 (1980)
10 HOLM, N.J.: Acta. Orthop. Scand. 52, 135 (1981)
11 WEISL, H.: Acta Anat. 20, 201 (1954)
12 WEISL, H.: Acta Anat. 22, 1 (1954)
13 RICH, E.A.: Cong. Rec. 110, 5157 (1964)
14 WEISL, H.: Acta Anat. 23, 80 (1955)
15 FRIGERIO, N.A., R.R. STEWE, J.W. HOWE: Clin. Orthop. 100, 370 (1974)
16 GOEL, V.K., S. VALLIAPAN, N.L. SVENSSON: Comput. Biol. Med. 8, 91 (1978)
17 GOEL, V.K., N.L. SVENSSON: J. Biomech. 10, 195 (1977)
18 CAULFIELD, H.J.: In *Optical Data Processing*, ed. by D. Casasent, Topics in Applied Physics, Vol. 23 (Springer, Berlin, Heidelberg, New York 1978)
19 BASMAJIAN, J.V.: Muscles Alive, Their Function Revealed by Electromyography (Williams Wilkins, Baltimore 1974)
20 CARMEN, D.J., P.L. BLANTON, N.L. BIGGS: Am. J. Phys. Med. 51, 113 (1972)
21 VUKIČEVIĆ, D., V. NIKOLIĆ, S. VUKIČEVIĆ, J. HANČEVIĆ, Z. ŠUČUR: Holography in Medicine and Biology, ed. by G. von Bally, Springer Series in Optical Sciences, Vol. 18 (Springer, Berlin, Heidelberg, New York 1979)
22. VUKICEVIC, S., R. STERN-PADOVAN, D. VUKICEVIC, P. KEROS: Clin. Orthop. 151, 210 (1980)
23 VUKIČEVIĆ, S., R. ŠTERN-PADOVAN, D. VUKIČEVIĆ, J. HANČEVIĆ: Coll. Antropol. 4, 201 (1980)
24 JANTA, J., M. MILER: J. Opt. 8, 301 (1977)

Part IV

Speckle – Techniques and Spectroscopy

Speckle Techniques for Use in Biology and Medicine

O.J. Løkberg

Physics Department, The Norwegian Institute of Technology
Trondheim, Norway

1. Introduction

By formal definition, a speckle pattern is a pattern made up of irregularly spaced dots (specs). In optics, the speckle pattern is (usually) created by interference between a great number of wavelets with random phase values. These wavelets are formed when coherent light is reflected from a rough surface or transmitted through a diffusing media.

Among workers in the field of coherent optics speckle is generally considered to be "optical noise", which is a rather misleading notion. The speckle is really trying to describe to us in microscopic details the surface structure of the object. We will in this paper describe methods which utilize the speckle to measure various characteristics of the object and its behaviour. In that context the speckle should be considered as the information carrier wave on which the more coarse information (fringes) is encoded.

To a holographer, the techniques to be described here may appear rather crude with coarsely looking fringe patterns as the end result. However, these techniques are able to give information essential for deformation analysis and which hologram interferometry has failed to provide. A typical example is measurement of in-plane movements for stress calculations.

Within the scope of this paper, we are able to cover only lightly the bare essentials regarding the speckle behaviour and related techniques. For a deeper understanding, the reader should consult (1-6), while bibliographies covering parts of the subject are provided by (7-8).

2. Speckle Characteristics

As already stated, the optical speckle pattern is an interference phenomena, although the term white light speckle is sometimes used to describe man-made or natural textures on object surfaces.

Although mainly associated with the highly coherent laser light, speckle patterns can also be observed with thermal light sources under certain conditions (the first speckle paper was published in 1877 (9)). For example, look closely at the sunlight reflected from a fingernail for a finely coloured pattern. The speckle pattern created by a Xe arc is shown on Fig.1a, where the black and white rendition does not reveal the dispersive colour effects in the individual speckles.

However, speckle patterns created by laser light are by far most important and will be described in some detail. A typical laser speckle pattern

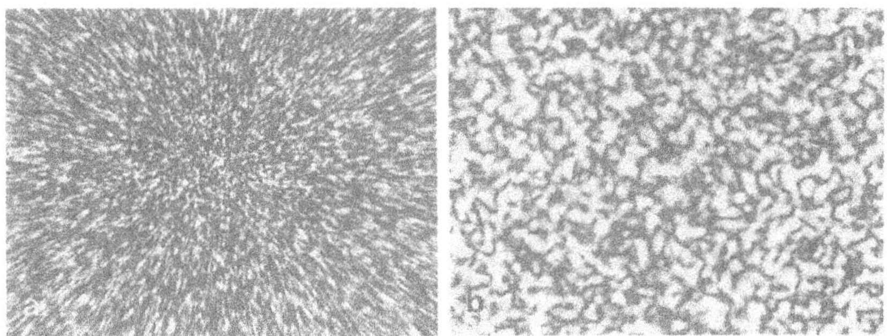

Fig.1a Speckle pattern produced by thermal light
b Laser speckle pattern

is shown on Fig.1b, where the reader may substitute the white areas with the colours of her/his laser (or heat if an infrared laser is used). The contrast for speckle patterns, defined as rms intensity variation divided by the average intensity is high. For a so-called fully developed pattern, created by reflection from a nonpolarizing surface, the contrast is unity. The contrast distribution resulting from combining speckle fields (and uniform waves) in different ways are important for many speckle applications and has been discussed by, e.g. BURCH (10).

The speckle size is another parameter of importance for speckle work. It is then common to distinguish between speckle formation under two different conditions.

The objective speckle wave is allowed to propagate freely from the scattering medium to the plane of observation. (Note, this speckle wave would be the object wave by holographic terms). The finest details in the pattern are created by interference between the wavelets scattered from the extremal parts of the object. But we are usually more concerned about the average size of a speckle, which for a circular, uniformly illuminated object of diameter D observed at a distance L will be $d_{o.av} = 1.2\ \lambda L/D$.

For the subjective speckle, the propagation of the wave is restricted by an aperture, e.g. the pupil of an eye looking at a laser illuminated screen, thus the term subjective. Of special interest is the speckle pattern in the image produced by a lens of diameter d_i and image-lens distance l_i. The average speckle size will then be given by an expression very similar to the subjective case, namely $d_{s.av} = 1.2\lambda l_i/d_i$.

If we compare expressions for the speckle size in the two cases, we find that for a given wavelength the size is related only to the angular extent of the interfering beams. In the objective case, coarser speckles are obtained by reducing object size and/or increasing observation distance, while stopping down the lens aperture enlarges the subjective speckles.

In both cases, we have referred to a plane transverse to the direction of propagation. In the longitudinal plane the speckles get elongated with propagation, looking somewhat like long cigars dangling in space.

The final question is how the speckle is affected by movements of the object in the illuminating wave. For a general answer the reader should consult e.g. Ch. 10 in (1). Two special cases should be mentioned. If an object moves in its own plane, the speckle pattern in the image plane moves in proportion. If the object light has been multiply scattered, e.g. from the interior of a translucent object, small movements of the object in the illumination produce a different or decorrelated speckle pattern. This is often the reason for the low fringe contrast obtained with biological objects.

3. Speckle Measurement Techniques

3.1. Surface Roughness Measurement

The problem of devising an optical instrument for characterizing the microstructure of an object surface has been actively pursued by many groups, see e.g. Ch. 3 in (1). For various reasons such an instrument would be of great practical value both for research and industry. Some success has been obtained e.g. by correlating the contrast of polychromatic speckle patterns to surface roughness. But today no general optical instrument exists which can do measurements on all types of rough surfaces.

3.2. Speckle Photography

In the introduction, we emphasized the importance of measuring the in-plane movement (or more correctly, the variation in in-plane movement) of a deforming object. This information can be determined by speckle photography - a very simple, but powerful technique.

In speckle photography, we simply use an ordinary camera to record the speckled image of a laser illuminated object on a high resolution film. The

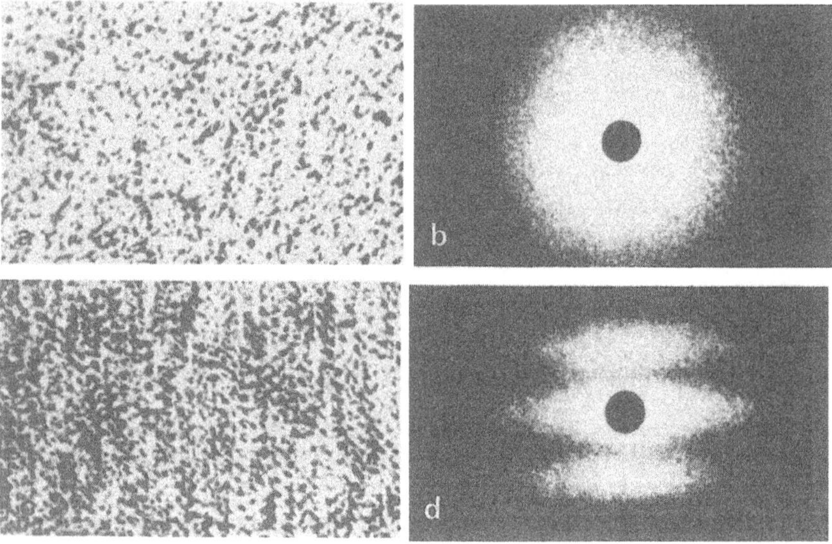

Fig.2 Speckle patterns (a and c) and their corresponding far field patterns (b and d) for (top) undisplaced and (bottom) displaced pattern (Courtesy: M. Fourney, Univ. of Cal. Los Angeles, USA)

lens is set at appr. f:4, which is a good balance between small speckle size and lens aberration. A double exposure is recorded before and after object deformation. The developed film contains two identical speckle patterns, which are displaced in proportion to the movement of the object. It is of course possible to measure this displacement directly on the film, but more accurate and elegant optical methods have been devised. We might for example shine a laser beam through such a recording and observed the resulting far field pattern. If no movement has taken place, we will simply observe a speckle pattern in the far field, Fig.2a-b. However, the displaced pattern results in a speckle pattern modulated by fringes, usually called Youngs fringes, Fig.2c-d. The spacing of these fringes is directly proportional to the object in-plane displacement, while the fringe direction is normal to the object displacement vector. By moving the speckle picture in the (thin) read-out beam we obtain a detailed map of the object displacement as shown on Fig.3. Using photoelectric read-out, the measurement can be made sufficiently accurate to compete with strain gages, except that one speckle recording equals numerous strain gages.

Although the point read-out is very simple to use and interpret, we would also like to have a full field view of the object's deformation. This is possible by using conventional optical filtering techniques on the speckle recording. A small aperture is used as off-axis filter in the frequency plane and the resulting image of the object will be covered by a fringe pattern. Each fringe represent a constant value and direction of the movement. By varying the position of the filtering aperture, we vary the sensitivity and/or the direction of fringe representation. This very useful feature is instructively shown by HUNG in (1).

A complete measurement of both the magnitude and direction of the deformation can be obtained by additional recordings of the speckle field in planes outside the image plane. For example by focusing on the illuminating source the displacement of the resulting speckle patterns will represent the object tilt.

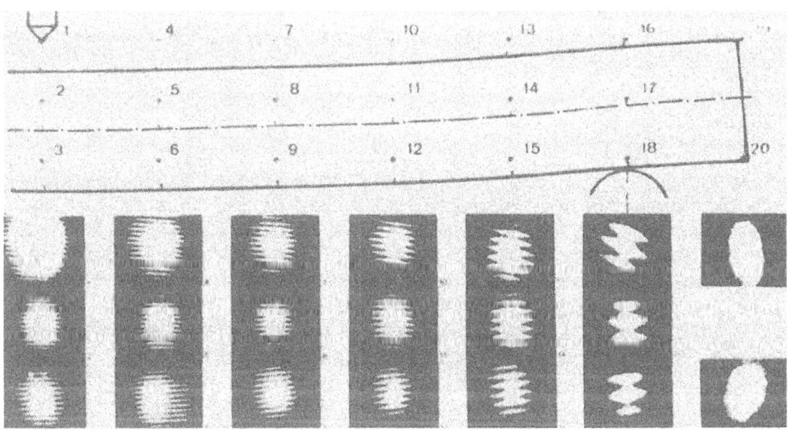

Fig.3 Far field fringes obtained with point-by-point analysis of a doubly exposed speckle pattern of bending lever (Courtesy: H.J. Tiziani, Univ. of Stuttgart, Stuttgart, West Germany)

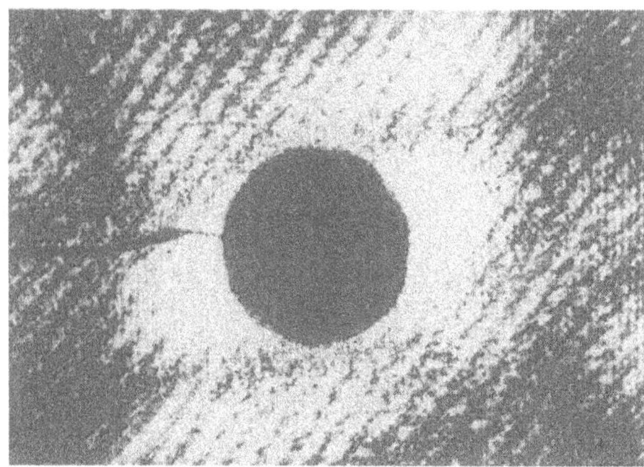

Fig.4 Fringes from "white-light" speckle recording of human skin movement (Courtesy: G. Cloud, Mich. St. Univ. East Lansing, USA)

The speckle photography method also records the object's movement during the exposure, e.g. harmonic vibrations. The fringe interpretation will be the same, except for the $\cos^2$-fringes being replaced by other fringe functions representative of the movement. When the movement is complex, the interpretation via Young's fringes will be rather difficult. But an interesting extension of the technique allows us to reconstruct the path of the object surface during the recording as described by WEIGELT in (1).

We should also note that the related problem of mapping fluid movement has been successfully attacked by speckle photography, see e.g. (6). The liquid has to be seeded with small particles and short laser pulses are used to arrest the motion.

The principle of speckle photography has also been used on "white light speckles" as defined earlier. Figure 4 shows a point read-out from such a recording where the fine details on the human skin have acted as speckles. The light sources were two ordinary flash lamps fired in succession. The surprisingly high fringe quality suggests that this technique deserves more attention also in the biomedical field.

3.3. Speckle Interferometry

While speckle photography was based on recording a single object speckle wave at each exposure, speckle interferometry records the interference between the object wave and a reference wave. The reference wave can be a uniform wave or another speckle wave. In the last case, the speckle wave may even come from the object itself but with the object illuminated from a different angle or with the object's image shifted (sheared) relative to each other. The border between speckle photography and speckle interferometry is rather vague in some cases and at least one technique has been put in both categories.

After we have brought in the reference wave in the appropriate way, the recording steps for the two methods are usually equal. But the speckle interferogram recordings are read out as full field fringe pattern, usually in an optical filtering set-up where the sensitivity is fixed. In certain cases the fringes may be observed directly in real-time.

As speckle interferometry is based upon interference between two separate waves, its sensitivity is generally higher than for speckle photography. But as a result, the speckle interferometry methods are more demanding on set-up stability and laser coherence.

Speckle interferometry techniques are used to measure various useful parameters of object deformations, like in-plane, out-of-plane and bending.

To illustrate speckle interferometry, we have chosen to describe shortly one method, whose versatility should make it interesting for biomedical measurements. In Fig.5a-b we illuminate the object surface with two plane waves, and record a double exposure of the deforming object. When the recorded image is examined in an optical filtering set-up, the interpretation of the resulting fringe pattern depends on how the illumination was arranged. Illumination symmetrical to the object surface normal gives sensitivity only to in-plane movement (Fig.5a-c), while a skew illumination also reveals the out-of-plane components of the movements (Fig.5b-d). As the absolute sensitivity mainly is dependent on the angle between the illumination waves we are able to vary the measuring range. (Note that the contour interval in Fig. 5d is 10 µm, which may be suitable for many biological experiments).

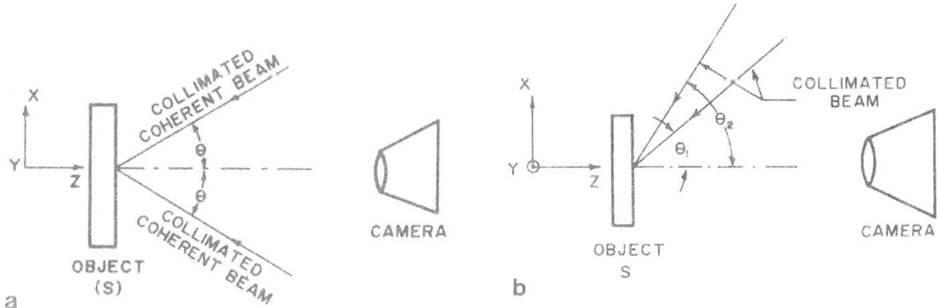

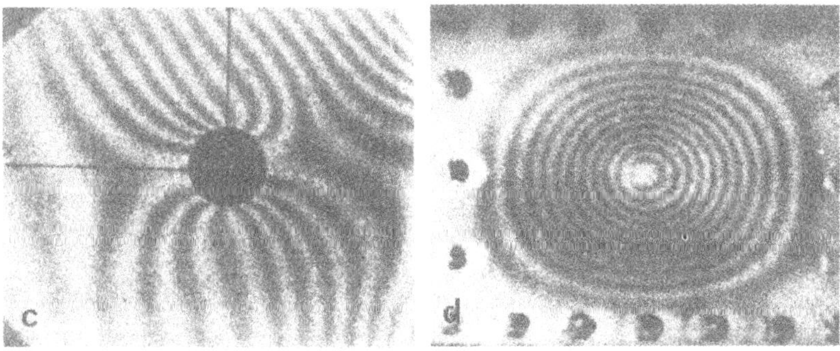

Fig.5 Speckle interferometry set-up for measuring in-plane (a) and desensitized out-of-plane movement (b). Resulting fringe patterns for in-plane (c) and out-of-plane movements (d). Fringe spacing in (d) repr. 10 µm displacement (from Ch. 4 in (1))

3.4. Electronic Speckle Pattern Interferometry - ESPI

ESPI can be explained both as a speckle and a holographic technique. Here we will follow the holography approach, which might be more familiar to most readers. Again, for a more detailed description the reader should consult e.g. (1,11).

Formally, ESPI can be defined as image holography with an in-line reference wave and where the recording and reconstruction steps are performed by videorecording and electronic processing. This slightly confusing accumulation of technical terms will be closer explained in reference to Fig.6 which shows the basic elements of a simple ESPI.

The illuminated testobject is imaged on to the TV target via mirror M_2. The aperture of the imaging system is small, about f:20 or less. To get interferometric sensitivity a reference wave is brought in through a small hole centered on the same mirror. The combination of low aperture imaging and in-line object/reference wave propagation results in a coarse speckle structure which can be resolved by the TV system. (Note, the resolution of a standard TV system is about 20-30 l/mm which is a factor of one hundred less than for holographic films).

The light exposure results in a charge distribution across the TV target which is subsequently transformed into a proportional current variation by the scanning action of the TV system. This process can be compared to the exposure and development of an ordinary hologram. The holographic reconstruction process is simulated by electronic processing as the video signal is high pass filtered and thereafter full wave rectified. Finally, this processed video signal is converted into a "reconstructed" image displayed on the TV monitor.

The optical part of most ESPI set-ups works in principle like the one depicted on Fig.6, but a radically new concept was recently introduced by SLETTEMOEN (12). He uses a speckle reference wave and a special beam combiner to get a system with many interesting properties.

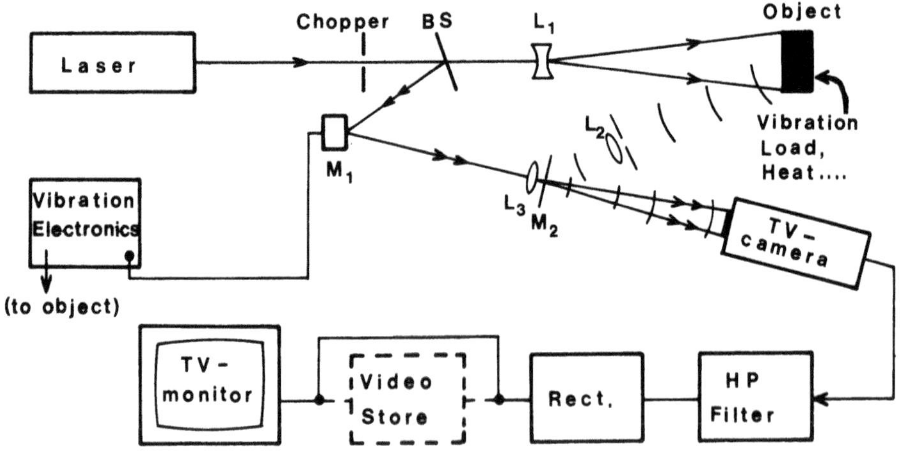

Fig.6 Basic ESPI set-up

ESPI produces fringe patterns which are identical to those obtained by similar hologram recordings. The only difference is a coarser speckle structure. If the object is vibrating during the exposure, the well known J_0^2-fringe function results, while a double pulse recording gives a $\cos^2$ pattern. If we want to measure slow changes in the object, the video store in Fig.6 is used to store and compare (subtract) speckle patterns from separate scans.

The photoelectric recording and electronic reconstruction of holograms give ESPI positive and negative features, some of which are listed and commented below:

The holograms are recorded and reconstructed in near real time. Even when the video store is used, the presentation is effected within a second.
The exposure time is short (40 msec. - European TV standard), and 25 new holograms are recorded and reconstructed each second (again European standard).
The combination of the last two properties makes ESPI a very stable interferometer, especially for vibration analysis. If the chopper in Fig.6 is used to further reduce the TV-exposure time, even extremely unstable objects can be investigated (13).
A properly designed ESPI set-up is easy to operate with almost nil running expenses. The operator(s) can observe and discuss the interferograms shown on a TV screen under ordinary room light levels.
Finally, on the positive side, ESPI may shorten the experimental time significantly. The real-time presentation speeds up the adjustment, while the experiment can be recorded in a video recorder for later analysis, maybe by a computer.

On the negative side, the ESPI interferograms are more speckled and coarse looking than hologram recordings. This difference in quality is however greatly reduced by speckle averaging in the speckle reference ESPI (12).
The combination of fixed, short exposure and small aperture imaging might create exposure problems, especially on uncoated objects. Using low light intensifier cameras solve that problem, however, usually the new class of vidicons (Chalnicon, Newvicon) will do the job.
The initial cost of an ESPI system might seem high, but this cost is saved by shorter experimental time and low running costs.
Finally, ESPI gives a two-dimensional representation of the object observed from only one direction (as does image holography).

3.4.1. Vibration Analysis by ESPI

If the object is vibrating sinusoidally during the exposure, the resulting fringe pattern will, as already mentioned, be the well-known $J_0^2(4\pi/\lambda\cdot u_0(x,y))$ function from which the amplitude distribution $u_0(x,y)$ across the object can be found, Fig.7a. Under optimal conditions at least 30 fringe orders can be detected. We should note that damped oscillations can be observed by ordinary time averaged ESPI.

If mirror M_1 in Fig.6 is also vibrating at the same or slightly different frequency, the fringe pattern represents the relative movement between M_1 and the object. This technique, usually called phase modulation, increases the practical use of ESPI for vibration studies. Amplitudes down to $2\cdot10^{-5}$ λ have been measured by photoelectric lock in detection, from approximately $2\cdot10^{-3}\lambda$ we can detect the vibration visually, while the vibration cycle can be seen as slow motion intensity variations from about $8\cdot10^{-3}$ λ.

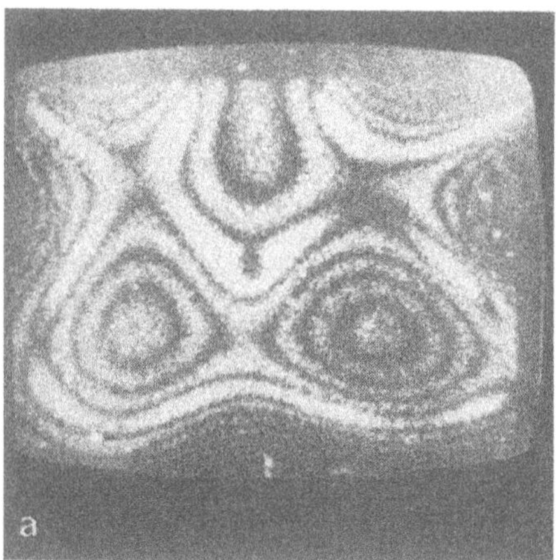

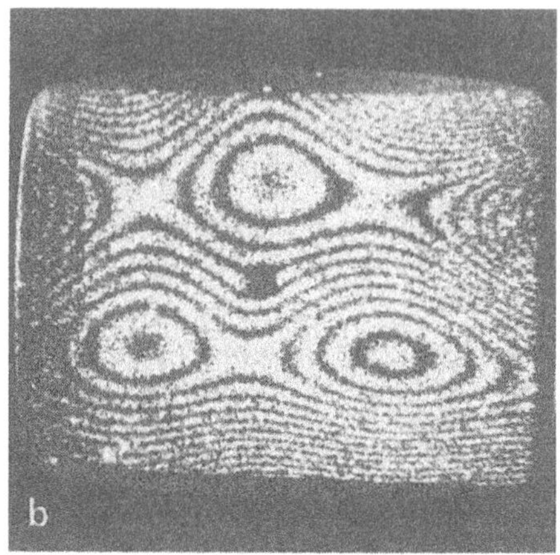

Fig.7 Time average (a) and stroboscopic (b) ESPI recordings of steel plate vibrating at 2100 Hz

Upwards, amplitudes to at least 20λ can be measured, which gives ESPI a total measuring range of 10^6 for vibration measurements. Phase modulation also can be used to provide contours of constant phase across the object, which is very helpful for interpreting complex vibrations.

Periodic vibrations of more general nature can be analyzed by stroboscopic ESPI, where the stroboscopic light pulses are placed at fixed parts of the vibration cycle. More random vibrations (or fast movements) can be analyzed by exposing with two pulses in each TV frame. In this case the experiment must be recorded on a video recorder for later replay in slow motion or still picture mode. Both techniques results in $\cos^2$ patterns as shown on Fig.7b.

3.4.2. Two Step or Slow Deformation

As mentioned, a video store must be used to compare (subtract) the speckle patterns from different TV frames. Commonly used video stores are storage tubes, videodiscs and lately digital stores. The signals can either both be recorded in the video store and read out simultaneously (double exposure) or the recorded signal can be continuously compared to the camera signal (real-time observation). In both cases the speed of the ESPI system enables us to make double exposures in rapid succession or to renew the reference state hardly without delay. The videorecorder should of course be used as a combination of notebook and camera during the experiment.

For measurement of out of plane deformations the standard set-up in Fig.6 can be used. The fringe pattern will follow the ususal $\cos^2$ function provided the object motion during the exposure is negligible. As the fringe contrast is constant some 50 to 60 fringes should be resolvable across the TV screen.

For in-plane measurements, the object is illuminated symmetrically like in Fig.5a. No reference wave is used. The frames representing the speckle patterns from the deformed and undeformed object are subtracted, resulting in a fringe pattern which represents contours of constant in-plane movement as in the speckle interferometry method described earlier.

3.4.3. Contouring

Using a two wavelength laser, absolute height contours can be obtained as in hologram interferometry. More interesting, however, is the fact that ESPI can directly provide the difference contours between the test object and its prototype.

4. Concluding Remarks

In this paper, we have described various speckle techniques in general terms without relating them to specific biomedical applications. So far speckle methods have had a rather small impact upon biomedical research. As usual industrial research has been faster to apply new measuring tools.

My personal opinion is that also biomedical holographers should consider speckle photography and interferometry as complementary or alternative solutions for many measuring problems. Both methods are simple to use and interpret. Their ability to measure in-plane movements at different sensitivity levels should help to solve strength related problems, for example in the construction and test of artificial limbs. The methods also have a sensitivity range which make them interesting as a bridge between the sensitivity of hologram interferometry and Moire methods. So far, most of the techniques have not been capable of real-time observation, but this drawback may be solved with the phase conjugating crystals as shown by TIZIANI and KLENK (14).

As for ESPI, this technique has already proved its role in biomedical research. Some specific applications will be described in the following paper. The practical application range will be mainly limited by the imagination and video-optical cleverness of the experimentalists.

References

1. Speckle Metrology. Ed. R.K. Erf. Academic Press 1978
2. J.C. Dainty (ed.): Laser Speckle and Related Phenomena, Topics in Applied Physics, Vol. 9 (Springer, Berlin, Heidelberg, New York 1975)
3. Laser Speckle and Applications in Optics. M. Francon. Academic Press 1979
4. A.E. Ennos, Progress in Optics XVI, 235, North Holland 1978
5. K.A. Stetson, Optical Engineering, 14, 482, 1975
6. Hologram Interferometry and Speckle Metrology. Technical Digest, Opt. Soc. of Am. Topical Meeting, 1980
7. K. Singh, Atti Fond. Giorgio Ronchi, 27, 197, 1972
8. K. Singh, Ind. Journ. of Opt., 18, 51, 1979
9. K. Exner, Sitzungsber. Kaiserl. Acad. Wiss (Wien), 76, 522 (1877)
10. J.M. Burch, Optical Instr. and Techn., Ed. J. Home Dickenson, 213, Oriel Press, 1979
11. O.J. Løkberg, Phys. in Technol., 11, 17, 1980
12. G.A. Slettemoen, Appl. Opt. 19, 616, 1980
13. O.J. Løkberg, Appl. Opt., 18, 2377, 1979
14. H.J. Tiziani and J. Klenk, Appl. Opt., 20, 1467, 1981

Bio-Medical Applications of ESPI

O.J. Løkberg, and P. Neiswander*

Physics Department, The Norwegian Institute of Technology
Trondheim, Norway

1. Introduction

In this paper, we will describe how electronic speckle pattern interferometry - ESPI - has been used to study and measure the movements of biological objects like the human body both in vitro and in vivo. The main part will deal with work on the human hearing mechanism, in particular recent experiments on vibration analysis of the basilar membrane.

2. Why ESPI?

To answer this rhetorical question we will first consider some of the difficulties encountered in interferometric work on biomedical objects. As a representative example we have chosen vibration analysis of the human hearing mechanism.

a) The objects are highly unstable. In vivo, blood pulsations, breathing and nervous trembling results in gross movements which makes the object interferometrically stable only for milliseconds. In vitro, gross movements are negligible, but changes in the object's microstructure result in an apparent stability in the seconds range (the value is highly dependent of the condition of the preparation).

b) The vibrating behaviour is often complicated and information about the amplitude and phase distribution is equally essential. The amplitude range is large, from μm in the ossicular chain to below 1 nm in the basilar membrane. Also, for a complete analysis, the frequency response should be sampled over the entire audio-range.

c) The experimental time is limited. In vivo, the temper and stress of the subject have to be considered, while post-mortem changes (drying, decomposition) can significantly alter the behaviour of a preparation. In addition, adjustments are time-consuming as the objects are small and difficult accessible.

d) Generally, fringe quality is low due to small random movements, low reflectivity, small size and (most important) multiple scattered light which decorrelates the speckle patterns.

As the virtues of ESPI were praised in the preceeding paper, we now leave it to the reader to prove the technique's ability to tackle the problems

* Present address: 1228 24th Str. # 3, Santa Monica, CA 90404, USA

above. One comment should however be made regarding d) - phasemodulation can be used to measure amplitudes corresponding to rather high fringe orders as described in (1,2).

3. Applications of ESPI

3.1. In vivo

Using double-exposures synchronized to the TV framing, the movement of the human skin have been studied at a rate of 25 recordings/sec. (3). The sampling rate can probably be extended to 50, may 75 rec./sec.. Large movements, e.g. due to finger movements, can be studied by single exposures giving so-called velocity fringes (3).

Vibration analysis of the human tympanic membrane (T.M) was performed by relative ease (2), except for rather prosaic problems like alignment and widening of the ear canal. To arrest gross bodily movements, the Ar laser was chopped to give TV exposures from 2 msec. and down. However, normal TV-exposure may be possible if a speckle reference ESPI set-up is used (4). The problem of accessibility can be solved by use of fiberoptic illumination and observation of the T.M (5). If some kind of speckle averaging (6) can be used, the fringe quality is likely to be good. The main problem is really to get reliable SPL measurements at the T.M. For example, the frequency-amplitude curves presented in (2) are rather dubious as the SPL refers to free field measurements.

As a whole, we feel optimistic about using ESPI for in vivo measurements on the T.M and may be also the ossicular chain during operations.

3.2. In vitro

ESPI has been used for routine measurements on the T.M and the middle ear components on human temporal bone preparations (7,8). Even at normal TV exposure, the stability properties of ESPI are so good that we can do measurement on very fresh or newly moistened preparations.

The basilar membrane (B.M) represents a more formidable challenge to the experimentalist and we will describe briefly the work done by one of us (P.N) on the B.M. A more detailed account will be published elsewhere (9).

The B.M is a small, nearly transparent membrane located between liquids with almost equal refractive index. The operative access to the B.M is rather difficult. The B.M itself is relatively stable, but the surrounding liquids create problems. Finally, the amplitudes are depressingly low, typically in the lower nm region at normal SPL.

The ESPI set-up constructed for measurements on the B.M, was in principle like the one shown on Fig.6 in the preceeding paper. The reader may therefore refer to this lay-out during the following description.

Fresh temporal bone preparations (24-48 hours post mortem) were used. The B.M was exposed through a 1.5 mm hole, which was sealed by a glass window. The position of the hole was 12 to 16 mm from the oval window and on the basal turn of the scala tympanic of the cochlea. The B.M. was excited by pure tones through the ear canal and the SPL was measured at the T.M.

The main experimental problem was to reduce stray light and optimize the light reflected from the membrane surface. This was achieved by a form of dark field illumination. The illumination beam went through the rim of the imaging lens and crossed the optic axis in the object plane and finally hit the scattering bony wall off the optic axis. In addition, polarizers were used to cut down on the unwanted scattered light.

The imaging lens was an ordinary camera lens (focal length 50 mm, N.A 0.24) used in reverse. The total magnification from the B.M to the monitor was 650 x giving a view of 220 x 280 μm^2 on the exposed part of the B.M. The speckle size, referred to the B.M, was 3 µm. When a 1 cm^2 PIN-diode was used for photoelectric detection, its size corresponded to a 17 µm probe on the B.M. A beamsplitter was also placed in the imaging branch to relay 20% of the object wave to an observing microscope.

The interferograms were stabilized by decreasing the TV exposure to 5 msec. by use of an acousto-optic chopper. Dynamic phase modulation (10) was extensively used during the experiments, both for visual observation ($\Delta f \sim 0.5$ Hz) and for photoelectric detection ($\Delta f \sim 5$ Hz). For post experiment analysis, the interferograms were recorded on the videotrack and the reference signal on the soundtrack of the videotape. The calibration procedure was rather

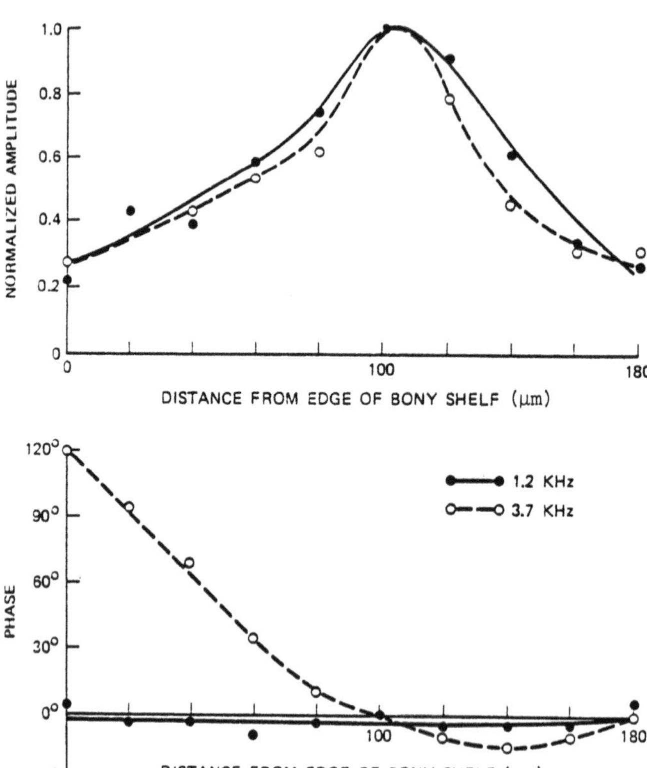

Fig.1 Transverse profiles of basilar membrane amplitude at 1.2 and 3.7 KHz. Distance from stapes footplate 13.3 mm.

complicated and will be described in (9). The lower limit of detection was 1 nm, but this value can probably be improved at least a decade.

During the experiments, successful measurements were done on 15 human temporal bone preparations. The behaviour of the B.M was at first scanned through visually using dynamic phase modulation. Thereafter dynamic fringe-patterns together with the reference signal were recorded in samples over the frequency range 400-7000 Hz (this range was determined by our present frequency translator and represented no real limit). From these recordings we could measure the amplitude/frequency response at a single point and the amplitude/phase distribution across the exposed membrane at constant frequencies. An example of the latter type of curves is shown on Fig.1. Note the large transverse phase variations across the B.M, an effect which was also observed in other samples.

The ESPI set-up was also used to observe how well tiny cobalt foil Mössbauer sources (size 60 x 60 μm^2) followed the motion of the B.M. The sources appeared to adhere to the membrane over the observed frequency range. However, they would often tilt depending on the frequency and their position. This makes the interpretation of Mössbauer data rather unreliable.

4. Concluding Remarks

We have described how the ESPI technique can provide valuable information about the motion of biomedical objects, both in vivo and in vitro. Future work should be on refinements of the method, especially stressing computerized data acquisition and analysis as the amount of information will rapidly outgrow the manual capacity of analysis.

References

1. O.J. Løkberg and K. Høgmoen. J. Phys. E. 9, 847, 1976
2. O.J. Løkberg et. al, Appl. Opt. 18, 763, 1979
3. O.J. Løkberg, Appl. Opt. 18, 2377, 1979
4. G.Å. Slettemoen and O.J. Løkberg, Appl. Opt., 20, 3467, 1981
5. O.J. Løkberg and K. Krakhella, Optics Comm. 38, 155, 1981
6. G.Å. Slettemoen, Appl. Opt., 19, 616, 1980
7. K. Burian et. al, Symp. 1976 on Electrocochleography and Holography in Medicine (Münster) A29
8. O.J. Løkberg and K. Høgmoen, Proc. SPIE 136, 222, 1977
9. P. Neiswander and G.Å. Slettemoen, to appear in Appl. Opt. 15 Dec. 1981
10. K. Høgmoen and O.J. Løkberg, Appl. Opt., 16 1869, 1977

Retinal Blood Flow Visualization by Means of Laser Speckle

J.D. Briers, and A.F. Fercher

Fachbereich Physik, Universität Essen, Postfach 103764
D-4300 Essen 1, Fed. Rep. of Germany

Introduction

Lasers have already found several diagnostic applications in ophthalmology. For example, the techniques of laser Doppler anemometry [1] have been used to measure retinal blood flow [2,3] and optic nerve-head compliance [4] , and the possibility of measuring eye dilatation by means of laser interferometry is presently under investigation [5, 6]. The properties of laser speckle [7] have also been exploited, in the measurement of eye refraction [8, 9].

The measurement of retinal blood flow is an important diagnostic tool in ophthalmology. A non-invasive technique is to be preferred, and several groups have investigated the use of laser Doppler anemometry [1-3]. However, this technique measures the velocity at only one point at a time, and many measurements are needed in order to build up an overall picture of retinal blood flow. It has been suggested that for some purposes - for example, the detection of reduced or restricted flow in certain blood vessels - a method which gives an instantaneous "map" of blood velocity might be a useful additional tool. We have attacked this problem by means of single-exposure laser speckle photography [1o-12] .

Theory

The basic principle of our technique lies in the phenomenon of speckle averaging. When a diffusing surface is illuminated with laser light, a random speckle pattern is observed which has well-defined statistical properties [13]. In particular a measure of the contrast of the speckle pattern is provided by the ratio of the standard deviation of the intensity to the mean intensity, i.e. $\sigma/\langle I\rangle$. Under ideal conditions of fully coherent light and a perfect diffuser, this ratio is equal to unity.

If the laser-illuminated diffuser is moving, then the speckle pattern recorded in a finite-time photograph will be blurred to an extent which will depend on the velocity of the movement and on the exposure time of the photograph. This blurring, or speckle-averaging, will result in a reduction in the speckle contrast, i.e. in the ratio $\sigma/\langle I\rangle$. The phenomenon is similar to that observed when botanical specimens are illuminated with laser light [14,15], and which proves to be a limiting factor in the measurement of plant growth by means of holographic interferometry [16-18].

In particular, if the subject is a flow field, then the speckle contrast in an area where flow is occurring will be reduced by an amount which will depend on the velocity of flow and on the exposure time of the photograph. The speckle pattern in an area of no flow, on the other hand, will remain of high contrast.

A mathematical model of the relationship between speckle contrast and flow velocity has been formulated [11]. This makes some rather drastic assumptions, but is useful as a starting point for the quantification of the speckle averaging technique. According to the model, the change from maximum to minimum speckle contrast should correspond to a velocity range of about 2 1/2 orders of magnitude. For example, an exposure time of 1 ms should allow the mapping of velocities between o.o2 and 5 mm s^{-1}. Changing the exposure time would, of course, change this range.

Experimental Technique

The experimental arrangement is illustrated in Fig.1.

Light from a helium-neon laser illuminates the retina via the lower half of the pupil of the eye, and the retina is photographed via the upper half of the pupil. In order to ensure that an image of the retina is focused on the film plane of the camera, the camera lens is focused on infinity and the subject is asked to focus and fixate his eye on an illuminated target located effectively in the film plane. The exact location of this target can also be used to determine which part of the retina is photographed.

Differences in speckle contrast are difficult to detect visually and certainly difficult to quantify. For this reason we have added a simple image-processing step to the technique. We use a standard high-pass optical filtering technique to convert speckle contrast variations to intensity variations. The argument is that areas of high-contrast speckle will diffract light into the usual diffraction halo, while areas with no speckle will direct all of the incident light into the central (zero-order) maximum in the Fourier plane. Hence a central obstruction in the Fourier plane, designed to filter out the zero-order component, should lead to a final image in which areas of high-contrast speckle are reconstructed with a much higher intensity than areas with no speckle (i.e.where the speckle has been blurred by the blood flow).

Preliminary Results

Early results have been encouraging. Laser photographs of a human retina do exhibit differences in speckle contrast between blood vessels and surrounding areas. However, as expected, their differences are difficult to observe visually (Fig.2a). In the high-pass filtered version of this photograph, however, the blood-vessels are easily identified (Fig.2b).

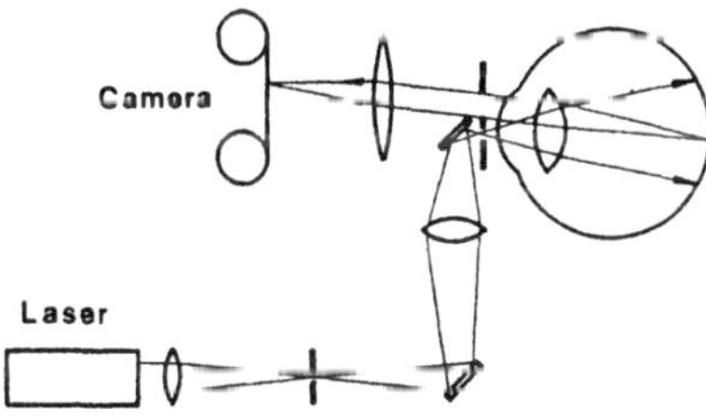

Fig.1 Experimental arrangement for laser photography of the retina

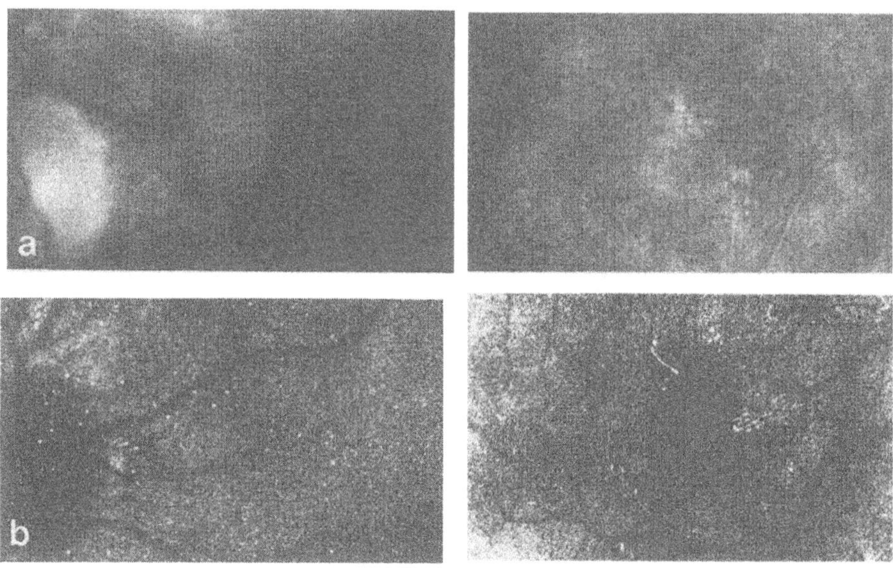

Fig.2 Laser photographs of a human retina: a. unfiltered; b. high-pass filtered

Although we have successfully obtained photographs like Fig.2b, in which areas of flow (blood vessels) are differentiated from areas of no flow, we have not yet been able to identify different velocities of flow. The main reason for this is that we suffer at the moment from a shortage of light. We are at present using a 5o mW helium-neon laser and Kodak 2415 film - the latter because of its high resolution. We find that with this combination we need exposure-times of about o.o2 s. This contrasts with the 1 ms mentioned earlier in this paper as necessary to resolve velocities around 1 mm s^{-1} which is the region of retinal blood flow velocities. Hence in our photographs the speckle is completely blurred out by the blood flow. (Even with the present arrangement, however, we should be able to diagnose blocked vessels, by comparing the filtered laser photograph with a standard fundus-camera photograph taken in conventional light.)

Conclusions

We have demonstrated the feasibility of differentiating between areas of flow and areas of no flow, but not, yet, of distinguishing between different velocities of flow. To this end we are presently working on improving and extending the technique. We are looking at the possibility of using a pulsed laser, in order to be able to use the shorter exposure times necessary to resolve the blood velocities encountered in the retina. We are also investigating alternative spatial filtering techniques and ways of improving the final photographs.

It is unlikely that this technique will be able to compete with laser Doppler anemometry [1-3] so far as precision is concerned, but we believe that the advantage of providing an overall "map" of blood velocities, even if it is only semiquantitative, makes the method attractive as an additional aid in ophthalmology.

The authors acknowledge the contributions of A. Pentrys and M. Peukert to this project, and also the financial support of the DFG (Deutsche Forschungsgemeinschaft).

References

1. J.B. Abbiss, T.W. Chubb, E.R. Pike: Opt.Laser Technol. 6, 249 (1974)
2. G.T. Feke, C.E. Riva: J.Opt.Soc.Am. 68, 526 (1978)
3. C.E. Riva, J.E. Grunwald, S.H. Sinclair, K. O'Keefe: Appl.Opt.2o,117(1981)
4. R. Zeimer, J.T. Wilensky, M.F. Goldberg, S.A. Solin: J.Opt.Soc.Am.71, 499(1981)
5. A.F. Fercher, P.F. Steeger, J.D. Briers: (to be published)
6. A.F. Fercher, P.F. Steeger: Paper S-6/4, ICO-12 Satellite Meeting Optics in Biomedical Sciences, Graz, September 1981
7. J.C. Dainty (ed.): Laser Speckle and Related Phenomena, Topics in Applied Physics Vol.9 (Springer, Berlin, Heidelberg, New York 1975)
8. H.A. Knoll: Arch.Am.Acad.Optom. 43, 7(1966)
9. E. Ingelstam, S.I. Ragnarsson: Vision Res. 12, 411(1972)
1o. J.D. Briers, A.F. Fercher: In Speckles, Procs. DFG Colloquium, Bad Lauterberg, 7 Nov. 198o, ed. by K.J. Ebeling, p.51
11. A.F. Fercher, J.D. Briers: Opt.Commun. 37, 326(1981)
12. J.D. Briers, A.F. Fercher: Paper B-3/2, ICO-12, Graz, September 1981
13. J.W. Goodman: In Ref.7, Ch.2
14. J.D. Briers: Opt. Commun. 13, 324(1975)
15. J.D. Briers: In Recent Advances in Optical Physics, Procs. ICO-1o, Prague, August 1975, ed. by B. Havelka and J. Blabla (Palacky University, Olomouc, and Society of Czechoslovak Mathematicians and Physicists, Prague 1976), p.145
16. J.D. Briers: ibid, p.5o1
17. J.D. Briers: J.Exp. Botany 28 (1o3), 493 (1977)
18. J.D. Briers: J.Exp. Botany 29 (1o9), 395 (1978)

Measurement of Interior Displacement of the Human Crystalline Lens by Using Speckle Pattern

H. Uozato, H. Itani, T. Matsuda, M. Saishin, and S. Nakao

Department of Ophthalmology, Nara Medical University
Kashihara, Nara 634, Japan

J. Okada, K. Iwata, and R. Nagata

Department of Mechanical Engineering, College of Engineering
University of Osaka Prefecture, Sakai, Osaka 591, Japan

Abstract

We propose a method for the precision measurement of interior displacement of three-dimensional transparent bodies by using speckle pattern technique. In this paper, we apply this method to displacement measurements in the interior of the human crystalline lens. The human crystalline lens (in vitro) is illuminated with a thin sheet-like beam of a laser light, and the speckle patterns created by the light scattering from the particles in the illuminating plane is recorded by double exposure with before and after deformation. In-plane displacement fringe patterns are obtained by optical spatial filtering techniques. This method is briefly discussed theoretically and the experimental results are presented to demonstrate the validity of this method.

1. Introduction

The crystalline lens of the human eye plays a major role in accommodation. The increase in curvature of the lens surfaces during accommodation is well established [1-5], but the interior behavior of the crystalline lens deformation has hardly been studied. Meanwhile, a number of new methods have been developed for measuring the displacement and the deformation of the surface of scattering objects by using speckle photography and speckle interferometry [6]. Unlike the conventional methods of measuring the displacement of scattering objects by using holographic interferometry [7], the speckle pattern techniques are very easy to perform and simple to construct. Recently, these speckle pattern methods are extended to interior displacement measurement of three-dimensional transparent bodies [8-11] for engineering structural problems. In this paper, we discuss a non-destructive and non-contact way to measure the interior displacement of 3-D transparent bodies, together with its application for measuring the interior behavior of the crystalline lens deformation.

2. Principle of the Method

To simplify the case, let us set up a recording optical system as shown in Fig.1. A thin sheet-like beam of a laser, which is made by two cylindrical lenses, is passing through a transparent 3-D body onto the plane of interest. A double-exposure photograph is then taken of the resulting scattered light speckle pattern, one exposure before and one after deformation with a camera normal to the

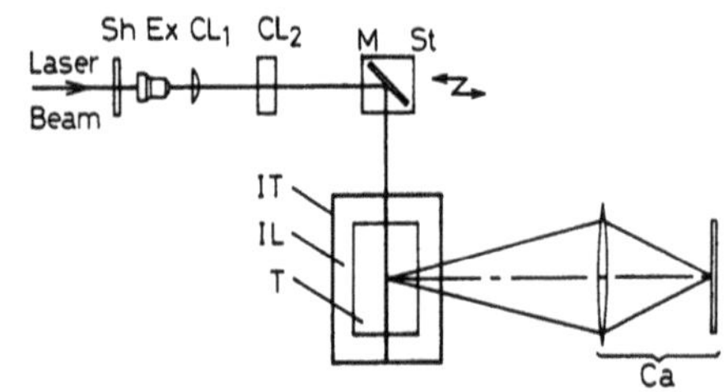

Fig. 1 Schematic diagram of optical system used for recording.

illuminated plane. The scattered light speckle pattern acts as a random grid embedded in this inner plane of the body. If the inner motion is large enough to cause a shift of the speckle pattern in the image plane of at least one speckle diameter, the processed double-exposed speckle negative will behave optically like a complex diffraction grating. There are two ways to acquire a displacement fringe from the resulting double-exposed negative based on the Fourier transformation (Fig.2). One is known as Young's fringes (pointwise spatial filtering) method and the other as isothetics (full-field spatial filtering) method. When a double-exposed negative is placed in a spatial filtering system as shown in Fig.2 and illuminated by a collimated beam, the resulting Fourier transformed image in the focal plane ($\mathbf{\nu}$) of the lens L_3 with a focal length f_3 is represented with the equation as follows [11]

$$<I(\mathbf{\nu})> = 2\cdot I'(\mathbf{\nu})[1+\cos(k\frac{\mathbf{\nu}\cdot\mathbf{d}_i}{f_3})], \qquad (1)$$

where $k=2\pi/\lambda$ (λ: wavelength of light), $\mathbf{d}_i$ is the displacement vector in the image plane, and < > stands for ensemble average. Here $I'(\mathbf{\nu})$ is an averaged Fourier transformed image without any speckle translation, and it is expressed by

$$I'(\mathbf{\nu}) = <|\int I_1(\mathbf{R})\cdot\exp(ik\frac{\mathbf{\nu}\cdot\mathbf{R}}{f_3})d\mathbf{R}\ |^2> , \qquad (2)$$

where $I_1(\mathbf{R})$ represents the intensity distribution in the recording plane before deformation. According to eq.(1), if $I'(\mathbf{\nu})$ spreads out widely enough, we can observe Young's fringes which cross the direction of the speckle translation $\mathbf{d}_i$ at a right angle. The fringe spacing here is

$$\rho = \frac{\lambda\cdot f_3}{|\mathbf{d}_i|}. \qquad (3)$$

Consequently, measuring the fringe spacing ρ, the displacement $\mathbf{d}_o$ (= $\mathbf{d}_i/M$, M: recording magnification) on the object space can be determined as

$$|\mathbf{d}_o| = \frac{\lambda\cdot f_3}{M\cdot\rho}. \qquad (4)$$

When the full-field spatial filtering is applied, the resulting isothetics are governed by

$$\mathbf{d}_i\cdot\mathbf{\nu}_o = N\cdot\lambda\cdot f_3, \qquad (5)$$

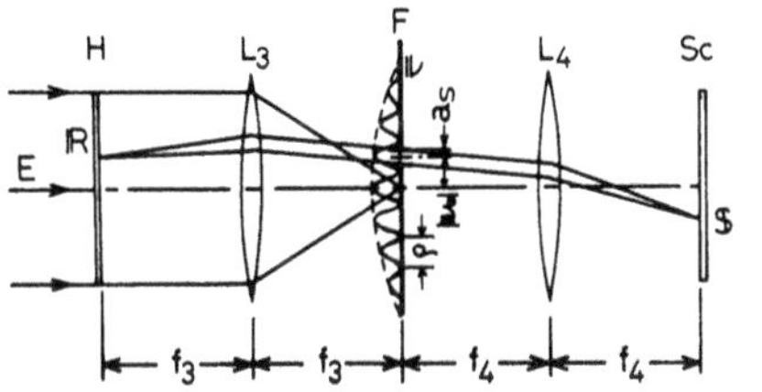

Fig. 2 Optical system used for spatial filtering. E: Illuminating laser beam, H: Double exposed negative, L_3 & L_4: Lenses, F: Aperture, Sc: Screen.

where N stands for the fringe order, $\mathbf{\nu}_o$ is the aperture vector. Rewriting this formula in order to find the displacement $\mathbf{d}_o$ of the object space, we obtain

$$|\mathbf{d}_o| = \frac{N\cdot\lambda\cdot f_3}{M\cdot|\mathbf{\nu}_o|}. \qquad (6)$$

When the displacement $|\mathbf{d}_i|$ is smaller than the average speckle size, no fringe appears, since the fringe spacing given by eq.(3) exceeds the width of the diffraction halo ($=\lambda\cdot f_3/\sigma_i$) caused by the speckles. We can find the diameter of a speckle σ_i in a recording system with a round aperture

$$\sigma_i = 1.22\lambda\cdot F_q = 0.61\lambda\cdot q/a_r. \qquad (7)$$

This diameter corresponds to the resolving power of the recording system.

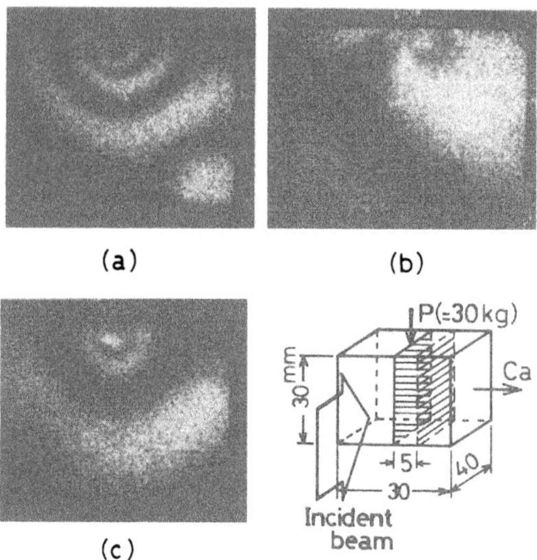

Fig. 3 Experimental results obtained from a 3-D model.

Both of the above-mentioned methods can work only when $|d_i|$ is larger than the speckle diameter σ_i.

3. Experimental Results and Discussions

Since the scattering light from the inside is generally weak, the laser beam must be made as powerful as possible. We used a He-Ne laser (λ: 633 nm) of 20 mW output for recording, and we employed Scientia 10E75 photographic plate or film, which have high resolving power. The Kodak D-19 process was used for development. In the experiment, we applied an immersion method to enable the

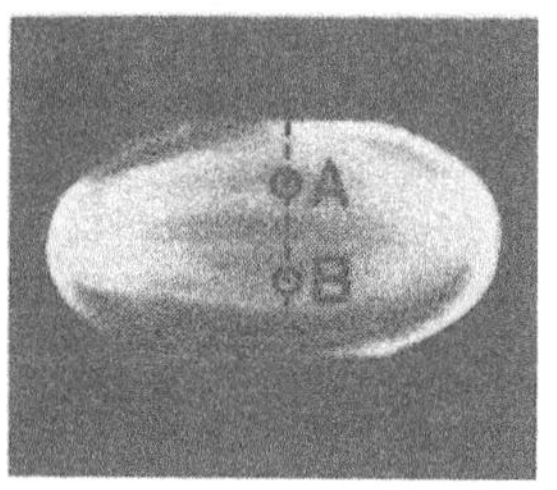

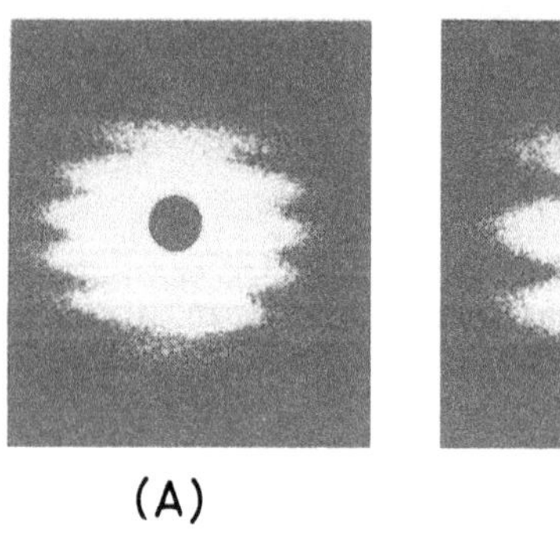

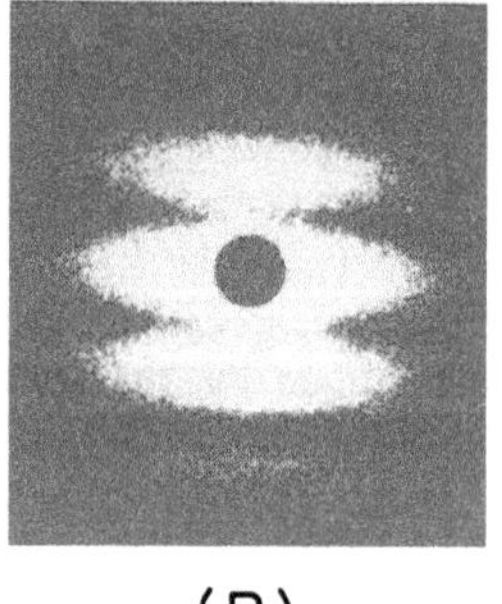

Fig. 4 An example of optical slice image of the human crystalline lens (in vitro), and Young's fringe patterns obtained at the position A and B.

sheet-like beam to go into the transparent body even when the boundary was not flat. The recording lenses used were Macro 50 mm/F:3.5 and Medical Nikkor 200 mm/F:5.6.

At first, to confirm the validity of this method, we performed a simulation experiment. Figure 3 shows an example of the simulation in which the load is concentrated on the center of the top plane of a 3-D block made by an epoxy resin. Figures 3(a) and (b) respectively indicate the vertical and horizontal displacement fringe patterns at the central section of the model, including the load point, and (c) shows the vertical displacement fringe pattern at the plane 5 mm in front of the central section. In each case, the displacement of a fringe is 8 μm.

Next, the application to the crystalline lens deformation in vitro is performed [12,13]. As a primary experiment in vitro, a crystalline lens was fixed by a cramp into an immersion liquid with a refractive index almost equal to that of the crystalline lens inside a glass vessel. As the displacement of the crystalline lens due to accommodation is normally larger at the anterior side, the posterior side of the crystalline lens was cramped to the datum plane, and the concentrated load was applied from the anterior side. In general, the crystalline lens behaves as a viscoelastic material [14]. Therefore, owing to decorrelation of speckles before and after deformation, it is very difficult to obtain good fringes. Making a double exposed recording before and after this load, we measure the interior displacement of each part by using a pointwise spatial filtering technique. Figure 4 is an example of the optical slice image of the human crystalline lens and the Young's fringe patterns obtained at the position A and B. Interior displacement distribution at the central section of the human crystalline lens is shown in Fig.5. In the experiment, a crystalline lens extracted after a cataract operation was used because it was easy to measure and had a greater scattering ability. Although the normal transparent lenses have an extremely weak scattering property, this method would be applicable to the normal lens due to accommodation (in vivo) with the aid of an image intensifier or an image orthicon and so on.

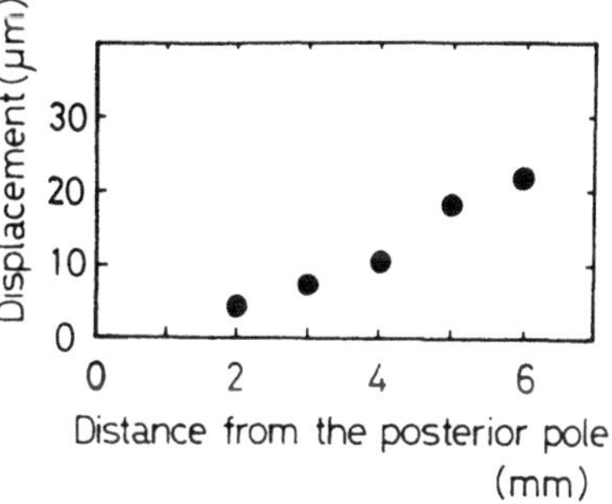

Fig. 5 Interior displacement distribution at the central section of the human crystalline lens.

4. Conclusion

We have discussed the method for measuring the interior displacement of a 3-D transparent body by using speckle patterns. The application of this method for measuring the crystalline lens deformation (in vitro) has been also demonstrated. The experimental results proved that the speckle pattern technique could measure its interior displacement without any single contact nor any physical destruction but with a remarkable accuracy. This method would contribute to the understanding of the accommodation mechanisms of the human eye and the biomechanical properties of the crystalline lens.

This work was partially suported by a Grant-in-Aid for Scientific Research from the Ministry of Education, Science and Culture (No. 56770802).

References

1. A. Nakajima and J. Nakagawa, Jpn. J. Clin. Ophthalmol. 20 (1966) 919.
2. B. Patnaik, Invest. Ophthalmol. 6 (1967) 601.
3. N. Brown, Exp. Eye Res. 15 (1973) 441.
4. V.Glezer, Fiziol. Zhur. 41 (1955) 830.
5. W. D. O'Neill and L. Stark, J. Opt. Soc. Amer. 58 (1968) 570.
6. A.E. Ennos, In: Laser Speckle and Related Phenomena, ed. by J.C. Dainty, Topics in Applied Physics, Vol. 9 (Springer, Berlin, Heidelberg, New York 1975) pp. 203-253
7. C. Vest, Holographic interferometry, John Wiley, 1979.
8. H. Uozato, K. Iwata and R. Nagata, Preprint of the meeting of the Jpn. Soc. Appl. Phys., 1976 Spring, p122, and 1976 Autumn, p168.
9. D. Barker and M. E. Fourney, Expl. Mech. 16 (1976) 209.
10. F. P. Chiang, In: The engineering uses of coherent optics (Ed. E. R. Robertson), Cambridge Univ. Press, p249-262, 1976.
11. H. Uozato, New measuring methods in photomechanics; Some new optical methods for use in experimental stress analysis, Doctoral Dissertation, Univ. of Osaka Prefecture (Feb. 1978) pp137.
12. H. Uozato, K. Iwata, R. Nagata and S. Nakao, J. Nara Med. Assoc. 31 (1980) 307.
13. H. Uozato, T. Ikuno, T. Matsuda and S. Nakao, J. Ophthalmol. Opt. Soc. Jpn. 2 (1981) 41.
14. Y. Kikkawa and K. Sato, Exp. Eye Res. 2 (1963) 210.

Categorization of Microparticles and Biological Cells by Laser Diffractometry

P.H. Kaye, I.K. Ludlow, and M.R. Milburn
Department of Physics, The Hatfield Polytechnic
Hatfield, Hertfordshire, England

1. Introduction

The potential of light-scatter measurements as a means of categorizing microparticles or biological cells has been demonstrated [1-5]. In particular, light-scatter data recorded from suspensions have proved useful for size distribution and refractive index determination of narrow distribution suspensions [2,3]. However, where a broad or bimodal size distribution exists the value of light-scatter measurements from suspensions is limited because of the loss of interpretable scattered intensity information. Recording data from individual particles is therefore preferable, though in such an approach adequate data for categorization must be recorded from each particle in a sufficiently brief interval to allow the examination of a large and statistically valid number of particles. This report describes instrumentation which approaches this aim.

2. Apparatus

Light from a vertically polarized Argon-ion laser is focused to ~100μm diameter at the centre of a scattering chamber (refer to Fig.1), where it intersects the liquid flow containing the specimen particles. The particle flow rate, up to $40 s^{-1}$, is determined by microprocessor controlled microlitre syringe pumps. A flow differential between specimen liquid input and output produces a constriction of the liquid flow such that at suitable particle concentrations $\sim 10^5 ml^{-1}$, multiple particle illumination is rare. Light scattered by each particle as it traverses the beam is received by a semi-circular scanned optical-fibre array of angular resolution 1.5°. The photomultiplier output is therefore a temporal variation corresponding to the angular scattered intensity distribution (scattering pattern) about the particle. Approximately three scans are completed whilst each particle is in the beam. The microprocessor accepts only the scan which corresponds to the particle in the central region of the gaussian beam.

2.1 Data Correction

The photomultiplier signal is corrected for both background scattering from the quartz scattering chamber, and for fluctuations caused by non-uniformities in the transmission of the optical-fibre array. Background is removed by adding an inverted, pre-recorded background signal at the photomultiplier anode synchronously with each scan. The additive signal is generated by the microprocessor via a 16-bit D/A converter. The transmission of the fibres within the optical array is normalized by attenuating the photomultiplier signal by predetermined correction-factors (one per fibre) within the correction

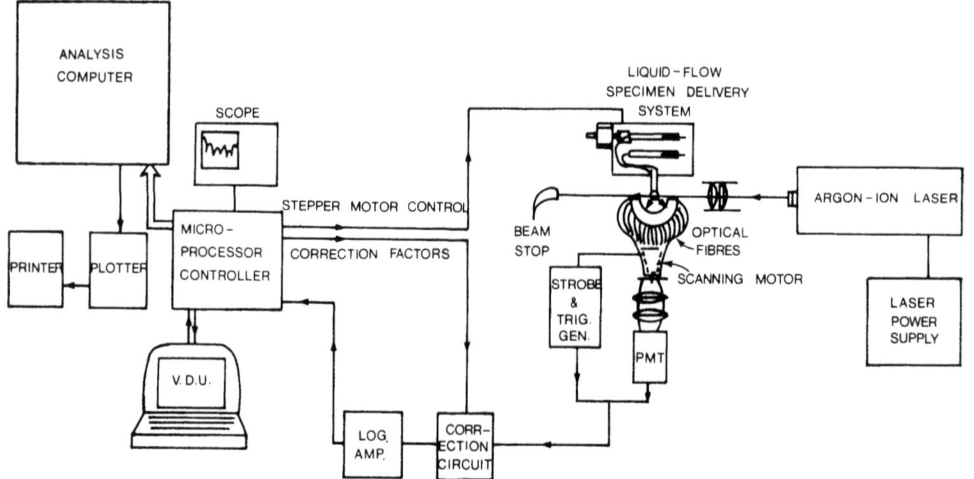

Fig.1 The Laser Diffractometer

circuit. The signal passed to the logarithmic amplifier is therefore an accurate representation of the radially scattered intensity from the particle alone. This signal is digitized and passed to the analysis computer.

2.2 Data Analysis

Currently, specimen particles are by necessity of spherical symmetry (section 4) and exact theoretical predictions of scattering patterns may be computed from Mie scattering theory [6]. It has been shown [7,8] that though scattered intensity maxima and minima at angles less that 90° are primarily of diffraction origin and thus relate to particle size, at greater angles they are more sensitive to internal dielectric structure. This suggests cells may be categorized on the basis of internal structure as well as size.

It is our intention that pattern analysis be carried out as near as possible in real-time. This requires the development of algorithms which may rapidly relate the information contained within the experimental scattering patterns to the theoretical models. A simple approach being investigated is to extract from the patterns the angular positions of intensity minima. This series of angles then becomes a code which may be matched to a library of theoretically derived codes in computer memory. Matching accuracy may be to within $\pm$ 2° for each minima. A record of the best theoretical match to each particle could then produce a distribution of size and dielectric structure within the population.

3. Results

Polystyrene latex microspheres have been used for instrument calibration. Fig. 2 shows the experimental pattern generated by a sphere from a suspension of quoted mean diameter 1.099µm. The incident wavelength was 0.4965µm. Theoretical matching by minima position indicated a sphere diameter of 1.20µm and refractive index 1.60. The corresponding theoretical curve is shown in Fig.2.

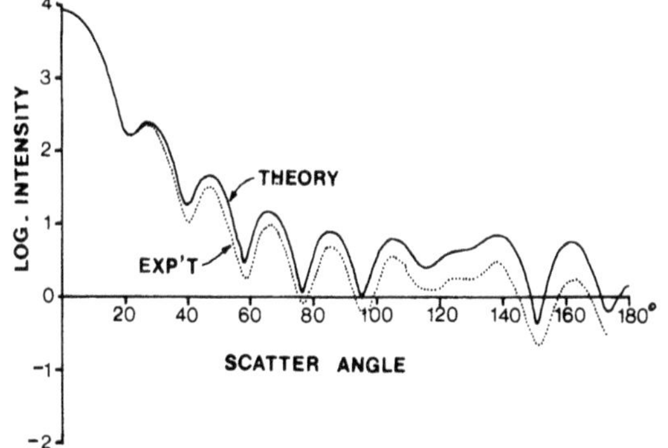

Fig.2 Experimental scattering pattern from single polystyrene latex sphere. Best-match theoretical curve corresponds to diameter 1.20μm, refractive index 1.60

From a series of such data the mean diameter was 1.21μm. This high value was confirmed by electron microscopy measurements and agrees with the results of Phillips et.al [9] on a similar latex.

3.1 Fungal Spores

Spores of the imperfect fungi Penicillium chrysogenum and Aspergillus niger have been examined. These spores were modelled on homogeneous spheres and gave values of mean diameter and mean refractive index of 3.46μm, 1.46 ± 0.02 and 2.47μm, 1.50 ± 0.02 for P. chrysogenum and A. niger respectively. Despite the use of an oversimplified theoretical model the results proved in good agreement with optical microscopy measurements [8].

3.2 Bacterial Spores

Bacterial spores of B. megaterium have been analysed using the theoretical model of a coated sphere. A typical scattering pattern and its theoretical best-matched parameter values derived from the minima positions are given in Fig.3. To achieve such matching entirely by computer requires some prior knowledge of the cell morphology to restrict the ranges of variables involved (refer to section 4).

4. Discussion

Despite potential for rapid particle categorization the technique suffers inherent limitations: (i) Specimens must be spherically symmetric such that scattering patterns are independent of particle orientation. We anticipate that minor cellular asphericity will not affect the mean values of size and dielectric structure derived from a population of such cells; (work is in progress at Hatfield to develop flow chambers which utilize hydrodynamic forces to orientate non-spherical cells axially along the flow. If successful, this facility would greatly increase the range of applicability of the technique); (ii) To restrict the library of theoretical data to a workable number ($\sim 5 \times 10^4$) the range of variables even in the simple coated-shpere model must be confined. This requires some a priori knowledge of cell morphology such as anticipated extremes of size, or typical wall thickness-to-radius ratio. For this reason the successful analysis of a totally heterogeneous population of cells is not foreseen. In matching experimental data to a theoretical library some ambiguity of matching is to be expected. Preliminary investigations suggest such

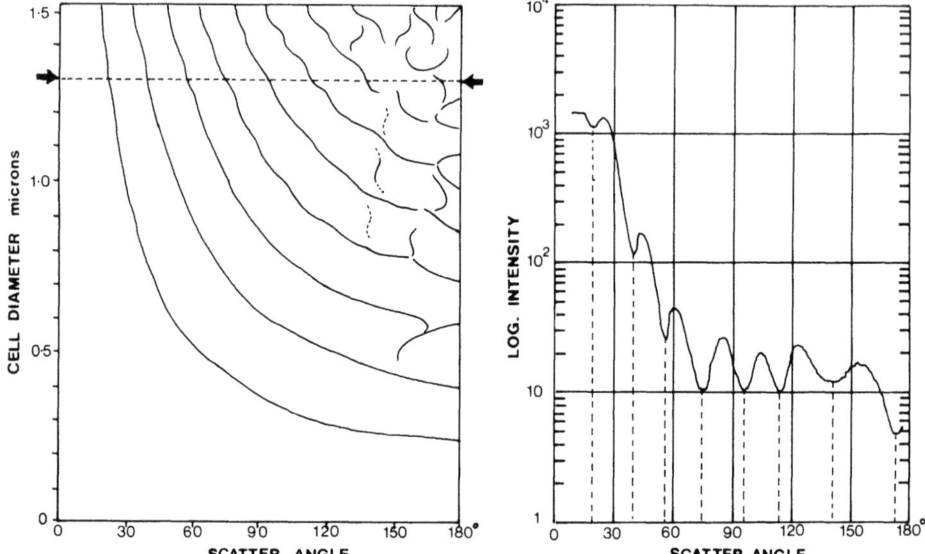

Fig.3 Experimental scatter pattern from single B. megaterium spore (right). The theoretical plot (left) shows the loci of scattered-intensity minima as a function of cell diameter for a coated cell having core and wall refractive indices 1.52 and 1.47. The ratio of wall-thickness to cell radius is 1 : 5 and the wavelength 0.4965 μm. By matching the positions of intensity minima the experimental pattern was attributed to a coated cell of diameter 1.3μm (arrowed line) having the above refractive indices.

ambiguous matching, when it occurs, has minimal effect on the mean values of variables determined from a statistically large number of experimental patterns; (iii) The determination of the optimum accuracy of matching, the presence of indeterminate experimental data, and the possibility of multiple particle scattering add complexity to the development of computer matching algorithms.

5. References

1 Wyatt, P. J., Applied Opt. 7,10 (1968)
2 Morris V. J., Coles H., and Jennings B., Nature(Lon) 249,240 (1974)
3 Wyatt, P. J., Berkman, R., and Phillips D., J. Bacteriol. 110,2 (1972)
4 Arndt-Jovin D. T.,and Jovin, T. M., Febs. Lett. 44,2, 247 (1974)
5 Eisert, W. G., Ostertag, R.,and Neimann, E., Rev. Sci. Instrum. 46,1021 (1975)
6 Kerker, M., in "The Scattering of Light", chaps.3-8, Academic Press, N.Y. 1969
7 Shimizu, K.,and Ishimaru A., 29th. ACEMB, Sheraton-Boston, USA. Nov. 1976
8 Ludlow, I. K.,and Kaye, P. H., J. Coll. Interface Sci., 69,3 (1979)
9 Phillips, D., Wyatt, P. J.,and Berkman, R., J.Coll. Interface Sci. 34,1 (1970)

Biomedical Applications of Molecular Luminescence

L. Szalay

Department of Biophysics, József Attila University
Szeged, 6720, Hungary

Image formation by light carries information about living organisms at many levels of organization, from the subcellular to the whole organism level. As a form of energy, light can be absorbed and may lead to luminescence of the absorbing molecules. Since the emitted light originates at a molecular level but depends on environmental factors, the luminescence carries information about structure and function at all levels of organization, including the molecular level. The scheme below shows the various roles of light in biomedical applications.

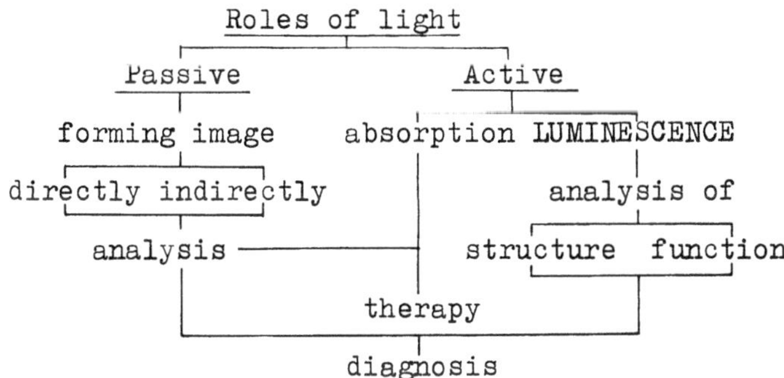

The experimental technique is generally known. Spectrofluorimeters of many types are commercially available both for steady-state and for time-dependent measurements. Many of them can be used for the determination of all luminescence characteristics: luminescence (fluorescence and phosphorescence) spectrum, excitation spectrum, yield, decay and polarization of luminescence. Since these parameters intrinsically depend on the structure of the emitting molecules, the molecular interactions and the structure of the molecular arrangement of the functional units, structural information can be obtained from the determination of these parameters. By changing the environmental physical and chemical factors (temperature, viscosity, pH, addition of drugs), additional functional information can be derived.

A few examples of the large range of applications are shown below, based mainly on the literature available in the last two years. In clinical chemistry antidepressants [1], enzymes, hormones [2] and antitumor agents [3] can be studied; immunoassay [4] and urea assay [5] can be performed. Carcinogenic polynuclear aromatic compounds can be simultaneously analysed, even in nanogram amounts [6]. At a cellular level multiparametric cell analysis, sorting and flow cytometry are widely applied [7]; the fluorescence of amniotic fluid provides information on respiratory distress of neonates [8], rheumatic factors [9], pulmonary neuro-endocrine cells in hypoxia [10]

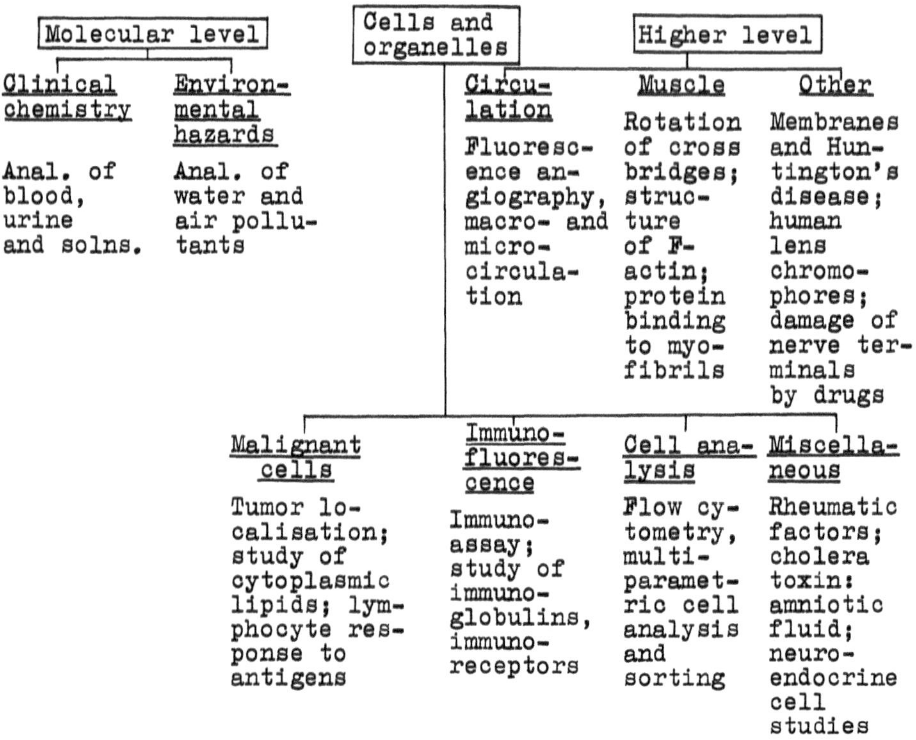

Molecular level

- Clinical chemistry: Anal. of blood, urine and solns.
- Environmental hazards: Anal. of water and air pollutants

Cells and organelles

- Malignant cells: Tumor localisation; study of cytoplasmic lipids; lymphocyte response to antigens
- Immunofluorescence: Immunoassay; study of immunoglobulins, immunoreceptors
- Cell analysis: Flow cytometry, multiparametric cell analysis and sorting
- Miscellaneous: Rheumatic factors; cholera toxin: amniotic fluid; neuroendocrine cell studies

Higher level

- Circulation: Fluorescence angiography, macro- and microcirculation
- Muscle: Rotation of cross bridges; structure of F-actin; protein binding to myofibrils
- Other: Membranes and Huntington's disease; human lens chromophores; damage of nerve terminals by drugs

and glucose in pancreatic cells [11] have been studied. At the level of organs and the whole organism micro- [12] and macro-circulation [13] studies and functional investigations of muscle [14] are examples of the versatility of the luminescence method. Some other clinically important applications: study of membranes in Huntington's disease [15]; damage caused to nerve terminals by amphetamine [16]; the appearance of chromophores in the human lens on aging.

Due to the relatively simple technique of the steady-state measurement, the extent of the biomedical applications is increasing: almost fifty publications now appear annually in the different journals. Though the technique of time-resolved measurements in the nanosecond and subnanosecond range is more sophisticated, the development of commercially available instruments has led to new modes of application.

In our laboratory we use molecular luminescence as a tool in the determination of some structural and functional characteristics of protein and photosynthetic membranes. To illustrate the wide range of the potentialities of the method we offer a few examples of our own research.

It is generally known that protein fluorescence originates almost entirely from tryptophanes because the electronic excitation energy of other amino acid residues is transferred to them. In peroxidase, tyrosines can transfer this energy only in part. If tyrosines are mainly excited (with 280 or 290 nm) overwhelming tyrosine fluorescence appears (peak at 305 nm), this is due to a steric hidrance effect in the energy transfer (Fig. 1. F = relative intensity).

If fluorescamine reacts with primary amines, intense fluorescence appears with maximum at 475 nm; fluorescamine and its hydrolysis products do not fluoresce. This reaction provides a rapid and sensitive assay for primary

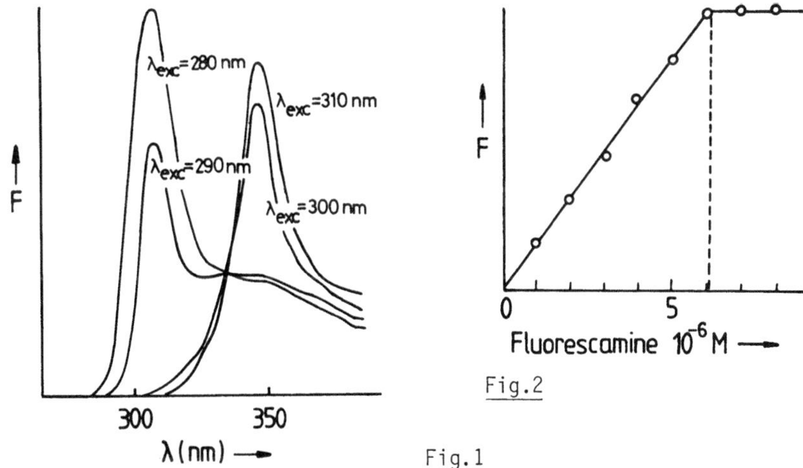

Fig.2

Fig.1

amines. The intensity of fluorescamine fluorescence of mixtures of $1 \cdot 10^{-6}$ M lysozyme and fluorescamine (from $1 \cdot 10^{-6}$ to $8 \cdot 10^{-6}$ M) solutions shows an increase of up to $6 \cdot 10^{-6}$ M fluorescamine and no change above (Fig. 2). It means that only 6 (of the 7) binding sites of lysozyme are accessible for fluorescamine molecules.

Very fine details of the contribution of amino acid residues to the total fluorescence of proteins can be obtained fluorometrically. 35-40% of the total fluorescence of lysozyme in aqueous solution is known to originate from the intensities I_A+I_B of fluorescence of tryptophan-62 and tryptophan-63 (TRP-62 and TRP-63). On account of their proximity I_A and I_B could not be determined separately with the chemical methods applied to date. We were able to measure I_A/I_B and $I_{AB}/(I_A + I_B)$, where I_{AB} is the intensity of fluorescence arising from the transfer of electronic excitation energy between TRP-63. $I_A/I_B = 0{,}53$ and $I_{AB}/(I_A + I_B) = 0{,}25$, in other words, the total intensity of the fluorescence of the pair originates from energy transfer. This is 9-10% of the total fluorescence of lysozyme, one third of the fluorescence intensity of the pair arising from TRP-63, and two thirds from TRP-62.

Cerulenin (a specific inhibitor of fatty acid biosynthesis) leads to changes in the fatty acid distribution of the thylakoid membrane of the green alga, Chlorella. As a consequence, the cell structure and function dramatically changes after cerulenin treatment. The fluorescence spectrum reflects the changes in chlorophyll organization (Fig. 3). In cerulenin-treated cells the long wave fluorescence is enhanced (broken line) compared to the fluorescence spectrum of intact cells (solid line). This indicates a profound change of chlorophyll integration associated with structural changes of the membranes.

The final example refers to an application of time resolved luminescence. The delayed luminescence of photosynthetic organisms originates from the back reactions in the electron transfer chain and starts already when the charge separation in the reaction center is completed. Due to the back reactions charge recombination occurs and the recombination is followed by a transition of the excited chlorophyll molecules into the ground state, i.e. by delayed emission. If a Chlorella suspension is illuminated with a laser flash (N_2-laser, 337,1 nm), the intensity of delayed luminescence (measured phosphoroscopically) falls about two orders of magnitude in the 40-500 μs time interval (Fig. 4). In this interval the intensity of light is very small. Photon counting and averaging of several thousand individual measurements

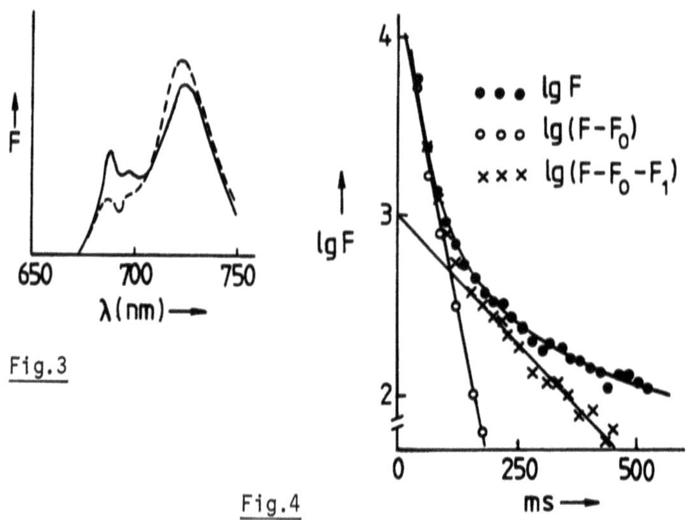

Fig.3

Fig.4

with a multichannel analyser should be applied. In logarithmic representation the decay of luminescence is nonlinear, therefore - following the usual procedure - the decay curve can be resolved into exponential components. A faster and a slower component (with average life time of 28 and 150 μs, respectively) can be obtained. The appearence of these components can be explained by attributing them to special donor-acceptor pairs present in the oxidation-reduction chain of the photosynthetic apparatus.

References

1 L.J. Cline Love, L.M. Upton: Anal. Chem. Acta *118*, 325 (1980); J.N. Miller, C.S. Lim, J.W. Bridges: Analyst *105*, 506 (1980)
2 K.W. Pearson et al.: Clin. Chem. *27*, 256 (1981); G.C. Pillai, R.C. Patel: Anal. Biochem. *106*, 506 (1980)
3 C.L. Richardson et al.: Res. Commun. Chem. Pathol. Pharmacol. *27*, 497 (1980); Biochim. Biophys. Acta *652*, 55 (1981)
4 H.T.T. Karnes et al.: Clin. Chem. *27*, 249 (1981); M.L. Dandliker et al.: In *Immunoassays*, Vol.4, ed. by Nakamura et al. (Liss, New York 1980) pp.65-89; I. Rédai et al.: Acta Biochim. Biophys. Acad. Sci. Hung. *15*, 177 (1980)
5 G. Jori et al.: 9th Annu. Meeting, Amer. Soc. Photobiol. June 14-18 (1981); E.A. Mroz, C. Lechene: Anal. Biochem. *102*, 90 (1980)
6 J.R. Maple, E.L. Wehry: Anal. Chem. *53*, 266 (1981); T. Vo Dinh et al.: Anal. Chem. *53*, 253 (1981)
7 T.M. Jovin: Flow Cytometry and Sorting, ed. by R.M. Melamed et al. (Wiley, New York 1979) p.137; J.P. Keene, B.W. Hodgson: Cytometry *1*, 118 (1980); W.G. Eisert, W. Beisker: Biophys. J. *31*, 97 (1980); J.W.M. Visser et al.: Exptl. Hematology *9*, 644 (1981)
8 D.O.E. Gebhardt: Clin. Chem. *26*, 1629 (1980); J.R. Lieberman et al.: Israel J. Med. Sci. *17*, 266 (1981); N.V. Simon et al.: Clin. Chem. *27*, 930 (1981)
9 A. Schulze, G. Geiler: Z. Rheumatol. *38*, 50 (1979)

10 I.M. Keith et al.: Cell Tissue Res. *214*, 201 (1981)
11 M. Deleers et al.: Biochem. Biophys. Res. Comm. *1*, 255 (1981)
12 P. Gaehtgens (ed.): Bibliotheca Anatomica, No.20, Karger, S AG, Basel (1981);
N. Thorball: Histochem. *71*, 209 (1981);
M. Oda et al.: in P. Gaehtgens (ed.) (see Ref.12)
13 F. Lund et al.: P. Gaehtgens (ed.) (see Ref.12)
14 K. Güth: Biophys. Struct. Mech. *6*, 81 (1980);
Yu.S. Borovikov: Biofizika (Moscow) *26*, 754 (1981);
A.L. Kirillov et al.: Biofizika (Moscow) *26*, 756 (1981);
C. Moos: J. Cell Biol. *90*, 25 (1981);
W.T. Stauber, Hui Ong Shu: J. Histochem. Cytochem. *29*, 866 (1981)
15 J.W. Pettegrew et al.: J. Neurochem. *36*, 1966 (1981)
16 H. Lorenz: Life Sci. *28*, 911 (1981)

A Measurement System for Otoscopic Spectroscopy in vivo

G. Heeke

University of Münster, Ear-Nose-Throat-Clinic
Medical Acoustics and Biophysics Laboratory, Kardinal-von-Galen-Ring 10
D-4400 Münster, Fed. Rep. of Germany

1. Introduction

In connection with holographic interferometry for investigating the vibration patterns of human tympanic membranes, spectroscopic properties of the eardrum are of practical interest. When using a ruby laser system for double-exposure holography, the wavelength is fixed. However, continous wave lasers with different wavelengths are used for real-time vibration analysis methods such as electronic speckle pattern interferometry. In this case it is possible to choose an optimum wavelength to maximize the energy reflected. A large amount of backscattered light is desirable for two reasons:

- the possibility of photobiological effects on the eardrum is minimized,
- the picture quality of interferograms increases with the strengths of the lighting.

Therefore, the reflection of light as a function of wavelength has to be determined.

This report describes a suitable system for such measurements and the results of some preliminary experiments. The possibility of using otoscopic spectroscopy for diagnosis of pathological disorders is discussed.

2. The Measurement System

Because the studies must be carried out in vivo, a special endoscopic spectrophotometer had to be developed. The main system design goals that influenced the photometer construction were measurement speed and ease of access to the eardrum with an instrument commonly available and familiar to a physician. The principle of the method is to illuminate the object with white light and analyse the backscattered light with a monochromator. The signal flow diagram of the spectrophotometer system is shown in Fig. 1. A 250 watt DC halogen lamp with a nominal colour temperature of about 6000 Kelvin is used as a light source. The condensor system which directs the light to a non-coherent glass fiber bundle includes a parabolic mirror with a special coating to suppress reflection of thermal radiation. By this means heating of the eardrum is reduced to acceptable levels.

The otoscope is a standard endoscopic instrument in clinical use. It allows direct observation of the eardrum through an optical system of rod lenses (Hopkins optic) contained in a metal shaft with an outer diameter of 4 mm, which is small enough in most cases. A simplified schematic diagram of the

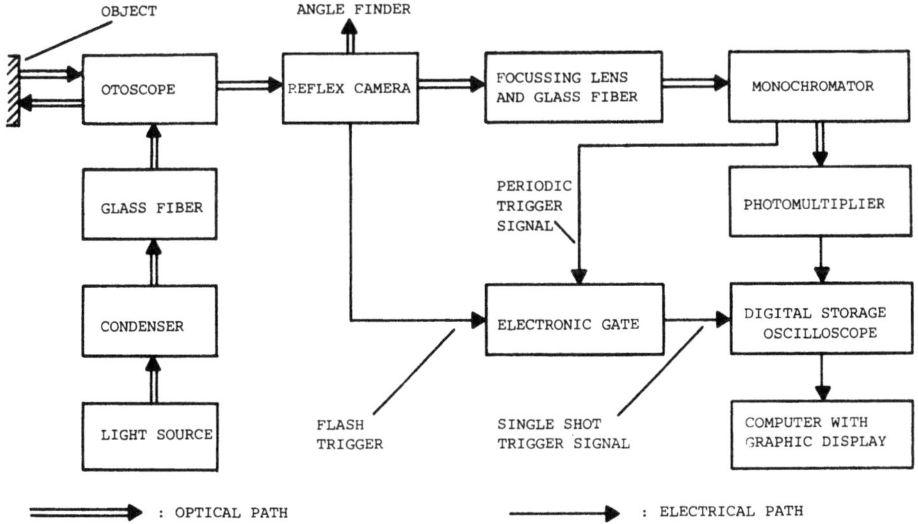

Fig.1 Signal flow diagram of the spectrophotometer system

otoscope is given in Fig.2. In addition to the lens system, the protective metal shaft contains a glass fiber bundle that guides the incident light to the tip of the instrument. From there a diffuse and nearly homogeneous object illumination is possible because of the coaxial arrangement of the fibers around the objective which has a nominal angular aperture of 80°.

A modified mirror reflex camera (Fig.3) follows next in the signal path. The standard backplane has been replaced by a lens holder containing a plano-convex short focal length lens. An angle finder is used during the positioning process. The actual measurement is initiated by a normal triggering procedure. While the mirror is in the upward position, light can pass through the camera housing, focussing lens and glass fiber bundle to the input slit of the monochromator with minimal loss of optical energy. At the same time an electronic gating circuit, with CMOS analog semiconductor switches, synchronizes a periodic trigger signal generated by the scanning monochromator with the flash trigger pulse. This produces a single shot write signal which triggers a digital storage oscilloscope so that the output waveform of the monochromator-photomultiplier combination can be stored for later analysis.

The monochromator (Fig.4) [1] uses a blazed diffraction grating (1200 l/mm) rotated by a small motor at 10 Hz as the dispersive element in a "side-by-side" Ebert type optical arrangement. The actual scan time of less than 5 ms for the wavelength range to be analysed is short enough to ensure that the object

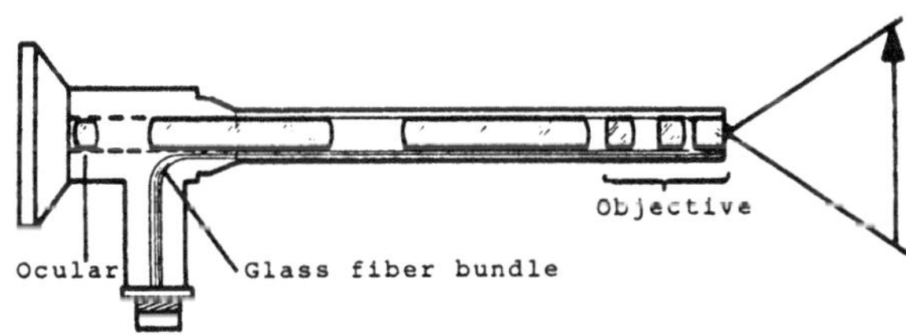

Fig.2 Schematic diagram of the otoscope

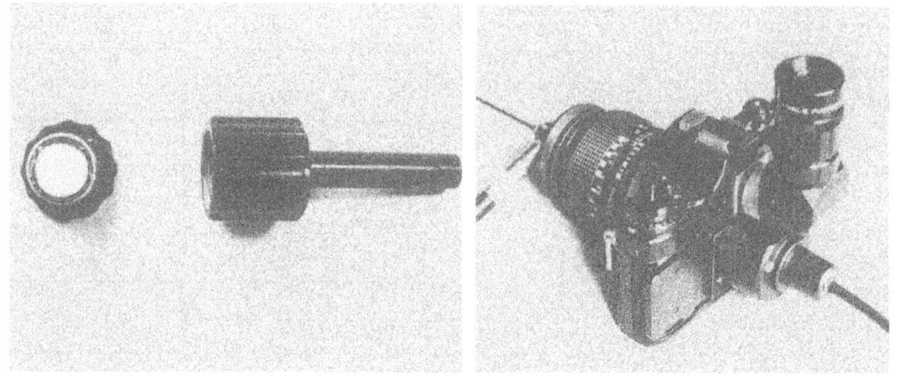

Fig.3 Modified mirror reflex camera with otoscope and calibration unit

is relatively stable. This short scan time, together with the weak reflection obtained from biological objects, results in an extremely low level of input energy to the monochromator and requires use of a detector with high quantum efficiency. A photomultiplier is used, but the output signal representing the spectral amplitude still has to be filtered. This can be done numerically following data acquisition using an online graphic computer system which also displays normalized data after a short processing time.

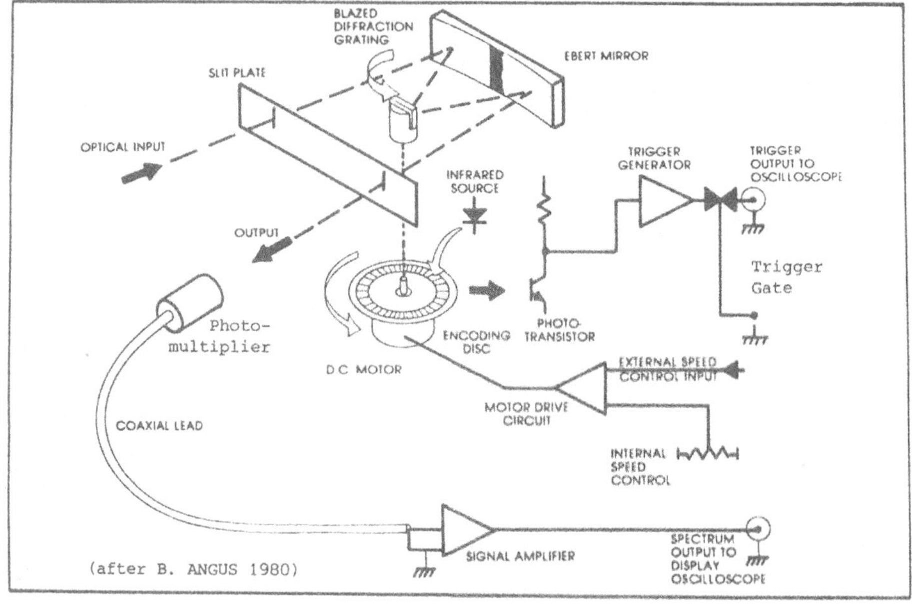

Fig. 4 Schematic diagram of the scanning monochromator

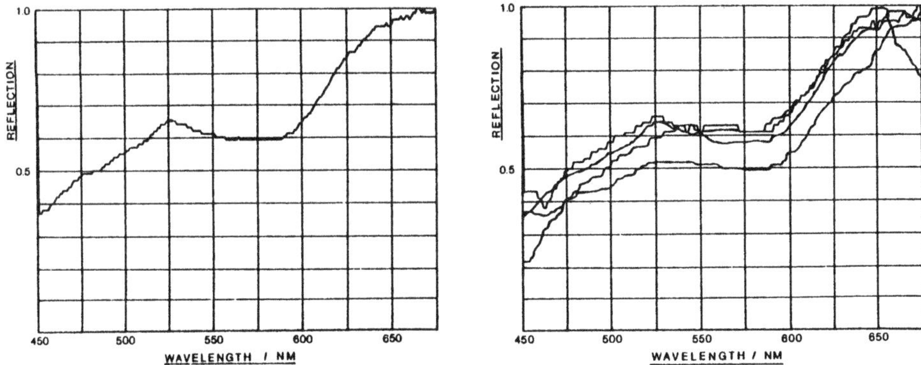

Fig. 5 Relative reflection measured at human tympanic membranes in vivo

As nearly all components of the measurement setup have wavelength dependent characteristics (transmission loss of fibers and lenses, energy distribution of the lamp, spectral sensitivity of the photomultiplier tube), the numerical data evaluation process has to compensate for those effects. As a secondary white standard magnesium oxide contained in a special adapter is used in the object plane during the calibration procedure.

3. Results

Preliminary investigations have been carried out with persons aged between 20 and 30 years. In order to ensure valid data, the spectral range to be analysed was limited to 450 - 700 nm, due to the photomultiplier characteristics. A single normalized reflection plot and a comparison of such measurements on four persons with regular middle ear status are given in Fig.5. These curves show a relative reflection maximum near 650 nm (red light) which is close to the wavelengths of the frequently used ruby (694 nm) and krypton (647 nm) lasers. At shorter wavelengths ($\sim$450 nm) the reflection value is only 20-40%, indicating higher losses either by absorption or transmission.

4. Discussion

Comparative investigations have to be performed on extracted specimen and special cadaver preparations in order to distinguish between reflections from the tympanic cavity and the membrane.

In pathological cases the visual appearance of tympanic membranes varies widely. Colours ranging from red (Otitis media) to dark blue (Haematotympanon, Otitis nigra) have been observed, and sometimes the colours are very intense. These observations lead one to ask whether otoscopic spectroscopy can be used to correlate middle ear diseases with the reflection spectrum measured at the tympanic membrane. Nevertheless, it should now be possible to take the first step in this direction since the photometer described allows rapid accquisition of data which can be quickly processed on a microcomputer.

Before more detailed studies are undertaken, the equipment should be improved in two respects. Firstly, the wavelength range should be extended to include invisible wavelengths. This could be accomplished by using a photomultiplier tube or avalanche photodiode with increased sensitivity as the detecting element. Secondly, the weight of hand-held system parts should be reduced for easier operation. The mirror reflex camera could be replaced by a small optical beamsplitter arrangement which allows simultaneous viewing and measurement. A disadvantage would be the further reduction of input energy to the photodetector.

5. Acknowledgement

The author gratefully acknowledges helpful discussions and assistance by W. Mette and Dr. Th. Wesendahl.

6. References

1 ANGUS, B.: A Scanning Spectrophotometer - Design and Applications, Optical Spectra, 8, (1980). Fig.4 reprinted with permission (modified).

This work was supported by grants of the Deutsche Forschungsgemeinschaft (SFB 88/B3).

Noninvasive Method for Evaluation of the Ratschow Test

V. Wienert[1], and V. Blazek[2]

[1]Department of Dermatology, University of Aachen
[2]Institute of High-Frequency Technology, University of Aachen
D-5100 Aachen, Fed. Rep. of Germany

1. Introduction

Occlusive arterial disease (OAD) is diagnosed on the basis of clinical symptoms and then assigned to stages I-IV according to FONTAINE [1]. In disorders of arterial blood flow, the color of the skin and filling of the veins is altered in a characteristic way by changing the position of the limb. According to the procedure practiced so far, it is diagnosed visually by the doctor (RATSCHOW test). This evaluation is subjective, since the human eye cannot detect color changes objectively. It is thus entirely possible to arrive at different evaluation criteria under different illumination.

OAD involves as a rule not only damage to the macrocirculation, i.e. a disturbance of arterial blood flow in the region of the large arteries, but also damage to the macrocirculation in the region of the terminal vascular bed. It is known today that disturbed hemodynamics, i.e. the reduced blood flow of the larger arteries, also affect the capillary network.

As early as 1938, HERTZMAN [2] found a correlation between the intensity of the light reflection of the skin and its blood content and termed his instrument a photoelectric plethysmograph. On the basis of experience with photoplethysmography, we have now developed the method of light-reflection rheography (LRR) in order to register the variations in cutaneous blood circulation by optical scanning of the skin by means of infrared reflection.

2. Method

The LRR method is based on the measurement of light reflection and scattering from the skin. Part of the IR radiation beamed onto the skin penetrates into the vascular bed. The scattered radiation thus depends on the vascular flow. It is measured by a photodetector. The skin temperature is registered simultaneously.

The LRR equipment is composed of an electronic measurement apparatus, of an x-y recorder and of the measuring head with opto-electronical components. The measuring head has small dimensions (ø 30 mm) and low weight (10 gr) and can be fixed on the limb to be investigated (e.g. back of the foot or lower leg 10 cm above the inside of the ankle) by means of commercially available foil rings sticking on both sides. The detector is in the middle of the sensor, three radiation emitters are arranged annularly around it at a distance of 5 mm.

Some physical experiments had to be carried out to determine optimal parameters (e.g. wavelength of the radiation) with respect to:

- the spectral location of the absorption band of the venous and arterial blood in the typical reflection spectra of the skin,
- the transmission quality of the horny layer and the depth of penetration as a function of the wavelength,
- the spectral extinction quality of the human skin and the ratio of absorbed to scattered radiation.

The radiation emitters we use are GaAs-IREDs with typically 6 mW radiation power at the wavelength of 940 nm and a spectral bandwidth of 50 nm. To eliminate the effect of ambient light, the emitters are modulated with a carrier of 10 kHz. Typical optical qualities of the skin in this spectral region are: extinction coefficient $K(\lambda)<10\ cm^{-1}$, $K/S \approx 0.5$ [3].

Because of the use of three "cold" emitters a nearly homogeneous illumination can be realized, without excessive heating of the skin. Both the three emitters and the radiation detector (Si phototransistor) are equipped with lenses ($2\varphi=10°$). Therefore, radiation is focussed at some distance in the skin, and the detector measures primarily those parts of the radiation scattered by the deeper vessels, and not the surface reflection of the skin.

After appropiate signal processing the arterial LRR curves are registered on the x-y recorder (the voltage (vertical axis) is proportional to the scattered radiation and is inversely dependent on the blood pressure as a function of time (horizontal axis) see Figs. 2-4).

It is of course possible to combine the equipment with a small computer in order to get an on-line analysis of the arterial LRR curve. A detailed description of the physical and technical system of the LRR method as well as the description of the procedure for calibration of the LRR curves in mmHg (or kPa) will be discussed in a subsequent paper.

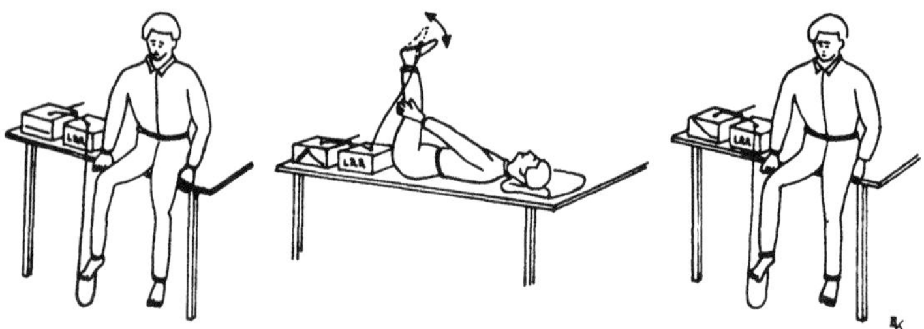

Fig. 1 Performance of the RATSCHOW test.

3. Results

In analogy with the familiar standard technique, we carry out the RATSCHOW test as follows:

First of all, the LRR measuring head is applied to the back of the foot in the manner specified above. The actual investigation then follows in four phases.

In phase I, the patient sits with pendent relaxed legs. In phase II, the patient lies on his back and stretches his legs vertically. He then performs dorsal flexures in both ankles (phase III). The skin becomes paler. In phase IV, the patient sits up again and allows the legs to hang down in the relaxed resting position. The skin becomes darker. During the course of the test, the degree of cutaneous reflection (change in blood flow) is continuously registered. This is otherwise usually observed only subjectively with the eye. Some selected results from healthy subjects and from patients will be presented below.

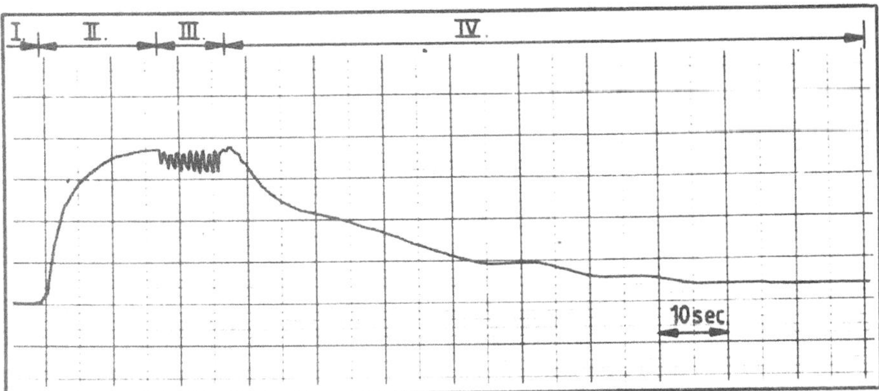

Fig. 2 Arterial LRR test in a 43 year old healthy male test subject.

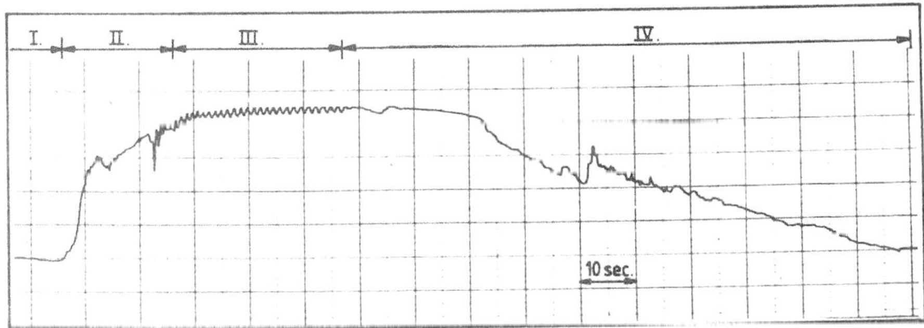

Fig. 3 Arterial LRR test in a diabetic with microangiopathy and macroangiopathy

In Figure 2, the arterial LRR curve of an arterially healthy patient is shown. Compared with this, the results of the same test in patients with arterial diseases are shown in Fig.3 and 4.

Soon after raising the legs to the vertical position, the skin becomes lighter in color. The movements of the foot then performed in the ankle joint hardly bring about an additional removal of blood from the limb (cf. the small rise of the LRR-curves within phase III in all figures). This movement program can thus be dropped; the maximum of the LRR-curve (maximum of skin scattered radiation) is already reached in most cases after about 20 sec elevation. After return to the initial position (phase IV), the radiation scattered by the skin of the back of the foot decreases practically immediately in healthy subjects ($t_a \approx 2$ sec, see Fig. 2), i.e. the blood flow rises.

In diabetic angiopathy, the cutaneous blood flow recommences only after several seconds (t_a 25 s in Fig. 3), in OAD of stage II, only after 35 s (see Fig. 4).

We regard the greatly differing time span (t_a) up to recommencement of cutaneous blood flow as a physical measure for objective quantitative evaluation of the severity of the disorder of blood flow. The greater this time span, the more severe is the disturbance of arterial blood flow.

4. Conclusions

A noninvasive method for quantitative evaluation of the hemodynamics of the arterial system of the lower limb which has already been successfully tested clinically is described. Phlebological syndromes can also be clarified with light-reflection rheography [4] .

The technique has the following advantages:

- experimental determination of physical parameters (refilling phase t_a)
- monitoring of the process of compensation after arterial occlusion
- painless technique free of complications
- noninvasive rapid and reproducible method.

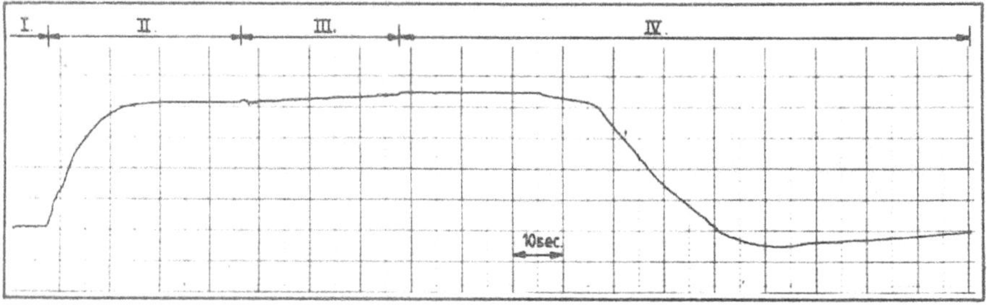

Fig. 4 Arterial LRR test in a 53 year old male patient with OAD (stage II)

References

1. WIDMER, L., WAIBEL, P.: Arterielle Durchblutungsstörungen in der Praxis. Huber Verlag Bern 1972.

2. HERTZMAN, A.B.: The blood supply of various skin areas estimated by photoelectric plethysmography. Am. J.Physiol. 124 (1938), 328-34o.

3. BLAZEK, V.: Absorptions- und Streuverhalten biologischer Gewebe bzgl. der monochromatischen, optischen Strahlung. Biomed. Techn. 24 (Erg.-Band) (1979), 84-85.

4. BLAZEK, V., WIENERT, V.: Licht-Reflexions-Rheographie: Eine nichtinvasive Technik zur Beurteilung der chronisch-venösen Insuffizienz. Wiss. Ber. 6. Jahrestag.Österr. Ges. f. Biomed. Techn. (1981), 267-271.

Part V

Optometry

Optical Techniques for Detecting and Improving Deficient Vision

H.C. Howland

Sections of Neurobiology and Behavior and Physiology, Cornell University
Ithaca, NY, USA

1. Introduction

The past decade has seen widespread development and application of new techniques for the study of vision and the diagnosing of optical maladies. It is impossible in this short space to make an exhaustive review, but the following studies may typify current trends and provide the reader with an introduction to the literature.

Contrast sensitivity measurements in adults have revealed that demylinating diseases of the optic nerve may reduce contrast sensitivity in mid frequency ranges without affecting the classical high contrast, high spatial frequency measurement of visual acuity [1]. Preferential looking techniques using grating stimuli have allowed us to trace (albeit indirectly) the neural development of the infant visual system [2], as have visual evoked response methods [3,4]. Laser speckle refraction has allowed us to make exquisitely precise measurements of meridional focus, thus permitting a new attack on the mechanisms of accommodation [5-7]. Double-pass linespread measurements continue to be used and improved upon for the measurement of the optical quality of the eye and are now finding use in developmental studies [8,9].

Computers and modern electro-optical-mechanical technology have opened up new possibilities for older techniques. Infrared optometers have been widely introduced in busier practices and their automated refractions found comparable to more conventional techniques [10,11].

On the side of prosthesis, the most promising outlook for the poorly sighted is the tremendous proliferation of microcomputers and computing equipment which promises the easy conversion of both spoken and written words into large and easy to read type.

In this multiplication of techniques and new opportunities I would like to discuss two optical techniques which I have helped develop together with some practical applications of them. The first concerns photorefraction and its application to one very prevelant type of visual deficiency, namely "amblyopia". I hope to show the relationship between this preventable malady and the more fundamental questions of the functioning of the human visual system.

The second technique concerns the measurement of monochromatic aberrations of the eye by a subjective aberroscopic method. While its application is not so far advanced as that of photorefraction, I hope to demonstrate its practical importance.

Lastly I want to discuss the applications of both these and other techniques to the problem of determining where and by what means the eye is focused.

2. Amblyopia

Amblyopia is called "lazy eye" in American lay terminology, doubtless because of its frequent association with strabismus. It generally designates a subnormal visual acuity in the face of a well focused image on the fovea. The prevalance of amblyopia varies with the definition of the criterion acuity, as has been shown by FLOM and NEUMAIER [12]. At a criterion of reduction of visual acuity of 20/40 or worse it affects approx. 1.5% of the U.S. school population.

As we learn more about the nature of amblyopia its classification has been revised, but it now seems certain that approximately half of the cases of amblyopia are due to the existence of an uncorrected refractive error in the first two years of life [13]. Here then is the challenge of amblyopia: How do we detect such refractive errors in early infancy?

2.1 Photorefraction

Some years ago my brother, BRADFORD HOWLAND of the Massachusetts Institute of Technology and I published an objective photographic method for refracting eyes from a distance [14], the principle of which is as follows. The method is similar to a linespread measurement wherein a point of light from a fiber optic light guide centered in a large aperture camera lens is directed towards the eyes of a subject some 1 1/2 meter distant. Light from an ordinary flash passes through the guide, into the eye of the subject, is reflected by the retina and returns to the camera where, with two pairs of cylinder lenses and the camera lens, the point spread of the returned light is concentrated into a star-arm pattern, the length of the star arms giving an approximation of the linespread along the corresponding meridian and hence a measure of the degree of defocus relative to the camera in that particular meridian (Fig.1).

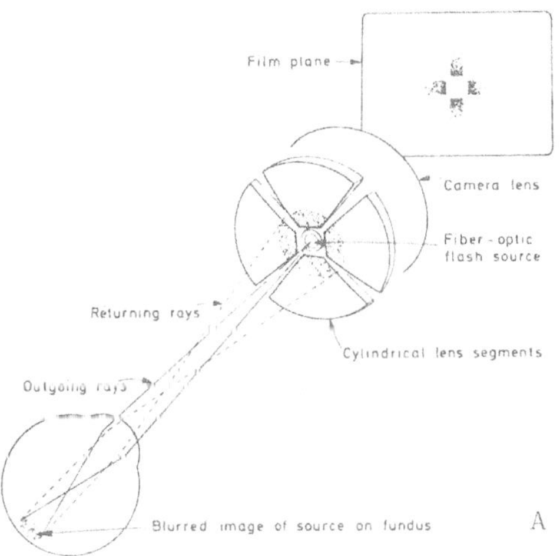

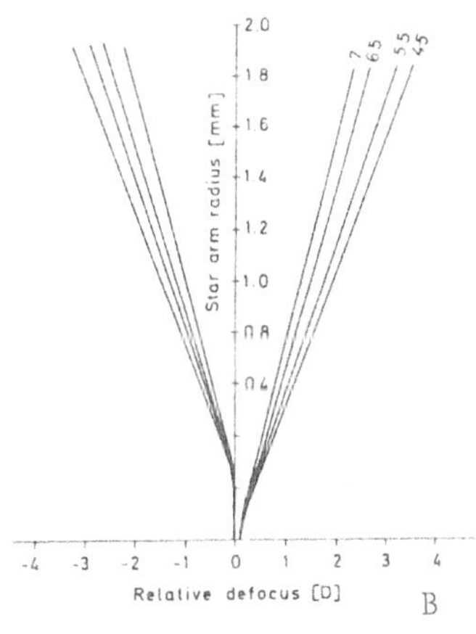

Fig.1.A. Optics of photorefraction. B. Radii of star arms (df/2) as a function of defocus and pupil size at 1.5 m. See ref. [17].

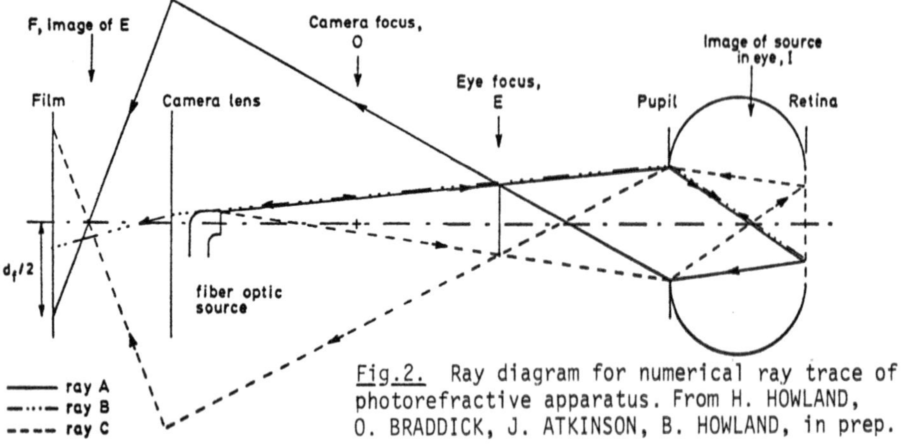

Fig.2. Ray diagram for numerical ray trace of photorefractive apparatus. From H. HOWLAND, O. BRADDICK, J. ATKINSON, B. HOWLAND, in prep.

It is possible to compute the concentration of light in these star-arm reflexes and hence to find the maximum intensity of light in the star arm as a function of defocus and pupil size. In practice, however, because the star arm is projected over the irregular intensities of the face, we have simply taken the maximum recorded extent of the star arm as an indication of the degree of relative defocus of the eye. This may be computed numerically in detail by geometric ray tracing (Fig.2) and for the main portion of Fig.1B solved analytically.

It may be seen that this technique tends to underestimate the degree of defocus in any meridian due to the cutoff characteristics of the photographic film (Fig.3).

There are, of course, two components of the linespread that may influence its magnitude other than defocus. These are the monochromatic aberrations of the eye, which are generally negligible compared to defocus, and those of longitudinal chromatic aberration, which in practical cases where one evaluates color film images may contribute the equivalent of 1/4 of a diopter of defocus to the linespread.

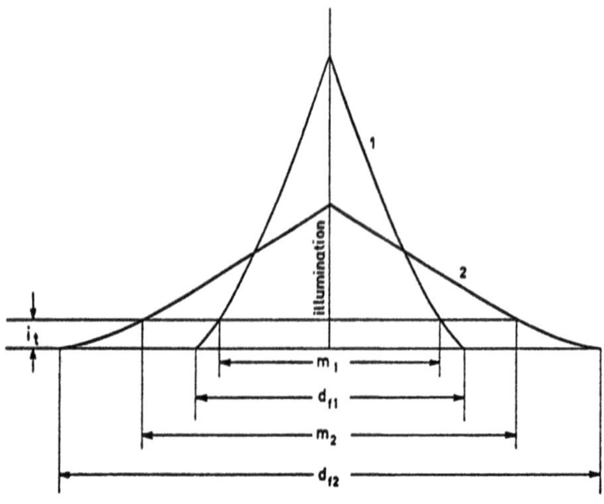

Fig.3 To show the effect of the illumination threshold, i_t, on apparent width (m_1, m_2) of defocused pupil in photorefraction.

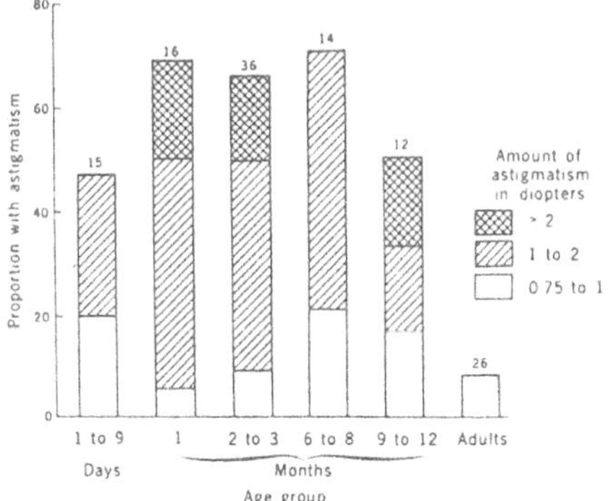

Fig.4. Summary of astigmatisms found by photorefraction for various age groups. See ref. [16].

2.2 Application of Photorefraction in Infant Vision

This method of refraction first found practical application in the study of infant vision with Drs. OLIVER BRADDICK and JAN ATKINSON of the University of Cambridge. Dr. INDRA MOHINDRA working with Prof. R. Held at M.I.T. had found by retinoscopic techniques that infants have a considerable amount of astigmatism [15]. In Cambridge, England, we were able to verify this observation by photorefraction [16] (Fig.4), and were able to trace the development of infants' focusing ability [17].

In the course of these studies we had ample opportunity to learn about the variation in fundal reflexes and the interpretation of photorefractive pictures. By accident we discovered that a modification of the photorefractive technique could yield not only the degree, but also the sign of defocus of the eye. We had constructed a simple centered fiber optic light guide for illuminating the pupils of our subjects so that we could easily measure pupil diameter. We discovered that by taking pictures with the camera when equipped with this attachment focused behind or in front of the subject we could a) determine the sign of defocus of the subject by comparing pictures and finding the smallest blurred pupil and b) easily determine the axis of astigmatism by finding the long axis of the blurred pupil (Fig.5A,B) [18]. Further, due to the chromatic aberration of the eye the negative cylinder axis of the subject is often distinctly reddish in tint (Fig.5C).

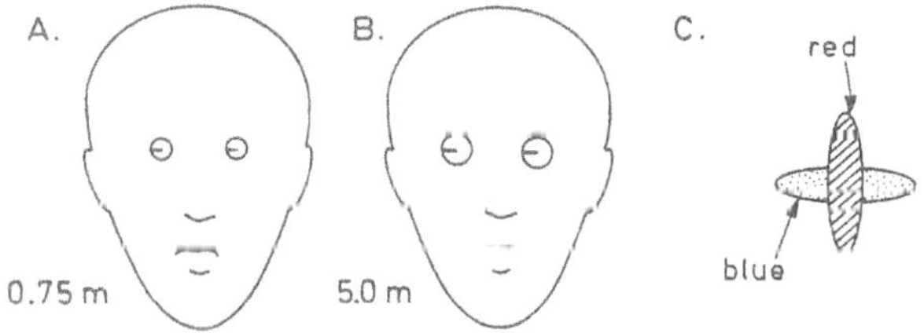

Fig.5. Appearances of pupil reflexes when myopic subject 1.5 m distant from camera is photographed with camera focused A) at 0.75 m and B) 5.0 meters. Shadow of fiber optic light guide centered in camera lens and used to illuminate pupils appears in pupil reflexes. C) Appearance of reflex of astigmatic eye with negative cylinder axis at 90 degrees.

In addition to the studies with my colleagues in Cambridge I have also photorefracted a large number of infants in Ithaca over the past two years and have conducted research with Drs. VELMA DOBSON, RONALD BOOTHE, AND MARTIN BANKS in Professor DAVIDA TELLER'S laboratory in Seattle. Collectively we have evolved the following picture of normal infant visual development.

2.3 Results of Infant Refractive Studies

A typical infant exhibits a diopter or more of against-the-rule astigmatism, and the refraction in both eyes is symmetrical. By three months of age infants are able to fixate objects accurately. Because the horizontal meridian of the eye is thus usually myopically focused relative to the vertical one, the infant at three months of age can focus vertical stripes in a preferential looking apparatus without making much accommodative effort [19]. The ability to focus rapidly improves and by six months the infant is able to focus accurately more distant objects at a meter and a half (Fig.6).

The magnitude of astigmatic error slowly decreases over the first years of life, but is still significantly above normal in the 1-2 year old age group.

By measuring corneal curvature in infants with photokeratometry it is possible to relate corneal astigmatism to total astigmatism of the eye and hence to show that much of the astigmatism of early infancy must be corneal (Fig.7).

While I am still gathering data on older infants, it appears from my preliminary results that, while astigmatism persists in the 1 to 2 year old group, less of this astigmatism is corneal and more lenticular than in the younger group.

It is fair to say that this picture of infant refractive state does not agree with classical descriptions of it [20], but that the presence of astigmatism in infants has been verified by both cycloplegic and non cycloplegic retinoscopy [21].

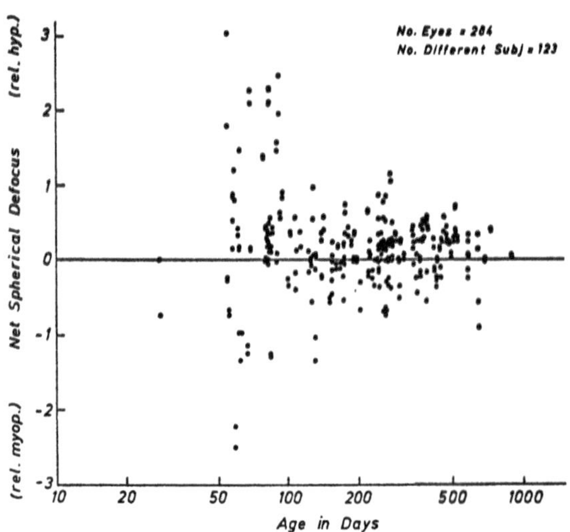

Fig.6. Net spherical defocus as a function of age for infants fixating photographer at 1.5 meters.

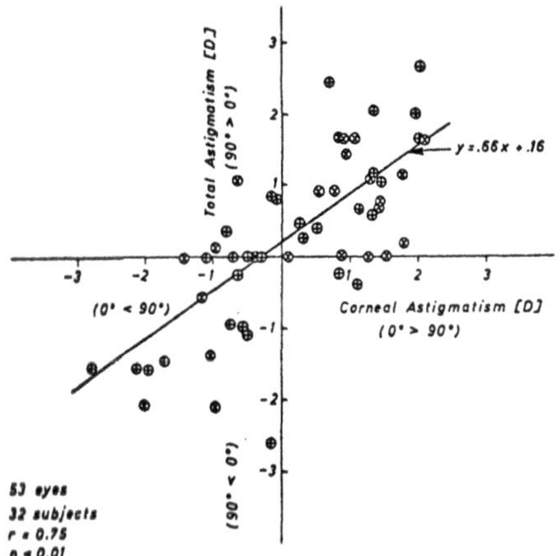

Fig.7. Total astigmatism as measured by photorefraction vs corneal astigmatism as measured by photokeratometry. Mean age of infants = 3.52 mos.

Now that we have a picture of normal infant refractive development we intend to follow those infants with more exceptional refractions with a view towards determining which refractive states are most likely to lead to amblyopia. Similar studies are underway in England.

3. Studies of the Monochromatic Aberrations of the Eye by a Crossed Cylinder Aberroscopic Method.

In the context of amblyopia the degradation of the optical image on the retina by the monochromatic aberrations of the eye must seem minor indeed. Nonetheless, good vision is a relative concept, and adequate sight for a university professor may be totally inadequate for an airline pilot. Before the aberrations of the eye can be corrected, they must be measured, and in this section I would like to discuss a practical measurement technique.

The concept of wave aberration of the eye embraces both refractive errors and monochromatic aberrations of the eye. Ideally, if the eye is focused at infinity, the wave front of light from a star converging on a point at the retina will form a perfectly spherical surface within the vitreous. Or, projecting in the other direction, light originating from a point on the retina, and passing through an aberration free eye would form a planar surface in object space. We may characterize the wave aberration of an acutal eye by describing the shape of this wave surface in object space. This is most easily accomplished by writing a Taylor polynomial for the surface in three dimensions [21].

$$W(x,y)=A+Bx+Cy+Dx^2+Exy+Fy^2+Gx^3+Hx^2y+Ixy^2+Jy^3+Kx^4+Lx^3y+Mx^2y^2+Nxy^3+Oy^4. \quad (1)$$

It may be seen that ordinary refraction of the eye deals with a determination of the prism, cylinder and spherical terms of the equation or constants B through F of the wave aberration equation. The constants G through O describe the "higher order" aberrations of the eye and may be grouped conveniently into what we will call here "third order" or coma-like aberrations: G,H,I,& J, and the "fourth order" or spherical-like aberrations, given by terms K through O.

Again, together with B. HOWLAND, using a method based on one of his inventions [22], we constructed a crossed cylinder aberroscope consisting of two

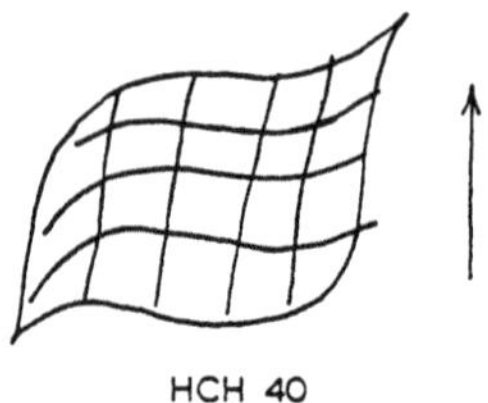

Fig.8. Drawing of grid seen through aberrsocope made by subject with "cylindrical" aberration (type o).

crossed cylinder +,- 5 D lenses and an intercalated grid of approx 1 mm spacing. With this instrument placed immediately in front of the eye like a spectacle lens the subject views a point source of light and sees projected on his retina the shadow of the grid. If the eye is focused at infinity and aberration free the grid will appear perfectly regular and rectangular. If an aberration free eye is focused at some other plane the grid will appear to scissor, its squares becoming rhomboids. Aberrations of the eye cause the lines of the grid to become nonparallel and/or curved as in Fig.8.

The distortion of the grid is quantitatively related to the wave abberation of the eye, the displacements of the intersections of the grid being proportional to the slope of the pupil. From an accurate rendition of the subjectively perceived grid it is possible with the use of orthogonal polynomials to totally reconstruct the wave aberration surface and hence to determine the values of all the coefficients of eq. (1).

The difficulty of the technique as so outlined lies in obtaining an accurate rendition of the perceived grid, since only the subject sees it. Again, with the advent of ubiquitous computers it is possible to generate comparison grids which the subject can then match to that of the aberroscope.

Despite these problems of rendition, because of the striking differences in the distortion of the grid introduced by the various aberration terms of equation (1) it is nonetheless possible to make statistical estimates of the importance of the various terms in the aberration equation in degrading the optical image. When this is done [23] we come to the surprising result that the major aberrations of the eye are not spherical aberration, as has so often been thought, but rather the third order, coma-like aberrations (terms G-J). Further, spherical aberration frequently occurs in one axis of the eye only, and is thus more properly the aberration of a cylinder than that of a sphere.

3.1 Possible Applications of Aberroscopic Wave Aberration Measurements.

Some of the occupations of our society make rather severe demands upon the visual system, and a practical method for wave aberration measurement may find use in association with screening appropriate personnel for these occupations.

In addition, some modern ophthalmological practices, such as radial keratometry where the cornea is deliberately mechanically weakened by a series of radial slits, must surely alter the aberration structure of the eye, and it seems a logical application of this aberroscopic technique to follow these changes.

4. Problems in the Study of the Focusing of the Eye: The Small Eye Error.

Above we took the view that perhaps half of all amblyopia is due to an uncorrected refractive error of infancy. But it may be that what is deficient in some infants is not their refractive state, but rather their ability to accommodate correctly. To study the focusing of infant eyes by objective means we must come to terms with the "small eye error" as described by GLICKSTEIN and MILLODOT [25]. They showed that many small eyes exhibit an apparent hyperopia due to the reflection of light from a layer anterior to the outer limiting membrane of the retina (presumably from the retinal-vitreous border). The difference may be 0.25 D in adult humans but can in theory increase to 0.5 D in infants, increasing inversely with the square of the focal length of the eye. While the small eye error does not affect retinoscopic or photorefractive estimates of astigmatism, since each meridian is mis-estimated by a constant amount, it is of significant magnitude when one is attempting to determine the absolute plane of focus relative to an astigmatic error.

The problem is that we do not know if such an error even exists in infants or in adults. The measurement in an adult is difficult enough and to my knowledge has never been correctly made. In the infant, the small eye error, if it exists, should be greater than in adults. It may be that with a clever experimental arrangement one could use laser speckle to induce optokinetic nystagmus in infants while simultaneously photorefracting their eyes. All objective reflectometric studies of focusing of eyes must be plagued to a greater or lesser degree by the uncertainty of the small eye error until it is correctly measured. In the meantime, there are still other uncertainties that make the study of focusing a difficult one which space does not permit us to discuss.

Portions of this work were supported by NIH grant EY-20994. Special thanks are due to M. Howland, J. Ballarino and T. Natoli for assistance on the project.

References

(1) J.A. Whitlock, T.J. Murray, & K.I. Beverley: Invest. Ophthalmol. Vis. Sci. 19, 324 (1980).

(2) V. Dobson: International Ophthalmology Clinics, 20, 233 (1979).

(3) V. Dobson & D. Teller: Vis. Res. 18, 1469 (1978).

(4) M. Millodot & I. Newton: Brit. J. Ophthalmol. 65, 294 (1981).

(5) E. Ingelstam & S.-I. Ragnarsson: Vis. Res. 12, 411 (1972).

(6) H.W. Leibowitz, & D.A. Owens: JOSA, October (1975).

(7) W.N. Charman & J. Tucker: Vis. Res. 17, 129 (1977).

(8) H. Howland & R. Rohler: J. Opt. Soc. Am. 10, 1 (1977).

(9) R. Williams & R. Boothe: Invest. Ophthalmol. Vis. Sci. (in press).

(10) H.A. Knoll & R. Mohrman: Amer. J. Optom. Arch. Amer. Acad. Optom. 49, 122 (1972).

(11) D.L. Guyton: Tr. Am. Acad. Ophthalmol. 79, 501 (1975).

(12) M.C. Flom & R.W. Neumaier: Public Health Reports, 81, 329 (1966).

(13) J.F. Amos: J. Am. Optom. Assoc. 48, 489 (1977).

(14) H.C. Howland & B. Howland: J. Opt. Soc. Am. 64, 240 (1974).

(15) I. Mohindra, R. Held, J. Gwiazda & S. Brill: Science 202, 329 (1978).

(16) H.C. Howland, J. Atkinson, O. Braddick & J. French: Science 202, 331 (1978).

(17) O. Braddick, J. Atkinson, J. French & H.C. Howland: Vis. Res. 19, 1319 (1979).

(18) H.C. Howland, J. Atkinson & O. Braddick: J. Opt. Soc. Am. 69, 1486 (1979).

(19) V. Dobson, H.C. Howland, C. Moss & M.S. Banks: Invest. Ophthalmol. Vis. Sci. 20, 146 (1981).

(20) I.M. Borish: Clinical Refraction, 3rd ed. pp. 9 (1970).

(21) A.B.V. Fulton, V. Dobson, B. Salem, C. Mar, R. Peterson & R. Hunson: Am. J. Ophthalmol. 90, 239 (1981).

(22) B. Howland & H.C. Howland: Science 193, 580 (1976).

(23) B. Howland: Applied Optics, 7, 1587 (1968).

(24) H.C. Howland & B. Howland: J. Opt. Soc. Am. 67, 1508 (1977).

(25) M. Glickstein & M. Millodot: Science 168, 605 (1970).

Computer-Assisted Objective Optometer

D. Bruneau[1], J. Corno, and J. Simon[2]

Institut d'Optique, Centre Universitaire, Bât. 503, B.P. 43
F-91406 Orsay-Cedex, France

1. Introduction

Apparatus capable of measuring ocular refraction automatically, called optometers or refractors, appeared during the last decade. Bausch and Lomb's OPHTHALMETRON seems to have been the first one of these instruments followed by Acuity Systems' AUTOREFRACTOR, and Coherent Radiation's DIOPTRON.

Automated refractors are interesting inasmuch as they allow a rapid approach to the correction needed by the patient; but their purchase and maintenance costs are very high, and they deprive the specialist of the ophthalmoscopic image, which allows detecting possible pathologies. For that reason we decided to construct a simpler instrument, associated with a microprocessor, which does not suffer from this drawback.

This instrument is a classical refractor with a small angular test, using white light. The specialist projects the test onto the patient's retina and observes the image there. He can adjust the test on the retina very easily, and thereby study the quality of the ophtalmoscopic image of the test. Automatic light data processing relieves the operator of fastidious tasks : presentation of a visual test in the dioptric space, reading and interpolation of dioptric and orientation scales, identification of the examined eye, writing out the prescription.

2. Principle of the instrument

The computer-assisted objective optometer has the following characteristics:

- refraction measurement through analysis of the image quality of a punctual test in white light;
- full pupil measurement - the pupil of the instrument is larger than that of the patient, so that measuring errors due to mis-centering are avoided;
- simultaneous presentation of a measurement test and a relaxing unfocused test;
- use of a microprocessor that controls the various measurement steps (centering, focusing, etc. ...), takes the focusing values into account, and avoids faulty operations, calculates refraction and displays it on indicators, before printing the results.

---

[1] Presently at the L.P.M.I.

[2] This paper was presented at the Int. Conference on Optics in Biomedical Sciences, Graz, Austria, by J. SIMON

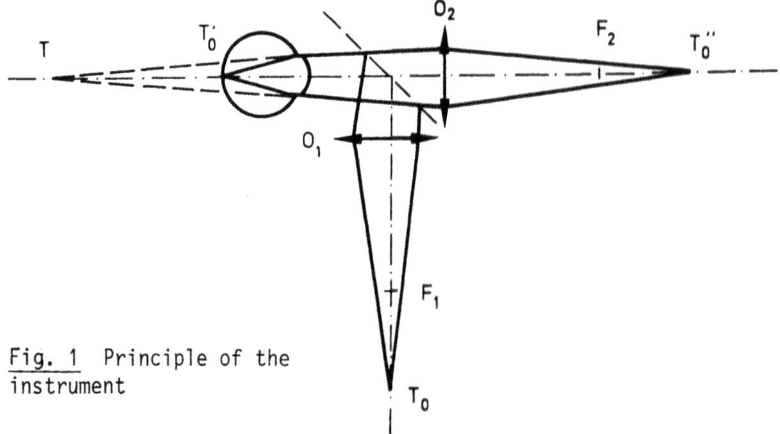

Fig. 1 Principle of the instrument

The measurement range is ± 20 dt, with an accuracy of 25 dt. Fig. 1 presents a simplified diagram of the optometer.

The test object, T_0, luminance 4×10^6 cd m^{-2}, is projected by the objective O_1 in T. The lens of the subject's eye forms an image on the retina in T_0'. The objective O_2 produces an image in T_0'', observed by the operator through an ocular, not shown in the figure. The instrument is constructed so that the points T_0 and T_0'' are shifted synchronically. A picture of a landscape, luminance 900 cd m^{-2}, is presented to the subject's eye in such a way that it is focused at infinity.

The microprocessor provides the following functions:

- starting of the procedure,
- real-time computation and display of orientation values and dioptric power according to one or two focusings necessary to determine ametropy
- result printout on a card,
- indication of faulty operations.

EXPERIMENTS

A first series of experiments was carried out in order to evaluate the accuracy of refraction measurements on a normal eye. A second series was performed to compare optometer-measured refractions with classical subjective refraction measurements.

- Accuracy of refraction measurements on a normal eye

Measuring accuracy was investigated by performing 50 focusing tests on two subjects: one myopic, mildly astigmatic, and one myopic, heavily astigmatic.

- Subject FC, myopic, astigmatic left eye (aged 40)

sphere	=	- 3.36 dt	σ = .14 dt
cylinder	=	+ 1.18 dt	σ = .18 dt
axis	=	177°	σ = 2°

- Subject DR, myopic, astigmatic right eye (aged 29)

sphere	=	- 8.7 dt	σ = .38 dt
cylinder	=	+ 5.88 dt	σ = .40 dt
axis	=	95°	σ = 3°

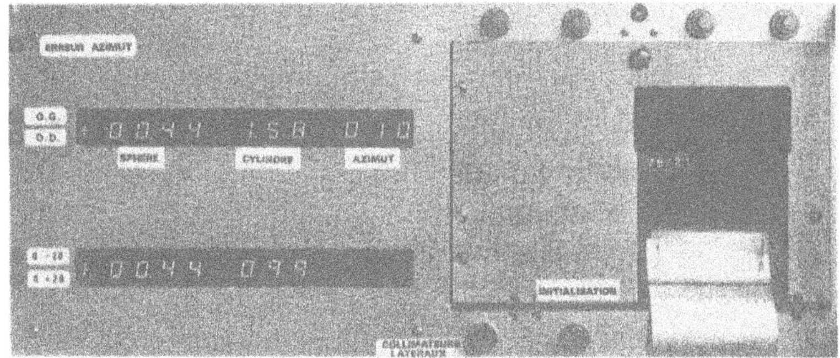

Fig. 2 Display and control panel of the computer-assisted objective optometer

- Comparison between subjective and optometer results

Comparative investigations were carried out on both eyes of 15 subjects. The subjective measurements were performed with an optotype table, contrasting sharply on a light background (luminance approximately 500 cd m^{-2}). The correction was finally tested in binocular vision. In the best cases, subjective correction for optimal visual comfort could be achieved with a ±.25 dt accuracy for spherical and cylinder lenses, and the axis could be determined with an accuracy of 5°.

Optometer measurements showed the following results:

sphere : 73% of results within ± .25 dt.
cylinder : 68% of results within ± .25 dt.
axis : 66% of results within ± 5° for cylinders over .5 dt.

The photographs in Fig. 2 and 3 show the instrument and its display and control panel, respectively.

Fig. 3 Computer-assisted objective optometer applied to a patient

This research work was carried out at the Institut d'Optique. The authors particulary wish to express their thanks to F. CORNO, P. FOURNET, J. MATHIAS, J.C. RODIER, and A. ROUSSEL, for their help in conceiving, setting up, adjusting, and carrying out experiments on the prototype, constructed under the DRET 78-329 contract.

Fundus Imaging by a Microprocessor Controlled Laser Scanning Device

U. Klingbeil, A. Plesch, and J. Bille

Institute of Applied Physics, University of Heidelberg
D-6900 Heidelberg, Fed. Rep. of Germany

Introduction

Imaging and documentation of the human retina is conventionally accomplished by a photographic fundus camera, a standard commercial product for many years. Exposure procedures require high flashlight intensities - usually 10 % to 100 % of the maximum just be accepted for a normal eye but can cause ocular damage in pathological cases. A different method of retinal imaging by laser scanning funduscopy will be presented that overcomes the intensity problem and promises further advantages.

Principles

In conventional funduscopy the whole retina is illuminated through the outer part of the enlarged pupil while a single point of it is observed by a small central part of ca.1/10 of the total exit pupil area. This reduces the intensity of the detectable light by a factor of 10 and limits resolution (Fig. 1).

Laser scanning funduscopy uses a collimated narrow laser beam of 1 mm diameter focused by the eye to ca.20 μm for illumination of a single point of the retina. The reflected light, normally 3-5 % of the incident light is collected through the outer 95 % of the pupil. Angular scanning of the illuminating laser beam results in sequential spotwise imaging of the retina.

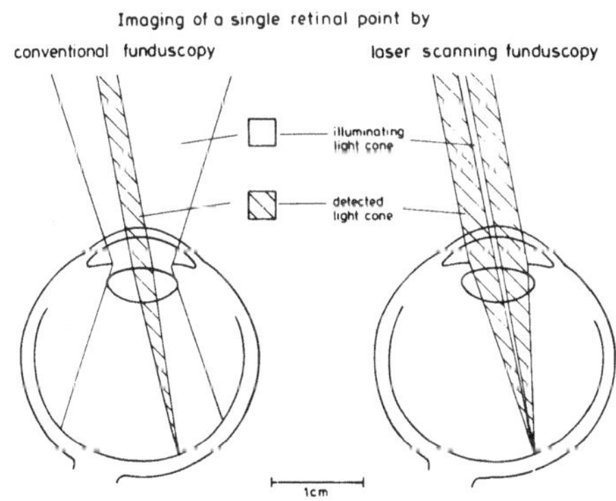

Fig. 1 Difference between conventional and laser scanning funduscopy

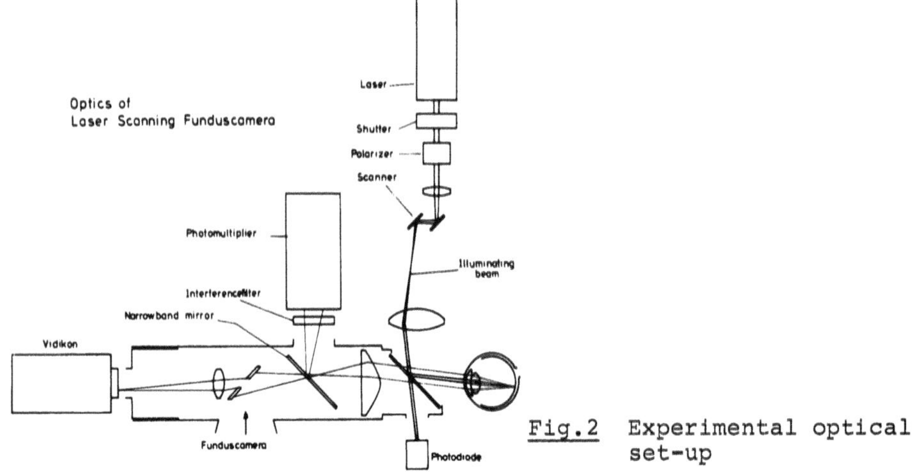

Fig.2 Experimental optical set-up

Optics

In the present experimental set-up (Fig. 2) a 1 mW He-Ne laser attenuated by a polarizer is used for illumination. Angular beam deflection is achieved by two perpendicular oriented linear galvanometer scanners. A rectangular field is scanned line by line with a line frequency up to 200 Hz and an opening angle up to 30°. The laser beam is projected into the eye by a low reflectivity semitransparent mirror cutting down laser power by a factor of 6. The entrance pupil of the eye is conjugate to the scanning plane in order to restrict the illuminating beam path to the center of the pupil. A photodiode detects the instantaneous laser power and triggers a security shutter. The laser scanner is adapted to a Zeiss fundus camera for ease of adjustment and simultaneous monitoring of the scanned eye. Radiation collected from the retina is focused by the aspheric front lens of the camera and separated from the common optical path-way by a narrow-band mirror. The laser radiation is passed through an interference filter and detected by a photomultiplier near the conjugate pupil plane of the eye.

Electronics

The laser scanner is driven by two 8-bit counter signal generators yielding two linear ramp signals for x and y deflection (Fig. 3). System timing control is accomplished by a 8080 CPU based microprocessor system.

The photomultiplier signal carries the image information. It is preamplified, digitized and stored on a discette. For image visualization the stored data are transferred to a PDP 11 minicomputer and read into the memory of a colour coding TV display system. Image resolution is limited by the TV system memory to 128 x 128 pixels/frame with 8 bits of depth.

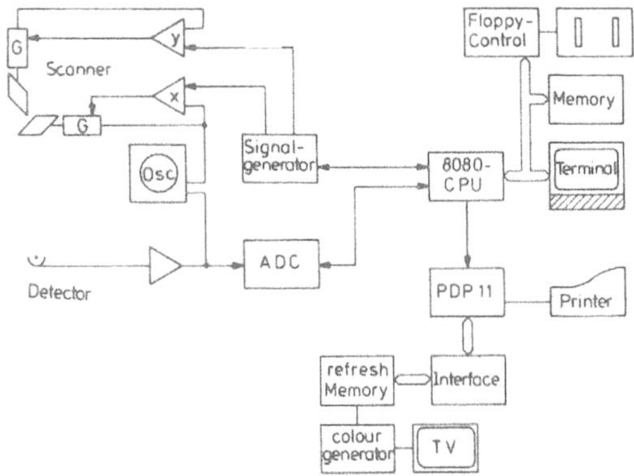

Fig. 3 Experimental electronical set-up

Results

For test measurements we used an artificial one-to-one model of the human eye. The machining procedure yields a surface structure of concentric rings at the artifical retina. Figure 4 shows a fundus photograph of the model. Fig. 5 represents a corresponding picture recorded by laser scanning funduscopy. Laser power was attenuated to ca. 1 µW at the eye which is three orders of magnitude below the maximum permissible exposure level. Signal-to-noise ratio was 50 - 100. Spatial resolution was ca. 12 µm/pixel.

Discussion

Laser scanning funduscopy has been proved to be a very sensitive and therefore safe imaging method. Its spatial resolution can compete with the resolution obtainable with conventional funduscopy. Image size is limited by the present digital TV-system memory. Memory expansion will improve the image size. Problems arising from the low scanning velocity will be overcome by a ro-

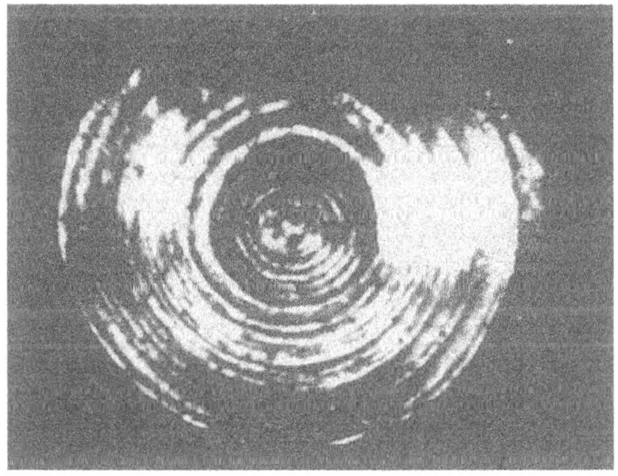

Fig. 4 Fundus photograph of an eye model

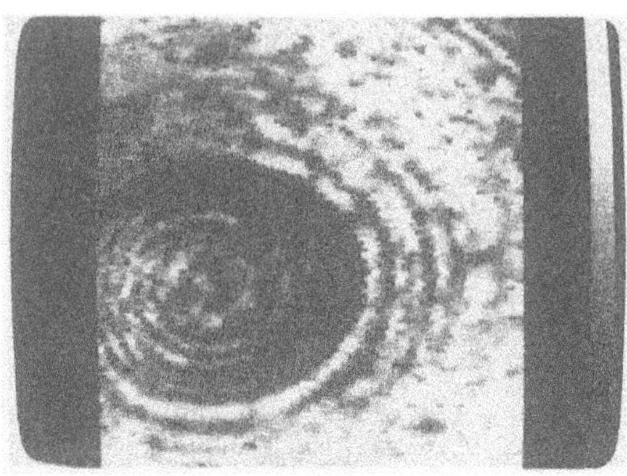

Fig. 5 Artifical retina of an eye model recorded by laser scanning funduscopy

tating polygon mirror scanner or a resonant TV-compatible galvanometer scanner. Both of them are designed to enable direct online fundus monitoring.

Laser scanning funduscopy promises an easy simultaneous detection of the two polarization components within the detected light. Thus, small local polarisation changes due to pathological ocular tissue variations should become accessable for diagnosis.

The presented experimental set-up for laser scanning funduscopy is based on digital data storage. Digital image processing methods such as filtering, differentiation, and segmentation algorithms can be applied immediatly to the images in order to extract and quantify retina description parameters or to stabilize the fundus image. Both procedures are suited for an analysis of the retinal vessel system.

Another interesting parameter is the velocity of blood in the retinal vessels. A differential Doppler technique analyzing the backscattered laser light by moving erythrocytes has been developed. By a fast Fourier analysis on a microcomputer system the frequency distribution of the observed particles can be calculated. The method needs constant fundus monitoring for precise adjustment.

This work was supported by the Deutsche Forschungsgemeinschaft.

Examination of Low-Luminance Myopia with a Laser Optometer

E. Ingelstam, K. Janson

Institute of Optical Research, The Royal Institute of Technology
S-100 44 Stockholm, Sweden

D. Epstein, B. Tengroth

Department of Ophthalmology, Karolinska Hospital
S-104 01 Stockholm, Sweden

1. Apparatus and Method

With a simple version of a laser optometer, Fig. 1, which enables testing of accommodation at photopic, mesopic and scotopic luminance levels (120 cd m^{-2}, 0.5 cd m^{-2} and 0.001 cd m^{-2}, respectively), low-luminance myopia studies were performed on about 400 eyes. This method has numerous advantages over previously used techniques. The most important are these: since the speckles produced by constructive and destructive interference of reflected laser light, the sharpness of its retinal image is independent of the focus of the eye; the pattern *per se* does not serve as an accommodative stimulus [1, 2, 3]; and the subject, being asked to evaluate the direction of the movement of the speckles, cannot relate this movement to the direction or amplitude of accommodation and is thus unable to influence the result. Also, the superimposing of the speckle pattern onto the subject's visual field causes only minimal interference with the ongoing task.

When the reflective surface moves, the speckles appear to move in a direction determined by the eye's refractive state. If the subject is accommodated to a plane nearer than the reflective surface (or rather the "plane of stationarity" [4]), the speckles appear to move with the relative motion of the surface. If the eye's focus corresponds to a plane beyond the surface, they appear to move in the opposite direction. When the retina is conjugate to the surface, the speckles appear stationary or "boiling", with no distinct direction of motion. For further details and a complete description of the

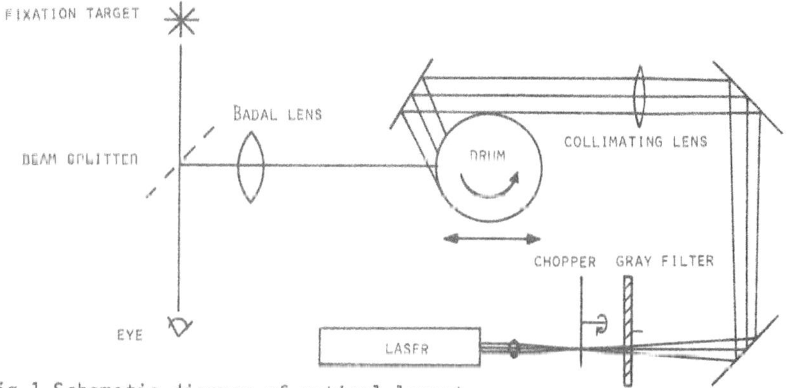

Fig.1 Schematic diagram of optical layout

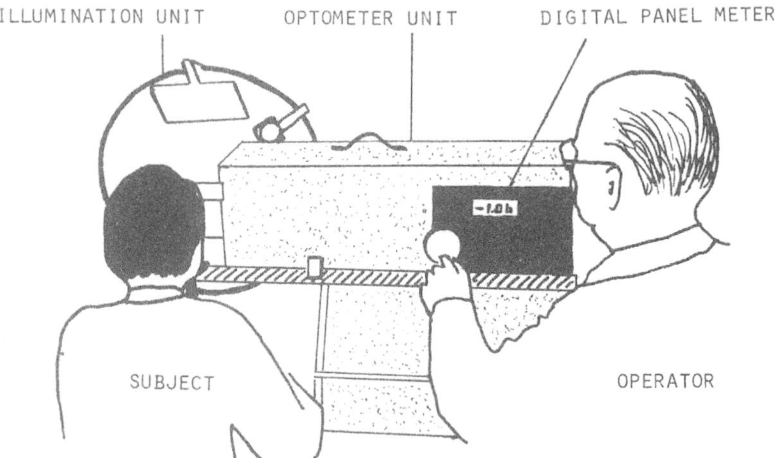

Fig.2 Overview of laser optometer. The refractive state of the subject's eye is registered on the digital panel meter

results obtained we refer to the full paper [5]. Fig. 2 shows the normal working procedure, which takes 15 - 30 mins. The subjects, refracted prior to the examination and fully corrected for distance, are tested monocularly. The difference between the mesopic and photopic values (M - P) is designated twilight myopia, the difference between the scotopic and photopic ones (S - P) is termed the actual night myopia.

2. Test Series

2.1 The reliability of the optometer was especially tested for three subjects. The intra-individual spread was found to lie within ± 0.50 D. Possible variations in the results obtained from individual subjects within one day, and the same subjects' twilight and night myopia as determined on three different days were also measured. In these tests, results did not vary appreciably; maximum variations were 0.32 D for (M - P) and 0.36 D for (S - P).

2.2 The main series were made with 100 recruits aged 18 - 24 and with 163 subjects aged 10 - 71, all with optimal correction for distant vision and a visual acuity of at least 20/25. The distribution of (M - P) values

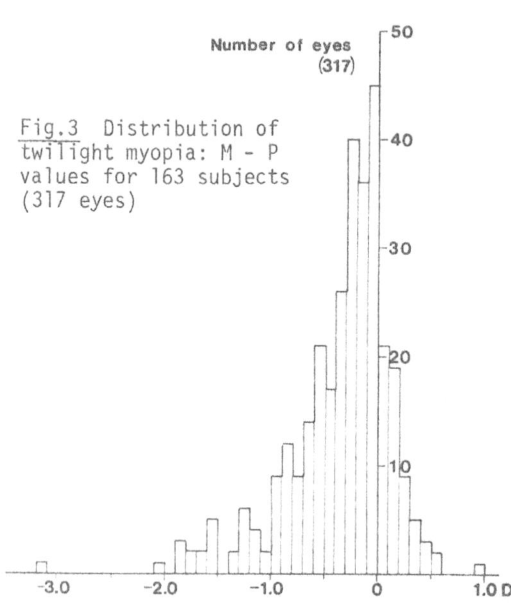

Fig.3 Distribution of twilight myopia: M - P values for 163 subjects (317 eyes)

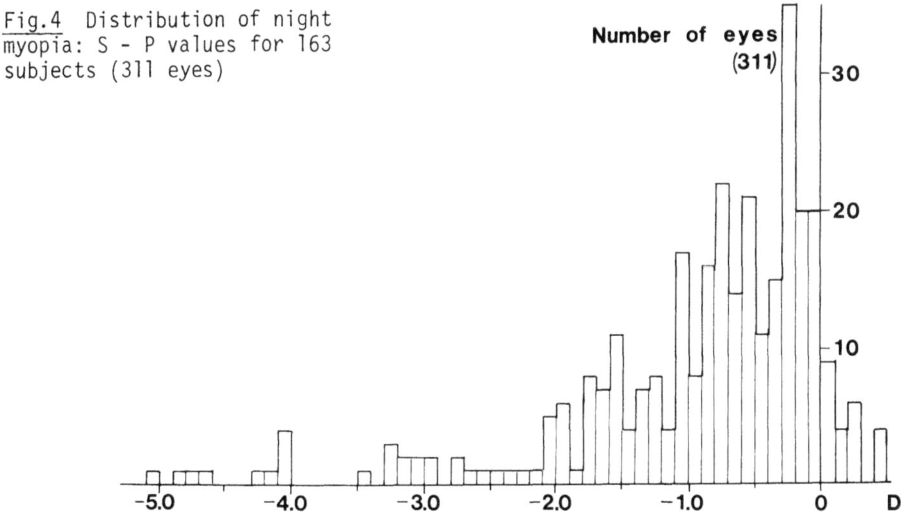

Fig.4 Distribution of night myopia: S - P values for 163 subjects (311 eyes)

for the 163 subjects is given in Fig. 3 and that of (S - P) values in Fig. 4. If one assumes that for practical purposes only a twilight myopia ≥ 0.75 D would require optical correction, 17 % fall into this category. Fig. 4 representing night myopia reflects a further shift toward myopia with a mean of - 1.01 D as compared to - 0.35 D seen in Fig. 3. Here 47 % of the eyes fall into the ≥ 0.75 D myopia group.

2.3 Finally, a series of tests was conducted using a model landscape, Fig. 5, under mesopic illumination to determine the relationship between the optometer myopia values and the optical correction required, as assessed by the subject. The landscape contained buildings, roads, cars and traffic signs. On top of the main building single rows of Snellen-chart prototypes (20/60 - 20/40) were mounted with the background luminance of these objects being 0.2 cd/m^2. Beginning with the subject's optimal daylight correction and using only subjects with an optometer twilight myopia ≥ 0.75 D in at least one eye, the greatest possible mesopic visual acuity was obtained by adding minus lenses as the subject viewed the Snellen prototypes from a distance of 5 m. The test was performed binocularly.

Fig. 6 gives the results. Of the 58 eyes tested, 83 % needed a lower negative correction (had less myopia) in the twilight landscape than that determined by the optometer, while 12 % required higher diopter values and 5 % preferred a correction that was identical to the values obtained. In all,

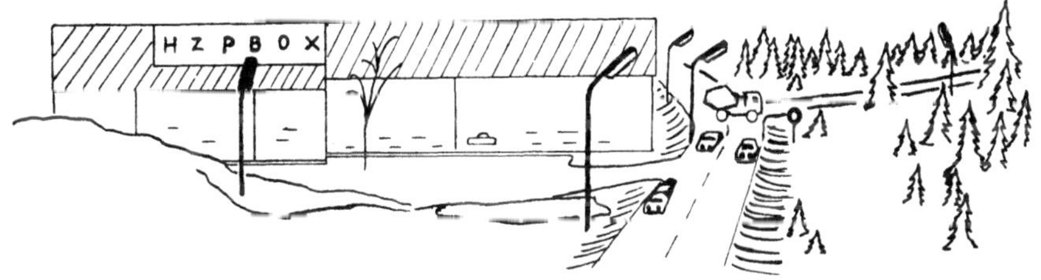

Fig.5 Sketch of twilight model landscape, see text

41 % of the eyes needed a minus correction ≧ 0.50 D in the landscape, and 21 % required minus correction ≧ 0.75 D.

3. Conclusions

This investigation has thus added new figures to the growing data bank of low-luminance myopia, and confirmed the dimensions of this refraction anomaly. Because of the large intersubject variability and the inability to predict these myopias on the basis of commonly used refraction procedures, measurements aiming at correction must be taken individually. We have found the laser optometer to be a valuable instrument in selecting those who may need special optical correction when functioning at mesopic luminance.

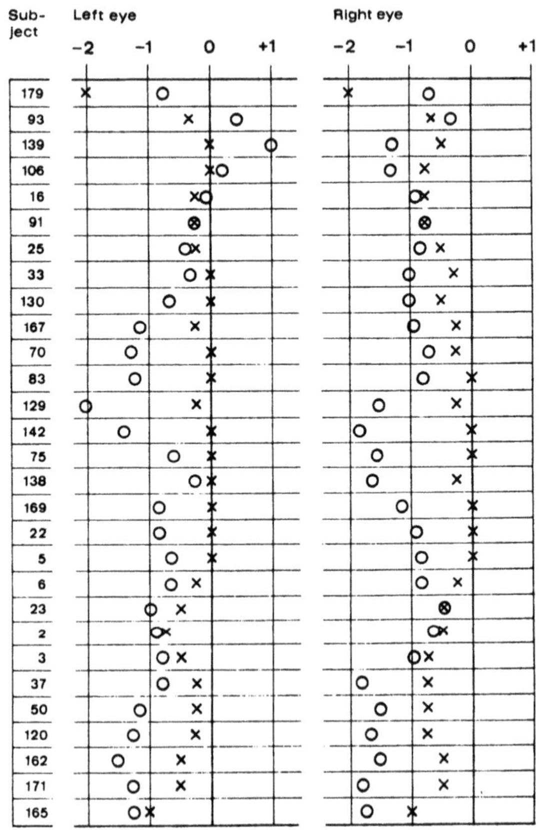

Fig.6 Diagram showing that with few exceptions, the correction required in the twilight landscape (×) was lower than the twilight myopia values obtained in the laser optometer (O). In order to clearly elucidate the relationship between optometer and landscape values, the subjects were not placed in numerical order.

Further clinical tests, as well as basic research on the nature of these myopias, are being conducted.

References

1 Hennesy R T and Leibowitz H W 1970. J. Opt. Soc. Am. 60, 1700-1701

2 Ingelstam E and Ragnarsson S I 1972. Vision Res. 12, 411-420

3 Leibowitz H W and Owens D A 1975. J. Opt. Soc. Am. 65, 1121-1128

4 Charman W N and Chapman D 1980. The Ophthalmic Optician 20, 41 - 51

5 Epstein D, Ingelstam E, Jansson K and Tengroth B. Acta Ophthalmologica, (in print)

Measuring the Shape of Soft Corneal Lenses

W.H. Steel, and C.H. Freund
CSIRO Division of Applied Physics, Sydney, Australia 2070

1. Introduction

The radius of curvature of a hard corneal lens in air is usually measured by the method proposed by DRYSDALE [1]. An autostigmatic microscope is used, the counterpart, for near images, of the autocollimating telescope. This is focused in turn on the surface and on its centre of curvature and the distance between these two positions is the radius. If the projected graticule consists of radial lines, different orientations can be focused separately to measure the two principal radii of a toric surface.

This instrument cannot be used directly with a soft corneal lens, which, since it is made of a hydrated hydrophilic material, must be kept in normal saline solution to retain its shape. The light reflected from the air-glass surfaces in the microscope objective completely swamps the images reflected from the corneal lens. However, we have shown [2] that the addition of crossed polarizers and a retarder to an autostigmatic microscope make possible its use to measure the central radius of curvature of soft lenses.

The surfaces of corneal lenses are not spherical and it is also important to know their whole shape. A simple method of obtaining an indication of the departure from sphericity is to measure the sagitta. The lens is placed with its concave side against a flat surface and the microscope focused first on this surface, then on the apex of the concave surface of the lens. In theory the value so obtained for the sagitta should be corrected for refraction in the lens when in water, but this correction is very small.

2. Tilting Microscope

To plot the shape of the whole surface of the lens, a new autostigmatic microscope has been made that tilts about its focus. This means that the microscope can focus on surfaces that are inclined at an angle; tilting the microscope away from the vertical to receive the reflected light does not change the focal setting. Such an instrument can be used in two ways. It can be focused on the surface and moved across it; the distances moved measured horizontally and vertically then give the surface directly. Or it can be focused on the centre of curvature and the variation in position of this plotted as a function of the angle by which the microscope is tilted. This plot is the involute of the surface and the surface shape can be computed from it.

Initial tests on a spheroidal surface of known shape showed the involute method to be the more accurate and reproducible. But when tried on actual corneal lenses, it would not give off-axis images of the centre, since the surface quality off axis was not good enough. The surface method, however, gave useful results.

3. Construction Details

The microscope is shown in Fig.1. The beam splitter is a coating cemented between two pieces of glass. One piece makes up a light trap to absorb the incident light transmitted by the beam splitter and the other a cone that dips into the saline solution. The only surface that can contribute an unwanted reflection is the front of this cone, and this is bloomed for water immersion.

The mounting of the microscope has adjustments in two directions to bring the focus to the axis of rotation. A lens with spherical surfaces has been made of black glass and the microscope is adjusted until the image reflected from this, under saline solution, does not move laterally nor change focus as the microscope is tilted. The image used is that at the centre of curvature.

Fig.1 Tilting autostigmatic microscope for measuring the full surface shape; n.a. = 0.1

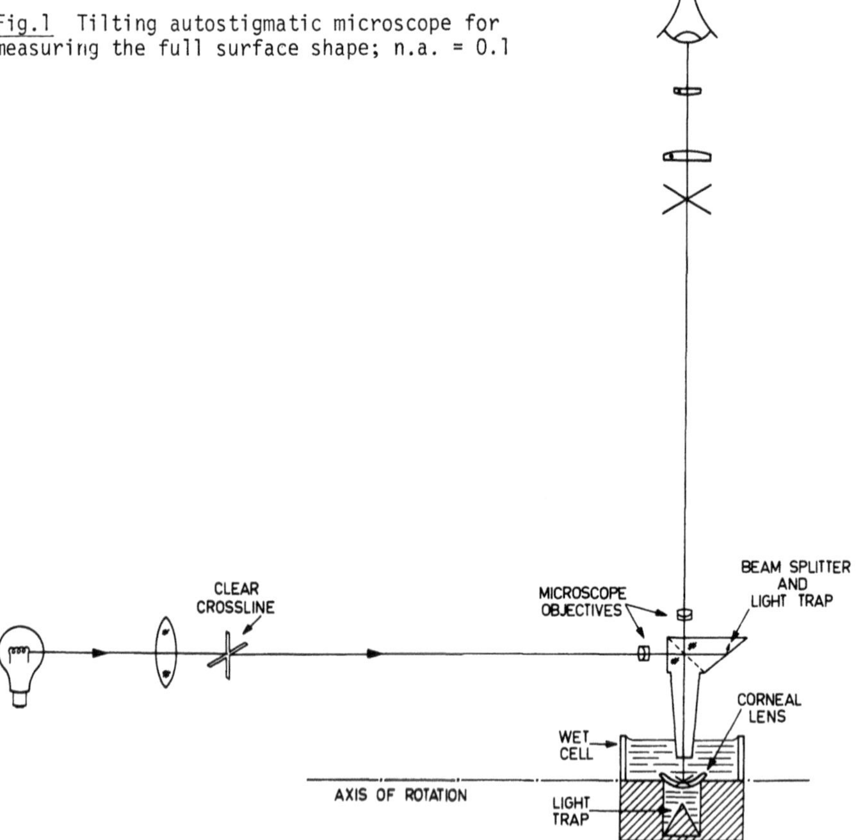

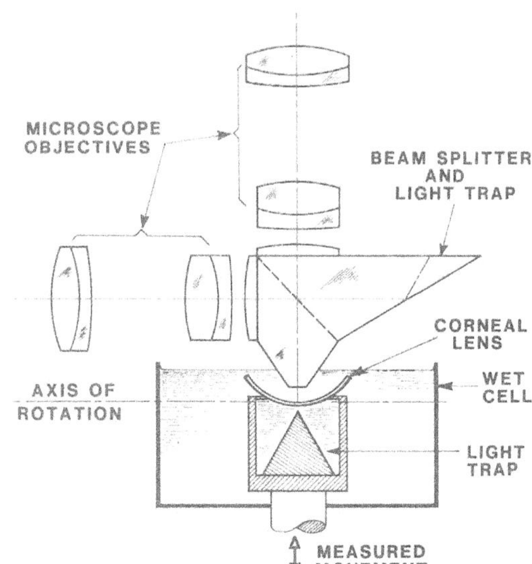

Fig.2 Front section and lens cell for the high-aperture microscope. This is based on 6.3 × objectives of n.a. = 0.2 and working distance 22 mm

The cell that holds the corneal lens is also shown in Fig.1. It can be rotated about a vertical axis to check that the lens is on centre and it is made of transparent plastic so that the lens can be seen from the side. Behind the lens is a light trap, a polished black cone in a black tube.

The microscope of Fig.1 has the low numerical aperture of 0.1. This was chosen for ease of construction and because, for the involute method, the centre needed to be found for small regions of the lens. Tests on setting on the surface have shown that this numerical aperture gives too great a depth of focus and hence insufficient accuracy. A new instrument has, therefore, been made with a numerical aperture of 0.27. This is probably greater than necessary but it will be useful for showing the limitations of the method. The beam-splitter arrangement and lens cell for this microscope are shown in Fig.2 and the microscope itself in Fig.3. The shorter cone means that measurements of the central radius must be made by lowering the corneal lens inside the cell, leaving the cell walls fixed.

4. Lenses with High Water Content

The recent introduction of hydrated corneal lenses of higher water content has reduced further the under-water reflectance of their surfaces. The standard autostigmatic microscope can no longer give an image, even when fitted with crossed polarizers and a retarder. But the design of the new microscope, with the beam splitter in front of paired objectives instead of in the usual position behind the objective, leads to an instrument with very low stray light. It is low enough for the reflected images from the new lens materials to be seen. It is possible that this use of the instrument for measuring the curvature and sagitta of such lenses will prove even more important than its ability to measure the full surface shape.

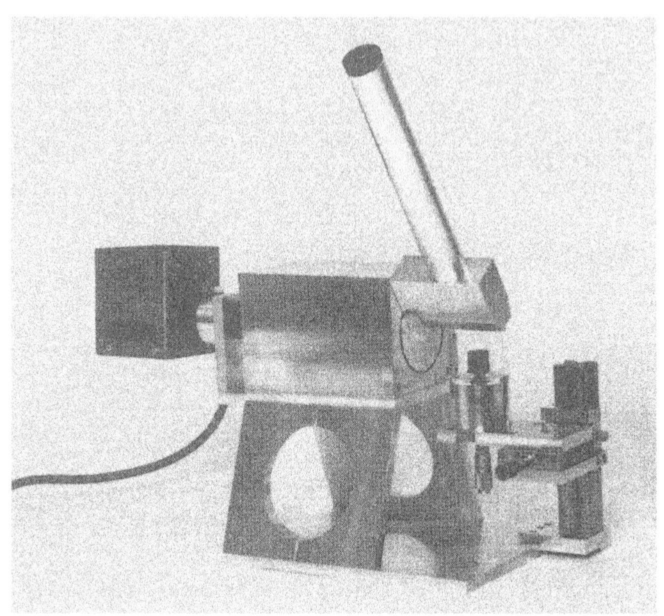

Fig.3 Photograph of high-aperture microscope

References

1. C. V. Drysdale, Trans.Opt.Soc.London **2**, 1 (1900).
2. W. H. Steel, and D. B. Noak, Appl.Opt. **16**, 778 (1977).

Topographical and Tear Film Thickness Measurements of the Cornea of the Human Eye

R. Buschner, D. Henrich, and J. Bille

Institute of Applied Physics, University of Heidelberg
D-6900 Heidelberg, Fed. Rep. of Germany

Introduction

A common procedure for documentation of the anterior segment of the eye is taking slit-lamp photographs. To analyse these images a film has to be developed first, before density distributions or geometric quantities, e.g. curvatures of the cornea or lens can be evaluated. For immediate analysis of these quantities a new technique has been developed using a TV-camera and digital image processing methods. In order to measure the tear film thickness a white-light interference method has been used.

Experimental set-up

The experimental arrangement consists of an optical and electronical part (Fig. 1). The anterior segment of the eye, including cornea, anterior chamber, and lens, is illuminated by a conventional slit-lamp, with 8.5 mm slit height and 100 µm slit width. The scattered light is focused on a TV camera at an angle of 40° corresponding to the optical axis of the slit lamp.

The problem of imaging an optical section through the anterior segment of the eye is that of limited depth of focus. This probl-

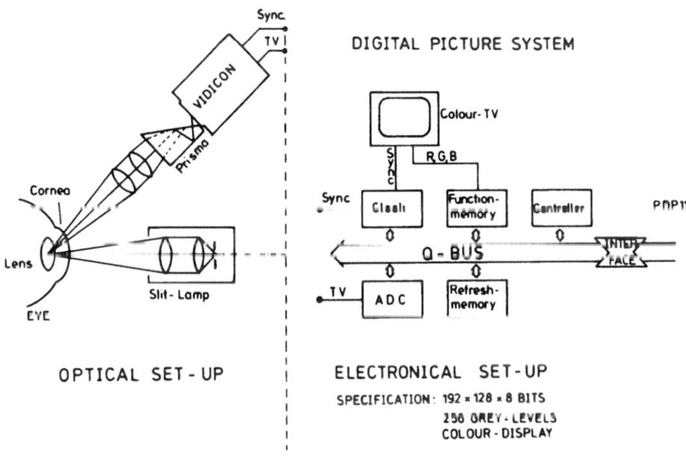

Fig. 1 Experimental set-up

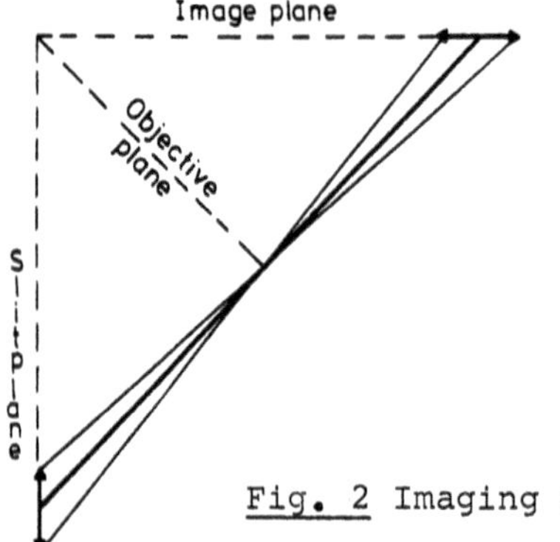

Fig. 2 Imaging according to Scheimpflug's principle

em can be overcome by using the principle of Scheimpflug [1,2]. This principle states that the image on an obliquely positioned object is formed such that the planes of object, image, and objective intersect (Fig. 2).

To utilize this principle one can either tilt the objective or the image or both of them. In the present arrangement the image plane has to be tilted to obtain the depth of focus. Instead of tilting the image plane a 60° prism was used (Fig. 1).

The analog TV signal is read into a digital picture system with a resolution of 192 x 128 pixels per frame with 8 bits depth, corresponding to 256 intensity levels. The data can be stored in the refresh memory and transferred to a PDP 11 minicomputer for further processing. To display the coloured data, the function memory assigns a special colour code to every intensity level.

Results

An image of the anterior segment of the human eye is presented in Fig. 3. The colour code is displayed on the right side. The color at the bottom of the color column corresponds to the lowest, that on top to the highest intensity level. The curvature and thickness of the cornea and the anterior lens capsule can easily be seen. As a result of the low signal-to-noise ratio it is difficult to resolve details of the structur of the lens, e.g. disjunction zone, anterior cortex or nucleus. This can be demonstrated by the intensity distribution of a single TV line. To obtain better data for evaluation of the curvature image processing techniques like thresholding and differentiation (Fig. 4) are used.

The curvature of the cornea is described by a least squares fit to a circle. The radius is computed to r_c = 8.4 mm. For comparison the radius of the same cornea was measured with a conventional instrument in an eye clinic. Using this method a radius r_h = 7.8 mm was found. The deviation from the computed radius is 7%.

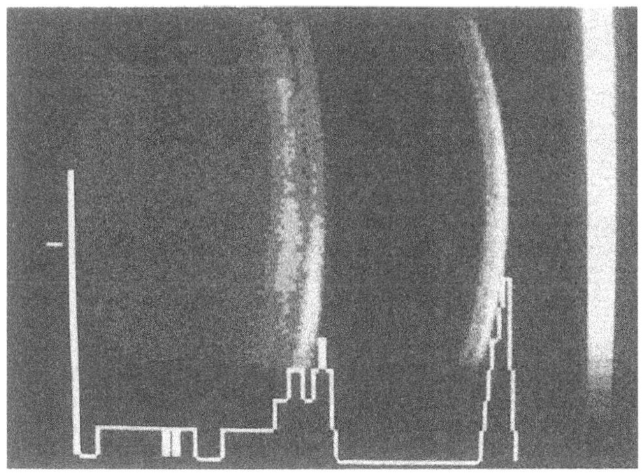

Fig. 3 Image of the anterior segment of the eye

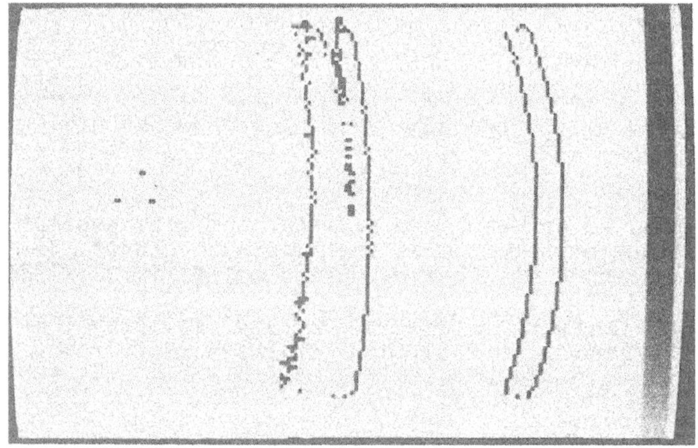

Fig. 4 Differentiated image of the anterier section of the eye

Discussion

The calibration error is about 3 % assuming an error of 2 pixels. The limited resolution can clearly be seen in Fig. 4. The central region of the corneal image is not curved but represented by a straight line. For this reason a principle error remains. These errors can be reduced by using a digital system with higher resolution. A system with 1024 x 625 pixels per frame and 1 bit depth is in preparation. A higher signal-to-noise ratio is very important for evaluation of the lens geometry and to obtain better contours of the anterior and backsurface of the cornea. The usual halogen lamp for slit lamp illumination is replaced by a XBO lamp. The slit illumination is performed by using a deflecting mirror scanner. With this system and further optical corrections of residual geometric errors of the vidicon target an error of less than 2% is expected.

Another scope is the measurement of the tear film thickness of the human eye for studying the dynamics of tear film ruptures - important for wearing contact lenses. The main problem is the small amount of reflected light due to a refractive index difference of about $\Delta n \approx 0.05$ between the cornea and the tear film.

The system cornea-tearfilm is simulated by 5 different single layer specimens with the same thickness of 0.8 μm and Δn ranging from 0.05 to 0.8. A white-light XBO lamp is used for illumination. The incident beam is reflected by the upper and lower surface of the layer. The resulting interference pattern is analysed in a grating spectrometer with a line array of 512 photodiodes [3]. For correcting the spectrum the data were normalized corresponding to the reflectivity of the glass substrate and fed into a microcomputer. The extrema of the modulation are used to evaluate the thickness of the layer with an accuracy of about 5 - 10%.

This work is supported by the Deutsche Forschungsgemeinschaft.

References

[1] Niesel P., Kräuchi H. und Bachmann E., "Der Abspaltungsstreifen in der Spaltlampenphotographie der alternden Linse", Albrecht v. Graefes Arch. klin., exp. Ophthal. 199, 11-20 (1976)

[2] Dragomierescu V., Hockwin O., Koch H.-R., Saki K., "Development of a New Equipment for Rotating Slit-Image Photography According to Scheimpflug's Principle", Interdisciplinary Topics in Gerontology 13, 118-130 (1978)

[3] Korth H.E., "Schichtdickenmessung mit rechnerunterstütztem Spektralphotometer", DGaO-Tagung 1979 in Bad Harzburg vom 5. - 9. 6. 1979

In situ Measurements of the Geometrical and Optical Properties of the Anterior Segment of the Eye

J. Politch, M. Oren, and M. Segal

Department of Physics, Technion City, Haifa 32000, Israel

S. Hyams

Department of Ophthalmology, "Ha-Carmel" Hospital, Haifa, Israel

Abstract An optical system for measuring in situ the geometrical and optical properties of the anterior segment of the eye is described. The measurements provide in one case the evaluation of the product of the thickness "t" by the index of refraction "n", and in the second case that of the ratio between these parameters. From each two corresponding equations, it was possible to obtain "t_i" and "n_i for every layer.

Introduction The problem of evaluating in situ the thickness "t" and the index of refraction "n" of every layer of the anterior segment to the eye, as well as the interpretation of the obtained results were found to be quite difficult tasks from the practical point of view.

In a multilayered system the reflected and the transmitted intensity are dependent on the incident angle, the indices of refraction n_i and the thicknesses t_i of every layer i. If the material is composed of N layers, there are 2N measurements of intensity (for 2N angles of incidence) in order to define the n_i and t_i parameters of every layer. There is another way to define these properties - achieved by measuring the angular distribution of the reflected intensity for a fixed angle of incidence. This type of solution is more complicated and is less accurate, since the mathematical treatment is derived from a geometrical-optics point of view, which does not take into consideration diffraction and scattering problems. Therefore,an integration over the whole beam is required. This measurement is limited to certain angles of incidence due to total reflection of the first layer, or between any two layers. There is a limit in the angular resolution of the reflected intensity, which limits therefore the accuracy of t_i and n_i.

Proposed here is a third way for obtaining these parameters, which seems to be simpler, applicable for in situ measurements,and with a reasonable accuracy. The system is based on a Michelson interferometer with variable imaging properties.

Measuring Flat Layers by Their Parameters Deformation of n_i and t_i must be accomplished through their parameters, defined by: $\eta = t \cdot n$ and $\xi = t/n$. The number of layers N will be defined by the number of times that a focus will be reached from the concentric circles by sliding the lens. This method also defines the number of interfaces (N+1) between the layers. The focusing regions are applied as well in order to determine the parameter ξ:

$$\xi_i = 4\Delta_i \frac{\Delta_i + [(L_1^{(i)} - 2f)^2 - 2f^2]^{1/2}}{2\Delta_i + L_1^{(i)} - 2f + [(L_1^{(i)} - 2f)^2 - 2f^2]^{1/2}} \,.$$

Since $L_1^{(2)} = L_1^{(1)} + \xi_1$, $L_1^{(3)} = L_1^{(2)} + \xi_2$, etc.

η_i was calculated from the optical path that the rays pass from x_{oj} and x_{oi} to the inclined beam splitter:

$$\eta_i = \frac{x_{o(i+1)} - x_{oi}}{2} - \frac{1}{2}[L_1^{(i+1)}a^{(i+1)2} - x(a^{(i+1)2} - \frac{1}{2}) - x_{o(i+1)}] \frac{\rho_{(i+1)m}}{x_{o(i+1)}}$$

$$+ \frac{1}{2}[L_1^{(i)}a^{(i)2} - x(a^{(i)2} - \frac{1}{2}) - x_{oi}] (\frac{\rho_{in}}{x_{oi}})^2 ,$$

where $a^{(i)} = \dfrac{f}{2[L_1^{(i)} - x - f]}$. In the case $\dfrac{\rho_{in}}{x_{oi}} \ll 1$, a simpler formulation can be obtained: $\eta_i = \frac{1}{2}[x_{o(i+1)} - x_{oi}]$.

Measuring Curved Layers by Their Parameters The image of a point source passing two curved surfaces is based on the images obtained from every layer. The analysis distinguishes between the case when the distance v from the point source to the lens is larger or smaller than the front focal length f_1 of the first separating layer Σ_1.

Case No.1 $v < f_1$. Fig.1 describes schematically the ray tracing for three layers.

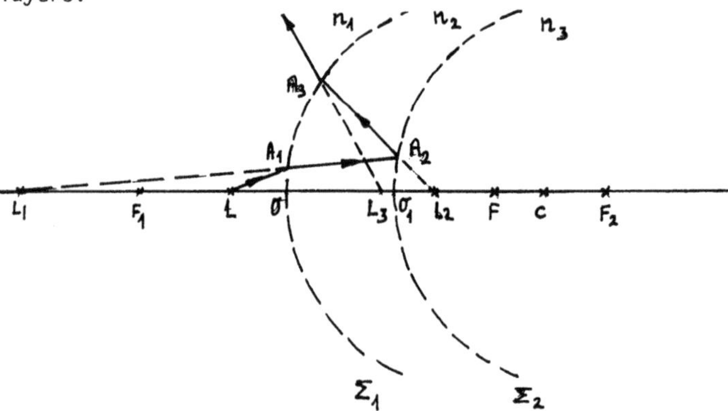

Figure 1: The image of a point source for 3 layers, when $v < f_1$.

L is a point source. Assuming $OF_1 = CF_2 = f_1$; $OF_2 = CF_1 = f_1$; $O_1F = FC = f = \frac{1}{2}R_2$; $CO = CA_1 = CA_3 = R_1$; $CO_1 = CA_2 = R_2$; $OO_1 = d = R_1 - R_2 = R_1 - 2f$; $LO = v$; $v < f_1$; $LC = R_L = R_1 + v$; one can get

$$R_{L_2} = \frac{(v+R_1)\, f_1 \cdot f}{(v+R_1)\, f_1 - (f_1 - v)f} ,$$

$$u_1 = f\, \frac{vf_2 + (f_1 - v)(R_1 - 2f)}{vf_2 + (f_1 - v)(R_1 - f)} ,$$

$$v_1 = \frac{vf_2 + (f_1 - v)(R_1 - 2f)}{f_1 - v} .$$

And in a similar way one can find

$$R_{L_3} = L_2C = \frac{(v+R_1)\ f_2 f}{(v+R_1)(f_1+f) - (f_1-v)f},$$

$$v_2 = u_1 + d = \frac{vf_2(R_1-f) + R_1(f_1-v)(R_1-2f)}{vf_2 + (f_1-v)(R_1-f)},$$

$$u_2 = \frac{v_2\ f_1}{f_2 - v_2}.$$

Case No.2 $v > f_1$, again 3 layers, and a point source located at L, so that $LO_1 = v$ from the 1-st surface Σ_1. Therefore

$$R_{L_1} = \frac{R_L \cdot f_1}{R_L - f_1} = \frac{(v+R_1)\ f_1}{v-f_1},$$

$$v_1 = u - d = \frac{(v+R_1)\ f_1 + 2f(v-f_1)}{v-f_1}, \quad \text{and } u_1 = \frac{v_1 \cdot f}{v_1 - f},$$

$$R_{L_2} = \frac{R_{L_1} \cdot f}{R_{L_1} + f};$$

and, similarly, $v_2 = u_1 + d = \dfrac{f_1(v+R_1)(R_1-f) + R_1 f(v-f_1)}{(v+R_1)\ f_1 + f(v-f_1)}$

and $u_2 = \dfrac{v_2 \cdot f_1}{f_2 - v_2}$; $R_{L_3} = \dfrac{(v+R_1)\ f_2 f}{(v+R_1)(f_1+f) + f(v-f_1)}$.

Measuring Procedure The measurable parameters are: a. Object-image distances ℓ_1 and ℓ. b. Distances between the images Δ. c. The height of the object h and its images h_1, h_2. From these parameters it is possible to calculate the magnifications m_1, m_2 and m according to:

$$m_1 = \frac{h_1}{h} = \frac{R_1}{2\sqrt{h^2+R_L^2}-R_1},$$

$$m_2 = \frac{h_2}{h} = \frac{R}{2\sqrt{h^2+R_L^2}-R},$$

$$m = \frac{h_2}{h_1} = \frac{R}{R_1} \cdot \frac{2\sqrt{h^2+R_L^2}-R_1}{2\sqrt{h^2+R_L^2}-R}.$$

R_1 can be solved by the following equations:

$$R_{L_1} = \frac{R_L R_1}{2R_L - R_1},$$

$$\ell_1 = R_L - R_{L_1} = 2R_L \frac{R_L - R_1}{2R_L - R_1},$$

$$m_1 = \frac{R_1}{2\sqrt{h^2+R_L^2}-R_1},$$

These three equations are solved in the case of $m_1 << 1$ (when the object is far from Σ_1), so that $m_1^3 << m_1^2$,

$$R_1 \simeq 2m_1 \frac{1i\ m_1 + \sqrt{2h^2 + \ell_1^2}}{1 + 4m_1 + 2m_1^2},$$

In a similar way the radius R was calculated:

$$R = \frac{n_{21} R_1 R_2}{R_1 + R_2 (n_{21} - 1)} \quad \text{where} \quad n_{21} = \frac{n_2}{n_1}, \quad R_2 = R_1 - d .$$

In order to define R_2 and therefore d and n_2, a "comparison measurement" with known material parameters has to be performed.

Measuring Results The first measurements with the technique proposed here were performed with three layers of known dielectric constants n_i and thicknesses t_i. The measurements showed a good agreement with the parameters of every layer; the differences in n_i were smaller than 0.1% and in t_i even better.

Analytic Recontruction of the Human Cornea by Means of a Slit Lamp and Digital Picture Processing

S. Fonda, B. Salvadori, and D. Vecchi

Clinica Oculistica Università di Modena, Italy
Industrie Ottiche Riunite, IOR, Porto Marghera, Venezia, Italy

1. Introduction

Photographic investigation of the human cornea and lens is widely used in ophthalmological clinical practice in order to achieve measurements of optical parameters of the anterior segment of the eye [1,2,3].

Some problems arise when dealing with lens parameters (power, curvature, size, width, etc.), mainly due to the poor focus depth of the ordinary version of the system used for ocular examinations, that is, the slit lamp. Duly modified according to the Scheimpflug's principle [2], the slit lamp photographic recording system produces good results. However, if only the geometrical shape on the corneal surface must be emphasized, no focus depth problem needs to be solved and direct measurements can be performed.

The aim of this paper is to present a procedure to reduce the number of trials and therefore the time necessary to design the optimal contact lens to be applied either to infants after cataract surgery, or to adult patients affected by keratoconus.

Both situations are critical. The first because it involves a still plastic and rapidly developing visual system; in this case sharply focussed images, with the same magnification in the two retinas are needed for the proper development of the binocular vision. The second one, because it often presents a very irregular corneal surface which doesn't match a sphere, as one would expect for a simple design of a contact lens.

In conclusion, both situations require a suitable design procedure for the clinical application of the optimal contact lens.

2. Methods and Results

The procedure we used is derived from a principal project, the aim of which was to reconstruct the optical features of a human eye by means of optical methods (the use of ultrasounds may be an alternative method , of course).

So, all the steps of the following description are largely biased by the original aim of the whole work.

The images of the anterior eye media (cornea,acqueous, lens and anterior part of vitreous) were taken by means of a slightly modified Zeiss photographic slit lamp, the most important part of which is reproduced in Fig.1a-b. A TV camera (high sensitivity NOCTICON tube, of 250 mA/lumen, spectral response

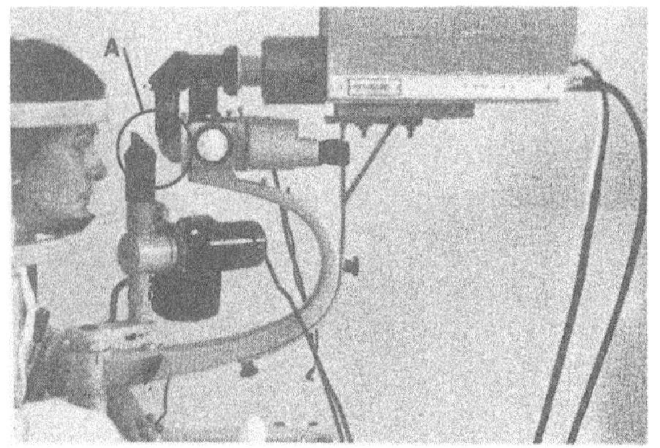

Fig.1a Slit lamp with TV camera; marked portion explained in Fig.1b

Fig.1b Frontal view,arrow indicates slit light exit

S20, Thomson) was mounted on the slit lamp and the TV signal fed into a TESAK VDC501 system for acquisition and image processing, connected to a PDP 11/03 computer. The definition of the images was 256x256 pixels, with 256 gray levels.

The camera's and subject's visual axes coincided as indicated in Fig.2.

Six images were acquired, normally at equally spaced angles during rotation of the slit light axis (±20, ±40, ±50 degrees), and were observed on a color TV terminal.

The constancy of the eye alignment was continuously controlled by means of another TV monitor in such a way that the corneal image acquisition was performed only when the subject's pupil image fell in a predetermined area

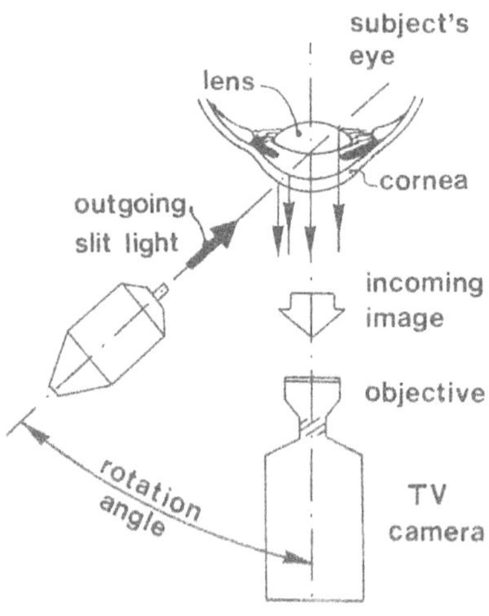

Fig.2 Schematic top view of the slit lamp TV system

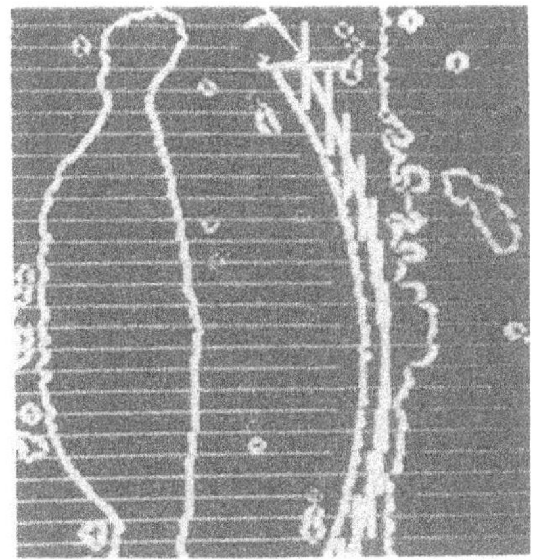

Fig.3 Lens (L) and cornea (C) of a human eye after picture processing for edge detection

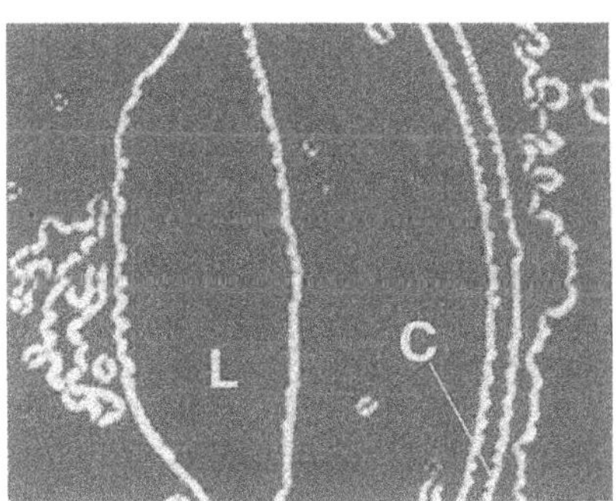

Fig.4 Fitting of the anterior corneal profile with a circle equation. Arrow indicates experimental points and the beginning of the theoretical profile.

on the screen. The acquisition time was 20 msec, which revealed to be a sufficiently short time interval to get images without motion blur. The contrast was enhanced by gray level histogram equalization, noise filtering and edge detection for improving the appearance of the features on the pictures (refer to Fig.3). For this reason, special purpose software has been developed in Fortran and Macro 11-Assembly languages.

When the corneal profile was clearly detected on the picture, it was fitted by a circle equation using chi-squared criterion (refer to Fig.4). The experimental points were selected among those indicated by the crossing between a 32stripes grating superposed on the picture and the profile itself.

The superposition effect was obtained exploiting the "graphic plane" option of the TESAK VDC501.
The choice of the experimental points was performed manually, moving a cursor on the screen through the TTY keyboard.
The theoretical profile was finally corrected according to the recording angle ("rotation angle")(refer to Fig.2) and referring it to Cartesian coordinates with the origin in the corneal vertex. In this way, a spherical surface interpolating the original cornea was obtained. Sufficiently good results are gained in normal adult corneas with circle equations.

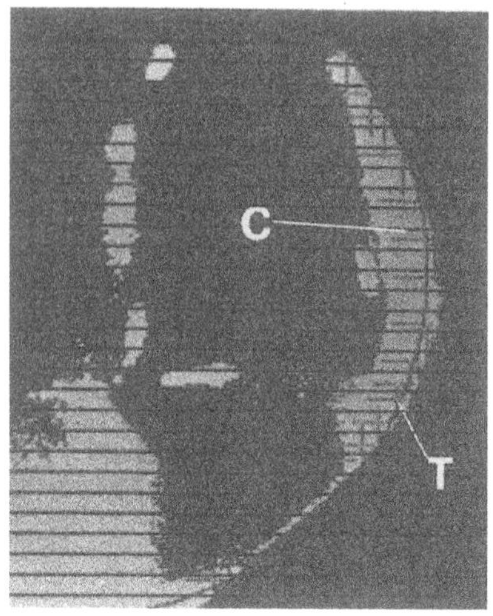

Fig.5 Processing of a keratoconus profile. C cornea, T theoretical fitting curve

It is clearly shown in Fig.5, however, that keratoconuses cannot be fitted by a spherical surface. In these pathological cases aspherical surfaces are the best solution for the contact lens design and we think that the procedure we are following may be suitable, time saving and precise enough to be applied to reach this aim.

3. Conclusions

We think that the TV fitting procedure offers the following advantages over photographic methods:

- the optimal corneal surface equation comes out very rapidly from the fitting procedure;
- it enables an easy control of the eye position through the observation of the pupil image;
- the images are directly stored on a storage device (we used floppy disks) or on video tape.

References

1 Brown N.:Photographic investigation of the human lens and cataract.Surv. Ophthalmol.23,307(1979)

2 Hockwin O.,Dragomirescu V.,Koch H.R.:Photographic documentation of disturbances of the lens trasparency during ageing with a Scheimpflug camera system. Ophthalmic Res. 11,405(1979)

3 Dragomirescu V.,Hockwin O.,Koch H.R.:Photo-cell device for slit-beam adjustment to the optical axis of the eye in Scheimpflug photography.Ophthalmic Res.12,78(1980)

A New Noninvasive Optical Method for Studying Binocular Vision by Infrared Image Processing

J. Bille, H. Helmle, and K. Müller

Institute of Applied Physics, University of Heidelberg
D-6900 Heidelberg, Fed. Rep. of Germany

For the diagnosis of irregularities in the movement of the horizontal eye muscle the z-transformation function of the measured saccadic eye movement is evaluated with the help of parameter estimation procedures. The constants of the transformation function are used as parameters for the classification of the saccades. These can unequivocally be described with high accuracy.

Introduction

The most frequent eye movements are the ones in horizontal direction in the range of +/-15° with regard to the primary position. They are controlled by the musculi rect. lateralis and medialis. If one takes into consideration that 20 - 40 % of all strabismus patients have to be operated a 2nd time [3] it becomes obvious that a better quantitative diagnostic method is necessary. Such measurements on the eye should be accomplished noncontactively and with as little inconvenience as possible for the patient. The mechanics of the muscles should be described with only a few parameters. In [2] it was suggested for the first time to use parameter estimation procedures for the eye muscle diagnosis. In this work [2] the theoretical pre-conditions are examined for gathering constants of the eye muscle from a mechanical excitation of the eye ball with the help of a pincette attacking at the limbus and the eye-movement resulting from it. We present a method with which evaluation of muscle parameters are possible just from the natural eye movement.

Control of Eye Movements

Figure 1 shows a principle model of the oculomotoric system [1]. The pursuit eye movement is a system which is regulated continuously [5]. A parameterisation of this movement discribes the regulation mechanisms but not the mechanics of the muscle. Saccades are induced when the eye cannot follow the fixation point with the pursuit movement system. These are jumping, very fast movements of about 20 - 50 msec duration and an angular amplitude of up to 50°. The saccades are figured in advance and proceed according to a certain scheme. Therefore, they depend only on the innervation structure and the mechanical characteristics of the muscle. Hence, it is possible to deduce mechanical characteristics of the muscle from saccades. The fixation movement con-

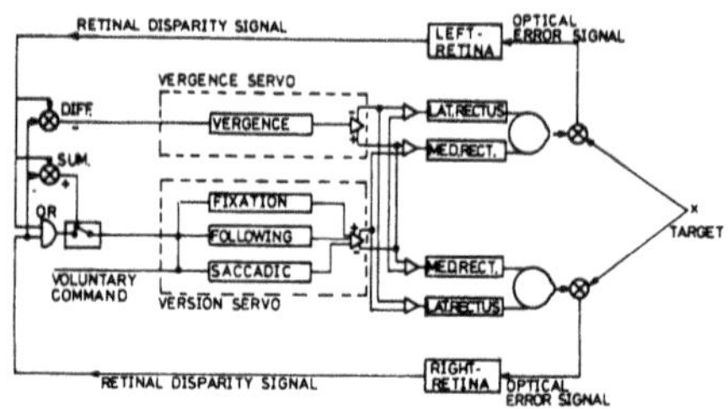

Fig. 1 The oculomotoric system [1]

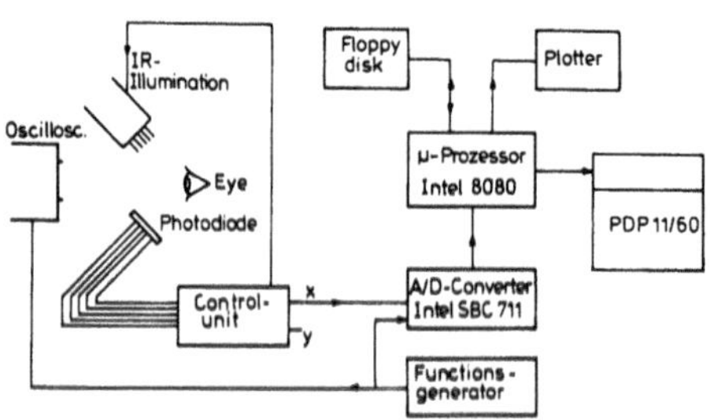

Fig. 2 Experimental set-up

sists of microsaccades with an angular amplitude of ca. 5' and a frequency of 80 Hz. This movement is too small and too fast to derive parameters from it.

Experimental Set-Up

Figure 2 shows the scheme of the experiment. The head of the test person is fixed by a holding device with a chin support and a holder for the forehead. The eye is illuminated from the top by infra-red light. The detector is fixed underneath the eye. The fixation target is a light point jumping into horizontal direction on an oscilloscope screen. The analog signal of the detector is stored on discs via an AD converter with the help of a microcomputer. For parameter estimation the data are transferred to a PDP 11/60.

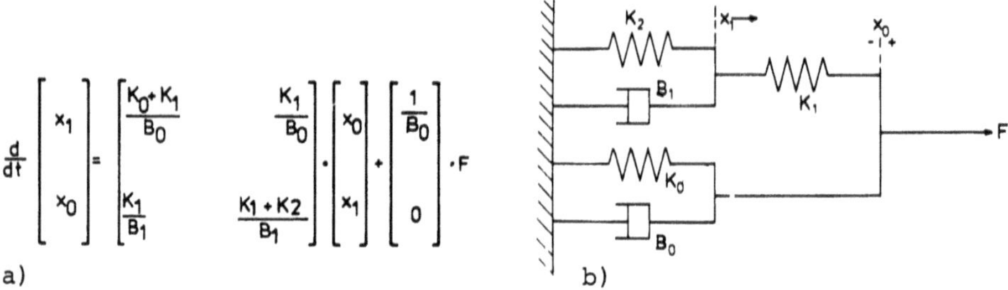

Fig. 3 a) Equation, b) simplified model of an eye muscle

Mechanical Model of an Eye Muscle

Figure 3 shows the mechanical model of an eye muscle [2]. For simplification it was assumed that the muscle is constantly innervated and that all constants are independent of time. The behaviour of this model can be described by a set of differential equations given in Fig.3a.

Parameter Estimation

By Laplace transformation the system of equations (1) can be linearized:

$$G(s) = \frac{x_o(s)}{F(s)} = \frac{s + A}{B(s^2 + Cs + D)}. \qquad (2)$$

After the insertion of a hold term of zero order, equation (2) can be written as a z transformation:

$$H \cdot G(z) = \frac{x_o(z)}{F(z)} \cdot H = \frac{b_1 z^{-1} + b_2 z^{-2}}{1 + a_1 z^{-1} + a_2 z^{-2}} \qquad (3)$$

$$\text{with } V = \frac{b_1 + b_2}{1 + a_1 + a_2}.$$

This is a transformation function of 2nd order with one pole and two zero positions. Together with the amplification 4 constants can be calculated from it if one presumes one constant of the muscle model. If $x_o(t)$ and $F(t)$ are known, the parameters of the z-transformation function can be estimated. Therefore, the following methods are used: method of the least squares, method of the instrumental variables, and correlation analysis. An exact description can be found in [4]. If a step function is used as an input signal the parameters of the z-transformation function can be estimated with an accuracy better than 3 %. The saccades can be reconstructed from the parameters calculated.

It is not possible to estimate the parameters directly from the measured data. For simplification we assume that the unknown innvervation and the resulting force of the muscle act in form of step functions. The force derived this way and the movement measured are then normalized to the same amplitude.

Results

Measurements and parameter estimations were performed on healthy and sick persons. The result from a patient with the Duane syndrome on the left eye is presented in Fig. 4. Table 1 shows the result of a discriminance analysis of the parameters estimated of 38 saccades and a following classification on the basis of discriminance functions. With the exception of group 3 all saccades can be classified well. Out of the three saccades of group 3 two are completely in the regular range in which no error innervation takes place.

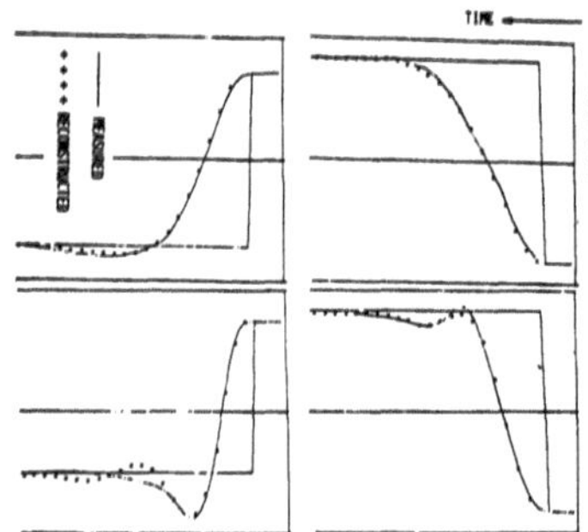

Fig. 4 Typical saccades of groups 1-4; top: left and right eye from right to left; bottom: left and right eye from left to right.

Table 1 Classification of the saccades

group	number of saccades	classification right	classification wrong	eye	direction
1	8	7	1	L	R → L
2	13	12	1	R	R → L
3	7	4	3	L	L → R
4	10	10	0	R	L → R

Discussion

With the procedure presented saccades can be described unequivocally and can unambiguously be classified for a diagnosis. A recalculation of the parameters estimated from the constants of the model is impossible so far since a real saccade in contradiction to the one assumed here is determined by 2 muscles. Besides this, the innervation structure of a muscle does not consist of one but of two jumps. But it only makes sense to take this into consideration if the transformation function of the two-muscle model [1], a system with 2 inputs and one output, can be estimated.

References

1 COLLINS, C.C.: in: G. Lennerstrand, P. Bachy-Rita (eds.): Basic Mechanism of Ocular Mobility: The Human Oculomotor Control System, Pergamon Press, New York, 1975.
2 GUPTA, N.K., PHATAK, A.V.: Identification of Human Oculomotor Parameters with Application to Strabismus, Proceedings of the Automatic Control Conference, San Francisco, 1975.
3 HELVESTON, E.M.: Reoperation in Strabismus, Abstract: Ophthalmology 85: 66-67, 1978.
4 ISERMANN, R.: Prozeßidentifikation, Springer-Verlag, Berlin, Heidelberg, New York, 1974.
5 ROBINSON, D.A.: The Mechanics of Human Saccadic Eyemovements, J. Physiol. 174, 1964.

Applications of the Nonlinear, Visual Operator "Lateral Inhibition" in Image Processing

A. Kriete, and S. Boseck

Fachbereich Physik der Universität Bremen, Postfach 330440
D-2800 Bremen, Fed. Rep. of Germany

1. Introduction

There have been some attempts to emulate a possible focussing mechanism of the human eye in technical optics, simulating lateral inhibition [1]. To study the basics of practical applications, this physiological principle was optically and information theoretically investigated in the human visual system (HVS).

2. Method and Material

A model for the HVS spatial frequency response was constructed by considering the system to be linear, isotropic, invariant in space and time, monocular and monochromatic. This paper discusses threshold vision for the object f and its image f' on the basis of the block diagrams as follows [5]:

$$f \rightarrow \boxed{MTF} \rightarrow \boxed{RTF} \rightarrow f' \quad \text{or} \quad f \rightarrow \boxed{MPF} \rightarrow f'. \tag{1}$$

The MTF represents the low pass of the physical eye as a spatial detector. The retinal transfer function (RTF) manifests the neural inhibition and a high pass gives rise to a subjective brightness perception, which differs from what one might physically expect, e.g. Mach band patterns. Following the multiplication theorem, a bandpass results. Here it can be called idealized modulation perception function (MPF), because it relates the reciprocal of the modulation threshold to detect sinusoidal gratings to the spatial frequency, but does not incorporate noise or experimental effects.

Such idealized functions are given in [2] and were chosen now as fundamentals, taking into account the system nonlinearities of different light levels (10 to 10^{-2} mL background luminance). The MTF low passes were obtained with the help of Lorentz functions. With the relation MPF/MTF=RTF, the RTF high passes could be found. Using a Fourier Transform (FT), the conditions in the spatial domain could be treated.

3. Results

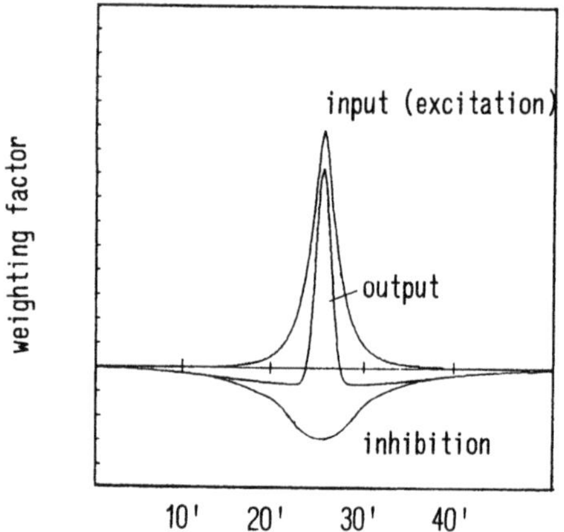

Fig. 1 Spatial domain (10^{-2} mL)
The inhibition function results by substraction of the input function (FT of MTF) from the output function (FT of MPF) or receptive field. This procedure follows the idea of unsharp masking.

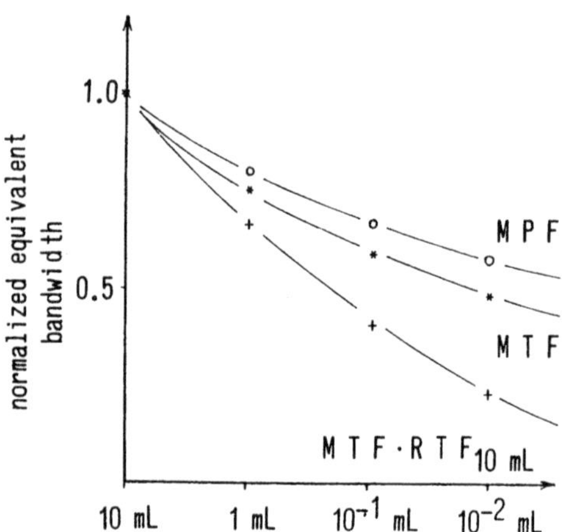

Fig. 2 Equivalent bandwidth
This half-logarithmic representation shows a falloff for the MPF, less than in the case of MTF, i.e. inhibition is reduced for lower luminances. As a defocussing simulation, the constant RTF gives the best focus criterium. Therefore, the data are normalized to 1.

4. Discussion

Photographic materials having a transfer function with adjacency effects give rise to "ghost structures". This phenomenon does not occur in the case of the "physiological adjacency effect", because the RTF's were found not to be greater than 1.

The information density in bits per area of one diffraction disc [3] with the aperture F is:

$$5.3\, g \iint \log_e \left(1 + (\mathrm{MTF} \cdot \mathrm{RTF})^2 R \right) du\, dv\ , \quad g = 2\, F^2 \lambda^2 / \pi . \tag{2}$$

The results are 2.6 (10 mL), 1.0 (1 mL), 0.42 (10^{-1} mL) and 0.2 (10^{-2} mL) for $\lambda = 550$ nm. R was chosen to 100, which means a good perception of fine details. In comparison to the MTF's alone, there is a loss of information density of about only 1%. Lateral inhibition therefore affects the ability of distinction surprisingly little. The equivalent bandwidths of the MPF's show a reduction with falling luminances, but this effect is less than in the case of the MTF's. Inhibition might cease at very low luminances. The qualitative defocussing simulation takes a constant RTF into account and gives the greatest gradient. If one considers the normalized bandwidth as a comparison of focus criteria (the maxima are 27.2 for the MTF and 11.4 for the MPF), the constant RTF high pass is the best method. This is also the equivalent explanation of the behaviour of the "Biofocus" [1] in the spatial domain. The results predict qualitatively experimental data given in [4]. According to (2) the information density, using different MTF's and a constant RTF, should come to a maximum in the focus as well.

When smoothed, the inhibition functions in the spatial domain (Fig. 1) have a structure similar to unsharp masking. RATLIFF (1965), RÖHLER (1967) and HALL (1979) combined lateral inhibition with this idea, MACH (1865 onwards), HUGGINS and LICKLIDER (1951) proposed differentiation methods as a model (for details see [5]). An algebraic procedure to simulate lateral inhibition could be found:

$$f_{out} = a \cdot f_{in} + (- 1 \cdot \Delta f_{in}); \tag{3}$$

f_{in} denotes the input-function, f_{out} the output function, a is a weighting factor and Δ the Laplacian operator.

Also the definition of acutance uses this operator [5], valuing the shape of an edge, and not resolution or contrast. It is no surprise that exactly this definition denotes the best correlation to the subjective sharpness impression as it works close to the HVS. A double slit simulation showed no improved resolution in the case of lateral inhibition, as the improvement of the central image sharpness suggests. This is clear from the frequency domain, because the high frequency part of the MTF and MPF was assumed to be equal.

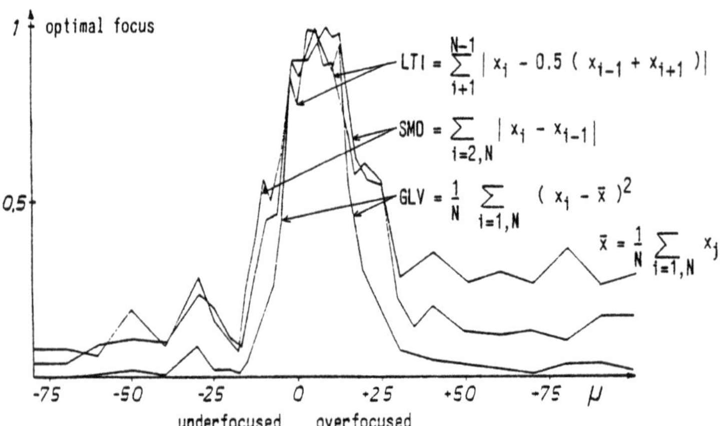

Fig. 3 Focal series

5. Practical Applications

Practical applications of lateral inhibition lie in the field of image enhancement. The subjective image interpretation can be improved, without the occurrence of "ghost-structures" and the information content is only slightly reduced. This is partly fullfilled in X-ray diagnostics by Logetronic. A further application of lateral inhibition is its use as a focus criterion. This can be achieved in the spatial domain, using simple algorithms, or in the frequency domain, using the equivalent bandwidth. Experiments with microscopic, photographic focal series showed that the criterion lateral inhibition gives the same focus point as other, e.g. sum modulus difference and gray level variance.

The result in Fig. 3 stands for a grating (object micrometer) and an objective PL FL 10/0.30. The human observer has an impression "sharp" at values greater 0.5.

6. References

1 Rechenberg, I; Koralewski, H.-E.; Bienert, P.: Deutsches Patent P 28 38 217.4-51, 1980
2 Kornfeld, G.H.; Lawson, W.R.: JOSA, 61 (1971) 811
3 Linfoot, E.H.: Qualitätsbewertung optischer Bilder, Verlag Vieweg und Sohn, Braunschweig, 1960
4 Campbell, F.W.; Green, D.G.: JOSA, 55 (1965) 1154
5 Hall, E.L.: Computer Image Processing and Recognition, Academic Press, New York, 1979

Visual Resolution of Colours in Overlapped Patterns of Two Colours with Sinusoidal Intensity Distribution

M. Pluta

Central Optical Laboratory, Warsaw, Poland

1. Introduction

For many applications in science and technology it is interesting to know the visual spatial resolution of two colours with sinusoidal intensity distribution. For simplicity such a colour pattern is outlined one dimensionally in Fig. 1. The colours are marked by λ' and λ'', and light intensity is denoted by I. The intensity oscillates spatially between a minimum and a maximum value resulting in a pattern of dark and bright fringes.

Suppose one pattern, e.g. λ'', is moving with respect to the other. Then it is interesting to evaluate the minimal displacement r_m, which is not perceived by the human eye. This displacement is named the visual spatial colour resolution or visual colour acuity. If $r<r_m$, then black fringes are perceived without any colouring, and, on the other hand, if r is slightly greater than r_m, these fringes are black in their middle, but they are slightly coloured on their sides. When r is considerably greater than r_m, the black fringes disappear.

It is difficult to produce optically two sine-wave patterns with different colours but the same period, as indicated in Fig. 1. However, it is easy to obtain such patterns with different periods p'and p", by using an interferometer equipped with monochromatic light sources emitting different wavelengths λ' and λ''. Such a procedure has been used in this work.

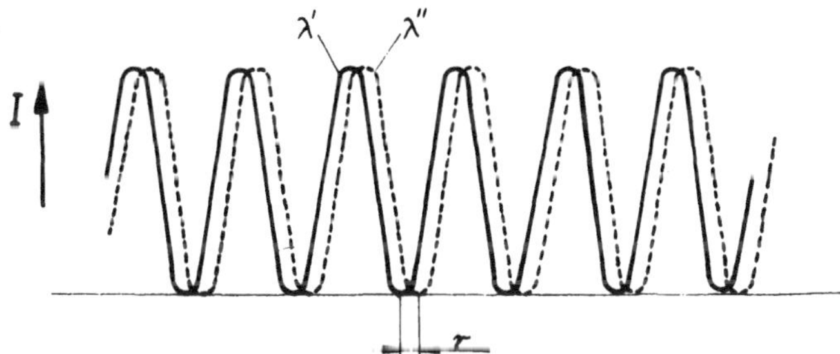

Fig. 1 Two overlapped patterns with different colours and sinusoidal distribution of light intensity

2. Experimental Apparatus

Patterns of two colours with sinusoidal, more exactly $\sin^2$-like, intensity distribution have been produced using a double-refracting interference microscope with variable wavefront shearing (Fig. 2). A detailed description of this system is given in a previous paper [1]. Its characteristic feature is a combination of two simultaneously acting birefringent prisms, W_1 and W_2, inserted between two crossed polarizers P and A. These are Wollaston type prisms made of quartz crystals. One of them, W_1, is located just above the microscope objective Ob, and is rotatable around the objective axis. The rotation enables the amount of wavefront shearing to be changed. The other prism, W_2, is placed in the microscope tube, and can be moved in a transverse direction t using a micrometer screw FS attached to the prism. This translation serves for shifting the phase between sheared light waves and, simultaneously, for measuring the period of interference fringes observed in the microscope image plane π'. Only the prism W_2 gives interference fringes in this plane, whereas the prism W_1 produces a homogeneous interference. The interference pattern is observed by means of an eyepiece, not shown in the figure, fitted with a cross-line. The period of fringes depends on the apex angle α_2 of the prism W_2: the greater this angle the smaller is the fringe period. The prism W_1 can be oriented at four characteristic positions relative to the prism W_2: additive, as shown in Fig. 2, crossed left-handed, crossed right-handed, and subtractive. For experiments presented here the subtractive position is most convenient, because the slit S of the diaphragm D, located in the front focal plane of the condenser K, can be relatively wide if the apex angle α_1 of the prism W_1 is comparable with α_2. In this case the slit S is oriented parallel to the apex angle wedges of the prism W_2, and it intersects at an angle of 45° the axes of the crossed polarizers P and A. The condenser slit is illuminated with two monochromatic light beams of wavelength λ' and λ''.

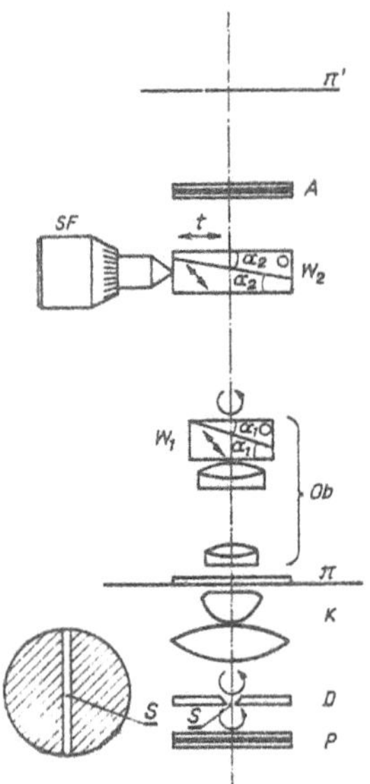

Fig. 2 Basic optical elements of a double-refracting interference microscope used in the experiments

Highly monochromatic light sources were used: He-Ne laser (λ = 632.8 nm), Ar laser (λ = 514.5, 488 and 457.9 nm), and He-Cd laser (λ = 441.6 nm). Additionally, a HBO lamp was applied, from which green light (λ = 546.1 nm) was extracted by using an interference filter. Simultaneously, two light beams with different wavelengths were directed to the condenser slit. Both beams produced independently two interference fringe patterns in the image plane of the microscope. These patterns were observed simultaneously as a mixed pattern, or alternately as single patterns if one of the beams was stopped. Both beams were adjusted so that they generated black interference fringes of zero order exactly at the same place of the image plane π'. Coherent noise produced by highly coherent laser light was suppressed by a rotating ground glass.

3. Experiments

For each combination of two wavelengths the fringe periods were measured with great accuracy. Inaccuracy has been estimated to 0.2 μm. It is not difficult to obtain such a result, because the highly monochromatic light used in the experiment gives a great number of identical interference fringes. Thus, one can measure not only single spacing between two neighbouring black fringes but considerably greater distances, e.g. between fringes of zero and 10-th or even 50-th interference order.

Next, the mixed interference pattern consisting of two systems of fringes produced by two light beams of wavelengths λ' and λ'', was observed. Characteristic feature of this pattern is tuning of adequate interference fringes. The tuning is not exact but runs with a smaller or greater disaccord which can be used for measuring the visual colour resolution, because this disaccord results in a smaller or greater colouring of the tuned black interference fringes of different orders. One can easily evaluate visually the orders tuned without and with a very small, just perceptable coloration. Knowing the fringe periods p' and p" one can determine the disaccord in tuning of the observed orders. This disaccord is exactly the measure for the visual spatial colour resolution r_m.

The measuring procedure is shown more in detail in Fig. 3 which represents two overlapped systems of interference fringes produced by red and blue light beams with wavelengths $\lambda' = 632.8$ nm and $\lambda'' = 441.6$ nm. For sake of simplicity, the upper and lower parts of this figure represent only the fringes for λ' and λ'', respectively. Fringe spacings for these wavelengths, as well as for the used interference system, are p' = 219 μm and p" = 147 μm, respectively. As seen, 3p" is only slightly greater than 2p', and the difference Δb = 3p"-2p' = 3 μm. If the interference system is adjusted so that the zero order interference fringes for λ' and λ'' are exactly in coincidence, then black fringes of the

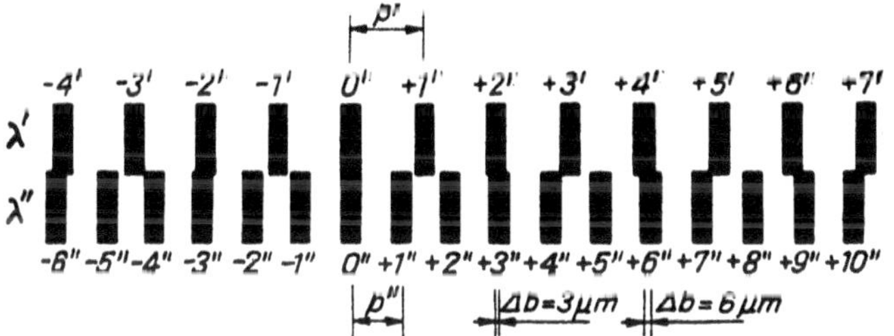

Fig. 3 Two systems of overlapped interference fringes

3-rd order for λ'' and 2-nd order for λ' are nearly in tuning. A sufficient tuning is also shown for black fringes of the 6-th order of λ'' and the 4-th order of λ', as well as for fringes of the 9-th order of λ'' and the 6-th order of λ', because respective disaccords $\Delta b = 3$ µm and $\Delta b = 12$ µm are not distinguished by the human eye in the case under consideration. The respective tuned black fringes are visible without any red and blue coloration. A noticeable colouring is observed for $\Delta b = 15$ µm. Thus, the threshold perception of colouring, and therefore, the visual spatial colour resolution r_m, ranges between 12 µm and 15 µm. The interference patterns were observed under 12.5 x magnification. For the naked eye r_m has in this case a value between 0.15 and 0.19 mm.

4. Results

The results for different combinations of two wavelengths λ' and λ'' are presented in Tab. 1. The values of r_m are averaged data obtained from a number of people of the author's laboratory. The differences between the individuals have been quite small, in general not greater than 0.001 to 0.002 mm. For the naked eye r_m, as well as p' and p'', given in Tab. 1 should be multiplied by 12.5.

Table 1 Visual spatial resolution r_m of different colours in overlapped interference patterns

Number of two wavelengths combination	λ' [nm]	p' [µm]	λ'' [nm]	p'' [µm]	r_m [µm]
1	632.8	219	441.6	147	12
2	632.8	219	457.9	153	10
3	632.8	219	488.0	165	6
4	632.8	219	514.5	175	7
5	632.8	219	546.1	187	8
6	546.1	187	441.6	147	10
7	514.5	175	441.6	147	9

From these results it can be gathered that for each combination of two colours, containing a wavelength in the deep blue spectral range (λ = 441.6 and 457.9 nm) the visual spatial resolution r_m is worse than for any other pair of colours.

Reference

PLUTA, M.: Optica Acta 18, (1971), 661-675.

Speckle Pattern Stimulator for Visual Evoked Potentials

H. Uozato, J. Fukuhara, M. Saishin, and S. Nakao

Department of Ophthalmology, Nara Medical University
Kashihara, Nara 634, Japan

Abstract

A new method to stimulate visual evoked potentials (VEPs) using laser speckle patterns is described. The principle of producing the speckle patterns is theoretically analyzed and an optimal optical system for stimulation of the VEPs is considered. The feasibility of this method is demonstrated by simple experiments carried out on a group of normal subjects. Application of this method to the assessment of the visual function of cataractous eyes is investigated.

1. Introduction

Visual evoked potentials (VEPs)are gross electric signals generated in the occipital cortex in response to visual stimulation. Analysis of VEPs is becoming a common clinical test for objective diagnosis of the function of the visual sense [1]. Especially, pattern VEPs has been developed remarkably with the aid of micro-computer systems and commerical TV monitors [2]. However, these conventional pattern VEPs are strongly affected by refractive, accommodative, and fixation states of the subject's eyes. Although these methods are applicable for objective determination of refraction [3], it is necessary to correct fully these factors for the assessment of visual functions. In infants and patients with refractive or accommodative anomalies which are extremely difficult to correct, the conventional pattern VEPs are of no use in clinical diagnosis.

We propose a new method for the stimulation of VEPs using laser speckle patterns [4,5] that overcomes these drawbacks. The speckle patterns are presented to the subject's retina regardless of the refractive conditions of the eye [6-8]. Therefore, it is possible to obtain VEPs directly in the state of uncorrected eyes. This method is also applicable to the assessment of visual functions of cataractous eyes, which could hardly be performed by conventional methods [4,5,9].

2. Principle and Method

When an optically rough surface is illuminated with highly coherent light, such as laser light, the scattered radiation possesses a marked granularity, or speckled appearance, and images formed by using this radiation also contain annoying speckle patterns. These speckle patterns are statistical in nature and their statistical properties are intimately connected with both the statistical characteristics of the diffusing object and the coherence condition of the illuminating light [10-12]. The optical system of a laser speckle pattern stimulator using a double diffraction system and Maxwellian view optics [13] is shown schematically in Fig. 1. When a diffuser D having a circular aperture (diameter:w) is illuminated with coherent laser light (wavelength:λ), average speckle diameter at the subject's retina is given by

$$\sigma = \lambda \cdot f_1 \cdot f_e / w \cdot f_2, \quad (1)$$

where f_1, f_2 are the focal lengths of double-diffraction lenses and f_e is the distance from the 2nd nodal point of the eye to the retina. Then the average angular size of the speckles can be expressed by

$$\theta = \sigma / f_e = \lambda / w \cdot M, \quad (2)$$

where M $(=f_2/f_1)$ is the lateral magnification of the double-diffraction system. Rewriting eq. (2) the spatial frequency (cycles/deg. or cpd) can be expressed by

$$\nu = 0.873 \cdot 10^{-2} \cdot w \cdot M / \lambda. \quad (3)$$

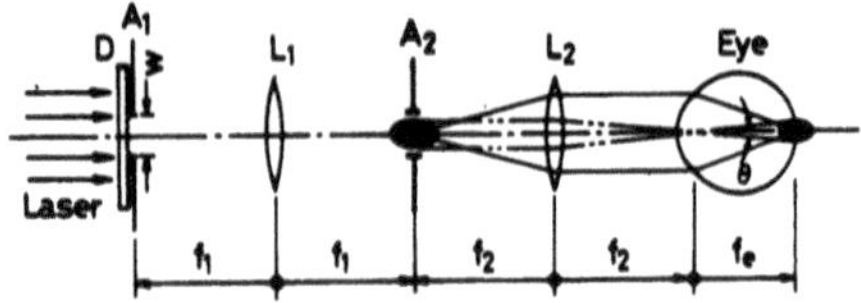

Fig. 1 Schematic diagram of the laser speckle pattern stimulator.

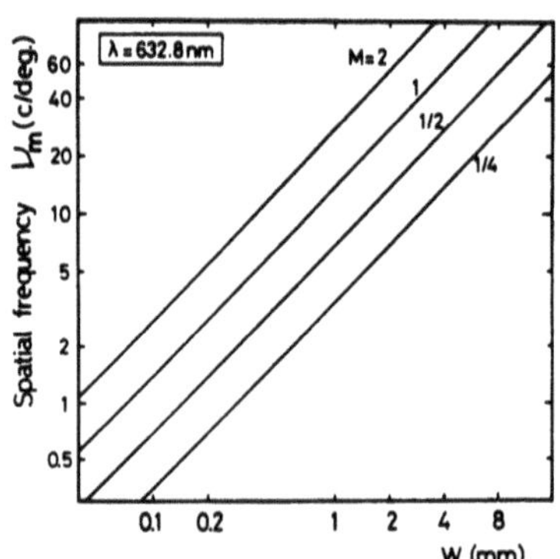

Fig. 2 Relation between the averaged spatial frequency of the speckle pattern and the diameter of the illuminated area (w).

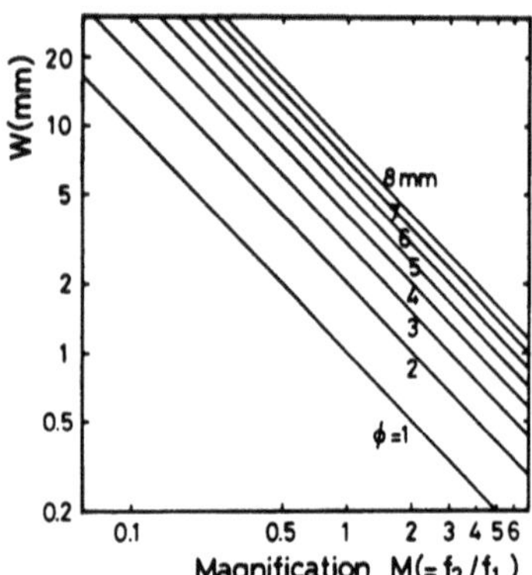

Fig. 3 Upper limit of the diameter of illuminating are (w) for various pupil sizes.

From eqs. (2) and (3), it can be deduced that the average speckle size or its spatial frequency at the retinal plane is independent of the optical system of the eye. As the variation of the positions of nodal and principle points due to accommodation of the eye is very small we can neglect these effects. Thus, size at the retinal plane is independent of the refractive and accommodative conditions. Figure 2 shows the relation between the averaged spatial frequency of a speckle pattern and the diameter of the illuminated area (w) when the wavelength is 632.8 nm. In general, the cut-off frequency of the visual system is in the order of 50 cpd [14]. This range of spatial frequencies is easily covered by the variation of the illuminated area from 50 µm to 5 mm.

Another remarkable feature is the high contrast of speckle patterns. Theoretically, the averaged contrast of a speckle pattern is usually defined by a normalized standard deviation of speckle intensity variation at the observation plane [10-12]. When an ordinary diffuser is illuminated by laser light, the contrast of a speckle pattern is almost unity. Therefore, in our system we can observe a high contrast speckle pattern in spite of the refractive and accommodative states, and the opaqueness of the ocular medium.

The diameter of the stimulation field on the retina is limited by the aperture size (2a) in the A_2-plane as shown in Fig. 1, and is given by $1/F_2$, where $F_2 = f_2/2a$.

As mentioned above, the averaged spatial frequency of the speckles is controlled by the illuminated area of the diffuser (w). However, the diffuser plane and the pupil plane are almost conjugate. When w becomes large enough, the pupil size determines the size of the speckles. Therefore, it is neccessary to make the image of the illuminated area (w) in the pupil plane smaller than the pupil size. This condition is expressed by $w \cdot M < \phi$, where ϕ is the pupil size. The upper limit of the diameter of the illuminated area for various pupil sizes is shown in Fig. 3.

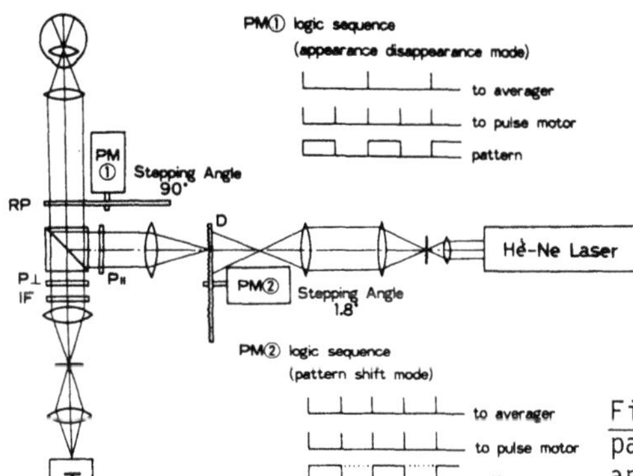

Fig. 4 Schematic diagram of a speckle pattern stimulator for appearance-disappearance and pattern shift modes.

3. Experimental Results and Discussions

Modulation without luminance change can be performed by two methods, pattern shift and appearance-disappearance. The system used is schematically shown in Fig. 4. A He-Ne laser and a tungsten light were used as coherent and incoherent light sources, respectively for contrast VEPs. A flow diagram for recording and processing is shown in Fig. 5. VEPs are recorded differentially by by two Ag-AgCl electrodes, one placed on the midline 3 cm above the inion and the other on the right earlobe, the left earlobe being earthed. The VEP signals are amplified by a factor of 10^4 before digital conversion. The signals are recorded as the average of 100 sweeps of 500 msec, and the waveforms are recorded with a X-Y recorder.

Effects of refraction and accommodative fluctuation on the VEPs are shown in Figs. 6 and 7, respectively. One of the most excellent applications of speckle pattern VEPs is the case of an opaque ocular medium such as a

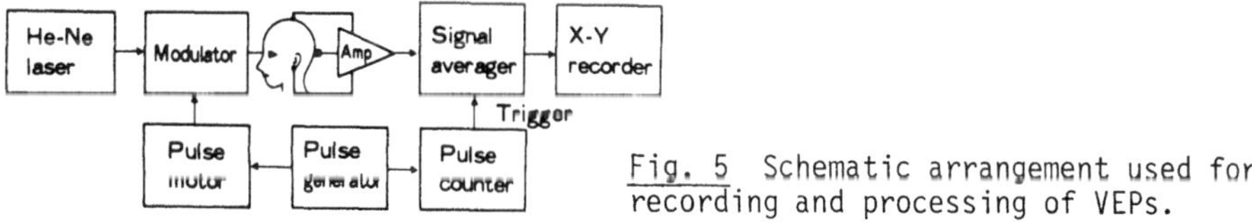

Fig. 5 Schematic arrangement used for recording and processing of VEPs.

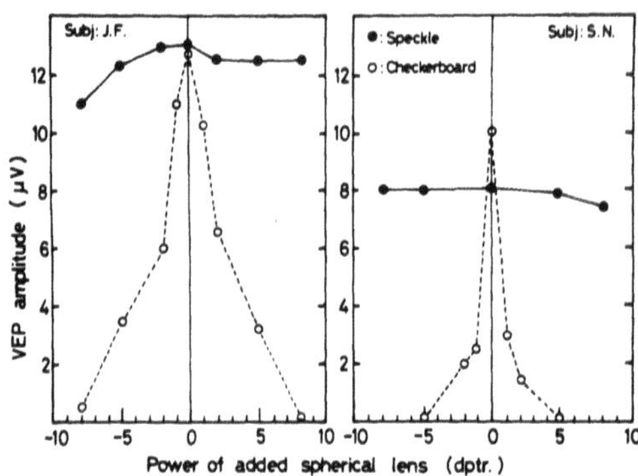

Fig. 6 VEP amplitudes as a function of spherical lens power.

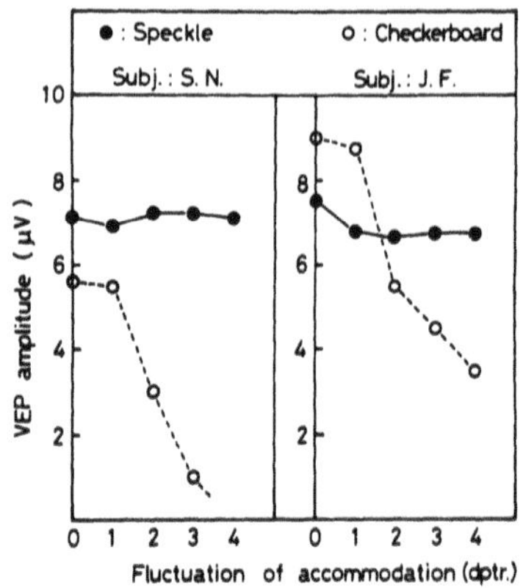

Fig. 7 VEP amplitudes as a function of accommodative fluctuation.

cataract [4,5,9]. A corresponding result is shown in Fig. 8. As known from Figs. 6-8, the conventional pattern VEPs (e.g. checkerboard) are strongly influenced by the optical abnormalities of the eye such as refractive or accommodative anomaly, keratoconus, corneal opacity and cataract. However the speckle pattern method is applicable even in the presence of those abnormalities [9]. It is difficult to obtain reliable VEPs in the case of a dense and diffuse opacity such as a mature cataract even when the interference fringe pattern VEP method [15,16] is adopted. In this case, the speckle pattern VEP technique can be applied, because the speckle pattern created with the cataractous lens itself is easily obtained by direct illumination with a laser beam as shown in Fig. 9. In general, the lens opacity is distributed spatially. It is suspected that the averaged speckle size and contrast is decreased by the effect of multiple scattering. Thus, a simulation experiment as shown in Fig. 10 was performed. Figure 11 shows the speckle patterns obtained with ground glasses (a,b,c) and a cataractous lens of a human eye in-vitro (d). This result indicates that the higher the number of ground glasses, the smaller is the average size of the speckles; the contrast is maintained, however, at a high value. Therefore, this technique may be useful for

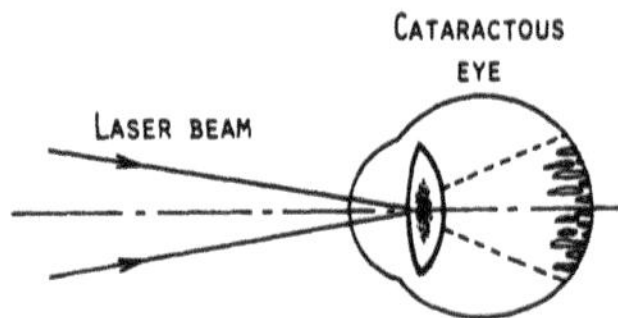

Fig. 8 VEP waveforms obtained from a cataractous patient. Checkerboard pattern reversal and speckle pattern shift VEPs are recorded from the same patient before and after operation, respectively.

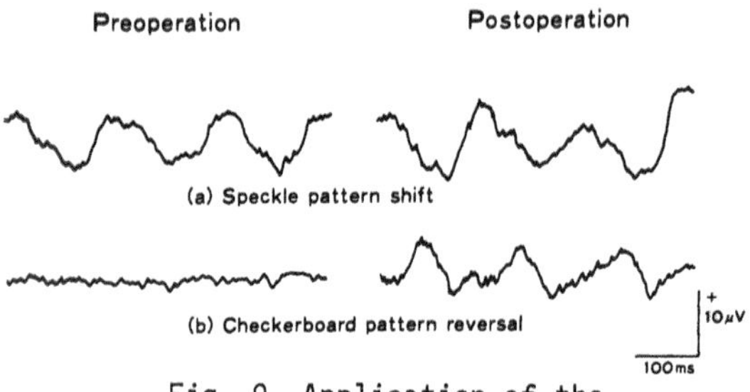

Fig. 9 Application of the laser speckle pattern visual stimulation to cataractous eyes.

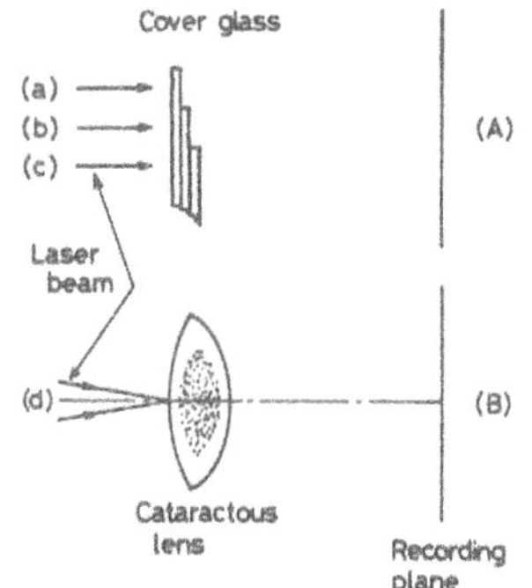

Fig. 10 Schemes of the arrangements used for recording diffraction-field speckle patterns created with ground glasses (A) and a cataractous lens of a human eye (B).

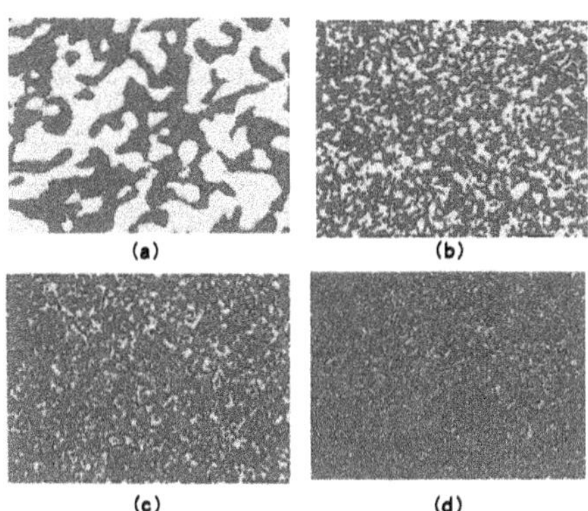

Fig. 11 Speckle patterns obtained with ground glasses (a,b,c) and a cataractous lens of a human eye in vitro (d).

a mature cataract whose examination is extremely difficult with conventional pattern VEPs, if we give the laser safety regulations and the speckle size our careful consideration.

4. Conclusion

A new method for stimulation of visual evoked potentials (VEPs) has been developed using laser speckle patterns. Experiments demonstrated the feasibility of the speckle pattern VEP method. It can be concluded that the speckle pattern VEPs are very useful in the assessment of visual functions regardless of the refractive and accommodative anomalies, and even in the presence of opaque ocular media such as a cataract or a corneal opacity, etc. Therefore, this method can be applied to infants and cataractous eyes whose evaluations are very difficult with conventional methods. Speckle pattern VEPs may become one of the most powerful tool in clinical tests.

The authors wish to thank A. Koike, M.D. and S. Nojima, M.D. for their technical assistance. This work was partially supported by grants of the Ministry of Education, Science and Culture (No. 577671 and No. 577673).

References

1 DESMEDT, J.E. (ed.): Visual evoked potentials in man: New developments, Clarendon Press,(1977).
2 ARDEN, G.B., I. BODIS-WOLLER et al.: in: J.E. Desmedt (ed.): Visual evoked potentials in man: New developments, Clarendon Press (1977) 3.
3 MILLODOT, M., L.A. LIGGS: Arch. Ophthalmol. 84 (1970) 272.
4 UOZATO, H., J. FUKUHARA, S. NAKAO: J. Nara Med. Assoc. 31 (1980) 354.
5 UOZATO, H., J. FUKUHARA, S. NAKAO: J. Ophthalmol. Opt. Soc. Jpn. 2 (1981) 127.
6 DAWSON, W.W., M.C. BARRIS: Invest. Ophthalmol. Visual Sci. 17 (1978) 1209.

7 FUKUHARA, J., H. UOZATO, A. KOIKE, M. SAISHIN, S. NAKAO: Folia Ophthalmol. Jpn. 31 (1980) 1864.
8 FUKUHARA, J., H. UOZATO, T. MATSUDA, A. KOIKE, S. NOJIMA, M. SAISHIN, S. NAKAO: J. Ophthalmol. Opt. Soc. Jpn. 2 (1981) 123.
9 FUKUHARA, J., H. UOZATO, A. KOIKE, S. NOJIMA, M. SAISHIN, S. NAKAO: Jpn. J. Clin. Ophthalmol. 35 (1981) 1009.
10 GOODMAN, J.W.: in: J.C. Dainty (ed.): Laser Speckle and Related Phenomena, Topics in Applied Physics, Vol. 9, Springer, Berlin, Heidelberg, New York (1975) p. 9
11 YAMAGUCHI, I.: Jpn. J. Optics (Kogaku), 3 (1974) 128.
12 ASAKURA, T.: in: R.K. Erf (ed.): Speckle metrology, Academic Press (1978) 11.
13 WESTHEIMER, G.: Vision Res. 6 (1966) 669.
14 KAWARA, T., H. OHZU: Oyo-Buturi, 41 (1977) 128.
15 SUGIMACHI, Y.: Acta Soc. Ophthalmol. Jpn. 82 (1978) 398.
16 ARDEN, G.B., U.B. SHEORY: in: J.E. Desmedt (ed.): Visual evoked potentials in man: New developments, Clarendon Press (1977) 381.

Part VI

Moiré Methods

Survey on Applications of Moiré-Techniques in Medicine and Biology

G. Windischbauer

Institut für Medizinische Physik, Veterinärmedizinische Universität Wien
Linke Bahngasse 11
A-1030 Wien, Austria

1. The Renaissance of Moiré Techniques

Moiré techniques are promising methods of applied optics with important contributions to medical and biological applications. Undoubtedly moiré topography is not the central part of optics in the biomedical sciences, but it is a method which produces images which are easy to recognize and in which the image processing is not too sophisticated. The result can be aesthetic in the case of a technical object and sometimes frightening in clinical use.

The term "moiré" itself has a biological origin. The shiny material made of the wool of angora goats was called "moiré" in French and "mohair" in English. Since the 12th century a special finishing of textiles produces dull and shiny waves. These materials are called Moiré Seide or watered silk.

The basic optical effect was noted by Lord RAYLEYGH as early as 1874 [1]. But within the following sixty years only little work was done, so that in 1964 STECHER concluded in a paper published in the American Journal of Physics: "It is odd that no recent optics text ... even mentions moiré fringes. Our engineering colleagues have been using the effect advantageously for about 20 years ..." [2].

In 1970 the renaissance of the optical moiré effect was initiated by three papers, which were published in the Journal of Applied Optics by TAKASAKI and by the American authors ALLEN, MEADOWS, and JOHNSON [3-5]. The greatest contribution came from Professor TAKASAKI, and it is my personal view that the "pretty biology" of TAKASAKIs object contributed a lot to the success of moiré topography in biology and medicine. But there are three other points concerning the explosive spread of moiré techniques.

1. Scientists were well prepared to accept interference-like patterns.
2. They had learned to process images.
3. They continued summarizing and defining the field of optical biometrics.

In 1960 the age of coherent optics began with the invention of the laser. Fourier optics and holography grew to new dimensions and the concepts of information theory were applied successfully to optical problems. Everyone used interfering waves and became accustomed to monochromatically illuminated surfaces and speckles. Holographic contouring methods were developed and I remember the surprise in studying the first paper of TAKASAKI, describing how to cover a human body with interference-like patterns only by simple incoherent optical means. It was the merit of this scientist to have developed an incoherent method similar to holographic contouring. And he expanded the moiré methods of strain analysis, which were limited to flat and small objects, to cover large three-dimensional bodies.

The development of digital image processing was the second condition for the world-wide acceptance of moiré techniques. Image processing with digital computers started in 1960, too. For the first time, computers with sufficient calculating speed and memories were available, but the processing was concentrated at great computing centres of universities or restricted to military goals. Image processing was troublesome and expensive; every user developed his own software. The rapid growth of image processing began with the use of the minicomputer. Large-scale integration, high-density memories, the development of the microcomputer chips, as well as a new generation of raster scan scopes, and last but not least, decreasing costs made image processing systems available for nearly everyone. Today this work is done decentralized by small dedicated systems and is used in every field of science. Thus, a photogrammetric interpretation of moiré topograms became possible at the same time.

In the past a great variety of optical methods were used for anthropometric measurements. In 1970 several excellent reviews were given of this widespread field, especially in the papers of HERRON [6,7]. He stated, that "although we have learned a great deal about human biology through the use of conventional anthropometric procedures, the continuing almost universal application of two-dimensional 'tape and caliper' methods is helping to perpetuate certain information gaps". In a similar manner KECK concluded in a review, that "the bodies of men and animals were measured with simple instruments compared with the expense of other physical methods. The lack of information on body form is due to these instruments and may be filled by optical methods" [8]. Summing up the different papers we can define the function of moiré topography in medicine and biology. Moiré topography is a modern scientific approach to the measurement, analysis, and mathematical description of the three-dimensional form of biological objects. It is based on the optical incoherent moiré effect and on the principles of analytical geometry.

2. Representation and Acceptance

Topographical methods and their results will hardly be accepted by physicians and biologists if the representation of the biological body form is very different from the conventional image. It is very easy to demonstrate this. We know that a distance between two points is a very insufficient measure to describe a whole body, but in combination with conventional photography it will be accepted. The information gap is not obvious. The digital representation of all points contains much more information, but in practice will not be accepted. In the field of stereophotogrammetry this problem is solved by calculating and drawing a contour map as the end product of the analytical process. Moiré topography produces the contour map as the first result of the physical process. Moiré topograms contain not only the moiré lines, but they simultaneously show the basic photographic image. There is no difficulty to become accustomed to the topograms.

But I must recall a critical remark made by HERRON: "Unfortunately, overemphasis on contour maps as end products of three-dimensional analysis has seriously hindered the adoption of stereometric procedures in physical anthropology and other fields. In practice three dimensional digital coordinates are so much more versatile, that the role of the contour map is now largely confined to that of a visual aid." Therefore every moiré topogram must be only the first step in the topographic procedure. In this sense moiré topography is a branch of photogrammetry, but a lot of the people doing moiré work are physicians and not photogrammetric engineers. Therefore some difficulties arose in the past about the importance and the role of the moiré topography, and some critical remarks were made about measuring accuracy and error handling. But no commercially available instruments meet photogrammetrical standards, so that moiré topography and stereophotogram-

Table 1

	Object Range [m]	Sensitivity [m]	Accuracy [%]
	10^3 1 10^{-3} 10^{-6}	1 10^{-3} 10^{-6}	
Stereophotogram.	xxxxxxxxxxxx	xxxxxxx	0.1
Moiré Topography	xxxxxxxxxxx	xxxxxxxx	1.
Holography	xxxxxxxxx	xxxxxxxxx	1.

metry are not competing, but complementarying methods. If you compare object range, sensitivity and measuring accuracy of these methods, you will see that in the field of close-range photogrammetry moiré topography meets best the demands of biological objects in regard to sensitivity and accuracy (Table 1).

3. Topographical Methods

As mentioned earlier, the theory of moiré figures was well known in 1970, but the moiré methods had to be expanded to large three-dimensional objects. Within the next few years, modified methods were developed simultaneously for different problems. Now we can distinguish

- the basic grating-shadow method,
- the grating-projection methods,
- the grating-TV methods and
- the synthetic grating methods [9].

In all the methods the moiré fringes are generated by the superposition of a periodical master grating and a grating image of the object. A first classification of the various types may be done by localizing the superposition plane, that means, whether the superposition takes place in the imaging light path, in the image plane, on a TV screen, or in a computer memory. A better way is to characterize the relationship between the two interfering gratings. The spatial binding of the transmission functions of the two gratings determines the precision of the gratings and the performance of the optical components as well as the overall accuracy.

The formation of the moiré figures may be understood as the correlation of the transmission functions of the two gratings. In developing these terms, profile and phase of each grating play an important role. Thus, the basic grating-shadow method offers the best accuracy and the simplest arrangement because projected grating and master grating are identical: they have the highest degree of binding. The method offers some further advantages like the moving-grating technique and high-contrast fringes, but the object size is limited by the grating area. Distinguishing between hills and valleys within object regions can be done by special gratings with asymmetrical profile functions [10]. The projection-type methods offer a lower degree of binding, larger object size, and more flexibility in adjusting the sensitivity, but there exist very rigid demands for the performance of the projection - and the master - grating. Distinguishing between concave and convex surface regions can be easily done. Promising efforts were done with adjustable master gratings, but with a loss in accuracy. If both gratings are connected rigidly, a binding similar to the basic method is obtained and moving-grating technique is possible. All methods where the master grating is generated by an electronic time varying signal or by a computational

Table 2

Moiré method	Superposition	Binding
Shadow grating	Light path	rigid
Project grating	Image plane	middle
Video grating	TV screen	low
Synth. grating	Memory	free

process offer the lowest degree of binding. This means complete independence of both gratings in amplitude and phase. The advantages are additional operations like phase detection, different types of superposition, and elevation detection. Their disadvantage is the limited accuracy of all opto-electronic devices (Table 2).

Most developments in moiré techniques were forced by the demands of the biological objects, for example, in measuring the corneal shape of an eye or the form of a horse. A lot of work was done simultaneously in Japan, in nearly all countries of Europe, and in the USA. It is impossible to mention all the important contributions or all the working groups, but two facts were important in the past:

1. Nearly every biological or medical problem needed a special solution, which had to meet the demands of the object in sensitivity and accuracy.
2. The papers were published too late or were so widely scattered, that parallel work and waisting of time could not be avoided.

The processing of moiré topograms may be done in various ways. For documentation of body form and analog interpretation, the calculation of simple measures by pocket calculator and/or nomograms may be sufficient. But the extraction of more than one cross-section through the object is very time consuming.

Analytical processing by digital computing can be done by all computers with graphic capabilities, but low-cost systems offer less comfort and consume more time [11]. The algorithms are straightforward. An image with central perspective has to be transformed point by point into an orthogonal projection. If image distortions have to be considered because of a low degree of binding of the gratings or by the optical systems, costs and time for the computation will increase. Therefore every application will have an economic part, were the costs and pay-back must be considered. But it seems to be necessary to make a distinction between topographical methods offering known sensitivity and accuracy in contrast to non-topographical contouring.

4. Biomedical Applications

In the following some typical moiré topographical applications in medicine and biology will be picked out. Some of the work was done in our institute and is included to demonstrate our research program and the range of sensitivity. But first two examples will be given, where moiré figures were used in non-topographical applications in order to have a glance at that moiré work not contained in this paper. In 1972 LOTMAR used moiré figures to test the retinal function [12]. He projected two superposed gratings into a human eye, thus producing high-contrast moiré fringes. By tuning the grating he changed the spatial period of the moiré lines down to the point where they could not further be resolved. SHOUP demonstrated a technique for determining joint instant centers of rotation with high accuracy [13]. Two fixed gratings interfered after motion and double exposition. By means of the moiré lines the rotation center could be determined.

In the field of moiré topography many demonstrations of possible biological applications have been given, such as the deformation of wood, the movement of a snail, the moving speed of a dogs heart wall, or the curvature of the ball of a joint [14-17].

A moiré topographical problem was solved by KAWARA with a very high sensitivity and accuracy [18]. In 1979 he published an apparatus for measuring the corneal curvature. Other authors did similar examinations, but KAWARA was successful by coating the eye ball with a fluorescent liquid and by using a projection method.

In our institute we have studied teeth, skeletal parts, and feet [19-21]. We proposed to plan individual protection filters in radiological treatment moiré topographically [22]. This work led to a commercially available moiré camera in the USA [23]. In veterinary medicine other groups as well as our own are trying to develop a method to objectify breeding criteria and the slaughter value of animals [24,25]. A lot of work is done in facial surgery for documentation and evaluation, in fitting artificial limbs, or in investigations for the clothing industry [26-28]. The most important application is the use of the moiré topography in scoliosis research [29]. In this field the potentials of moiré topography are obvious: "A non-invasive optical method, capable of minimizing the risks of illness and harm."

Acknowledgement. I gratefully acknowledge all the contributions of our moiré team, especially those of Professor Gertrud Keck, of my colleague A. Cabaj, as well as the valuable assistance of J. Gersthofer and Mrs. S. Domann.

Author's Note. When I presented this paper, I showed about sixty slides. Because it is impossible to include a reasonable number of them in this review, there are no figures in the text. I hope that the understandability has not been lost.

References

1 Lord Rayleigh (J.W. Strutt): "On the Manufacture and Theory of Diffraction Gratings", Scient. Papers, *1*, 209. Philos. Mag. *47*, 81 (1874)
2 M. Stecher: The moiré phenomenon. Am. J. Phys. *32*, 247 (1964)
3 D. Meadows, W. Johnson, J. Allen: Generation of surface contours by moiré patterns. Appl. Opt. *9*, 942 (1970)
4 H. Takasaki: Moiré topography. Appl. Opt. *9*, 1467 (1970)
5 J.B. Allen, D.M. Meadows: Removal of unwanted patterns from moiré contour maps by grid translation techniques. Appl. Opt. *10*, 210 (1971)
6 R.E. Herron: Biostereometric measurement of body form. Yearb. Phys. Anthropol. *16*, 80 (1972)
7 R.E. Herron: "Biostereometrics and the Communication of Biological Form", Am. Ass. for the Advancement of Science, 141st. Ann. Meet. Jan. 1975, p.12, 26-31
8 G. Keck et al.: "Moiré-topographische Verfahren für die Medizin: Problematik der Instrumentation", 76th Meeting of the DGaO, May 1975
9 H. Takasaki: "The Development and the Present Status of Moiré Topography", in *Holography in Medicine and Biology*, ed. by G. von Bally, Springer Series in Optical Sciences, Vol.18 (Springer, Berlin, Heidelberg, New York 1979);
G. Windischbauer: "Problems of Image Evaluation in Orthopedics Using Moiré Figures", in *Holography in Medicine and Biology*, ed. by G. von Bally,

Springer Series in Optical Sciences, Vol.18 (Springer, Berlin, Heidelberg, New York 1979)
10 G. Windischbauer: Ein Gitter zur Moiré-Topographie räumlicher Objekte mit Unterscheidung konkaver und konvexer Oberflächenbereiche. Optik *59*, 131 (1981)
11 G. Windischbauer: Digitale Bildverarbeitung von Moiré-Topogrammen mittels Kleinrechnern in der Skolioseforschung. EDV Med. Biol. *12*, 65 (1981)
12 W. Lotmar: Use of moiré fringes for testing visual acuity of the retina. Appl. Opt. *11*, 1266 (1972)
13 T.E. Shoup: Optical measurement of the center of rotation for human joints. J. Biomech. *9*, 241 (1976)
14 C. Sales: Application de la methode Moiré à la mesure des deformation sur materiaux bois. Bois Forets Trop. *174*, 41 (1977)
15 F. Heiniger, T. Tschudi, W. Künzler: Moiré-Tiefenkonturierung von bewegten biologischen Objekten. Optik *53*, 219 (1979)
16 T.R. Fenton, R. Vas: Application of moiré topography to the measurement of 3-dimensional body motion. Med. Biol. Eng. *12*, 569 (1974)
17 J.P. Duncan, J.P. Gofton, S. Sikka, D. Talapatra: A technique for the topographics survey of biological surfaces. Med. Biol. Eng. *8*, 425 (1970)
18 T. Kawara: Corneal topography using moiré contour fringes. Appl. Opt. *18*, 3675 (1979)
19 W. Holler: Vergleichende Untersuchung an den oberen ersten Molaren bei Wolf (canis lupus lupus), Haushund (Canis familiaris-Dobermann) und Goldschakal (Canis aureus) mit Hilfe der Moiré-Topographie. Säugetierkundl. Mitteilungen *26*, 131 (1978)
20 J. Respondek: "Beitrag zur Anwendung der Moiré-Topographie in der Medizin"; Dissertation Vet. Univ. Wien (1974)
21 G. Windischbauer, G. Keck, A. Cabaj: Moiré-topographische Verfahren in der Medizin: Ein Podometer zur Dokumentierung von krankhaften Fußstellungen. Orthopädie-Technik *28*, 7 (1977)
22 W. Binder, A. Cabaj, K.H. Kärcher, G. Windischbauer: Herstellung von Gewebeausgleichsfiltern für große Flächen bei Hodgkin-Bestrahlung mit Hilfe der Moiré-Topographie. Strahlentherapie *153*, 82 (1977)
23 A.L. Boyer, M. Goitein: Simulator mounted moiré topography camera for constructing compensating filters. Med. Phys. *7*, 19 (1980)
24 B.S. Speight, C.A. Miles, K. Moledina: Recording carcass shape by a moiré method. Med. Biol. Eng. *12*, 221 (1974)
25 G. Windischbauer, S. Zinner, G. Zinner, G. Keck: Moiré-Topographie von Großtieren. Wien. tierärztl. Mschr. *67*, 353 (1980)
26 M.C. Madden, M.S. Karlan: Moiré photography as a means of topographics mapping of the human face. Ann. Biom. Eng. *7*, 95 (1979)
27 J.P. Duncan, D.P. Dean, G.C. Pate: Moiré contourgraphy and computer-aided replication of human anatomy. IMechE *9*, 29 (1980)
28 K. Ashizawa, E. Tsutsumi: "An Experimental Study on the Application of Moiré Photogrammetry in Clothing Planning", Ann. Meet. Soc. Moiré Contourography, Tokyo, 1977
29 M.S. Moreland, M.H. Pope, G.W.D. Armstrong (eds.): Moiré Fringe Topography and Spinal Deformity (Pergamon, Oxford 1981)

Moiré Topography in Scoliosis Research

H. Neugebauer and G. Windischbauer

Abteilung für Wirbelsäulenerkrankungen und Haltungsschäden
Orthopädisches Krankenhaus der Stadt Wien-Gersthof, Wielemannsgasse 28
A-1180 Wien, Austria

Institut für Medizinische Physik, Veterinärmedizinische Universität Wien
Linke Bahngasse 11
A-1030 Wien, Austria

In a lateral view the human spine shows some harmonic curvatures called kyphosis or lordosis. Looking from the back, however, the spine should be straight. Every curvature in the frontal plane is pathological and it is called scoliosis. The history of conservative orthopedics is the history of scoliosis research.

There are many types of scoliosis, for example, neuropathic scoliosis after poliomylitis or fibropathic scoliosis following an operation of the breast, or osteopathic scoliosis, which is almost always congenital. Most frequent, however, is the so-called idiopathic scoliosis. We do not know the germ of this disease. But almost four percent of young girls are afflicted with this disease, and ten percent of them become very bad within a relatively short time. As a rule, idiopathic scoliosis is discovered when the child is about nine to twelve years old, and about three years later the scoliotic curvature may have become so much worse that the child is deformed like a cripple. We know many kinds of therapy, for instance, electric stimulation of the muscles, different types of braces like milwaukeebrace or bostonbrace, special gymnastics, and many methods of surgery.

But in any case we have to order a big number of X-ray pictures. Usually X-ray photographs of the whole spine, sometimes even in two planes, are ordered every six months or even more often. This results in a very high radiation exposure during the life of such a young girl. Therefore, it is necessary to find a new method to estimate the scoliosis without X raying, but by an optical method without any exposure to ionizing radiation.

Such a method seems to be moirê topography [1]. Since 1977 we are cooperating at the Orthopedic Hospital Gersthof in Vienna with the Institute of Medical Physics. We use the "asymmetrical grating" developed by one of us, permitting the optical distinction of "hills" and "valleys" within an object area [2]. The technical results are moirê topograms showing four shades in the black and white range of every moirê line for fringe interpretation and interpolation. The sequence of these four shades indicates whether the picture locally represents a convex or concave surface.

We wondered what criteria have to be fullfilled in order to produce well-defined and reproducible results. As far as analytical methods are concerned, the most decisive question to be asked is: Will it be possible to design a technique that permits repeated filming, yielding identical images and/or rather identical results?

We have done a great variety of studies and have tried to produce identical points on the nates (we call them "buttock points") in order to control a supposed rotation of the patient. Finally, we have tried to make moirê films of defined rotations of several plaster torsos. In doing so, we had to notice that the moirê topography is rather sensitive to rotations of the patient. If, however, it is correct that by means of moirê topography each surface point of an object can be defined in space, it should be possible to

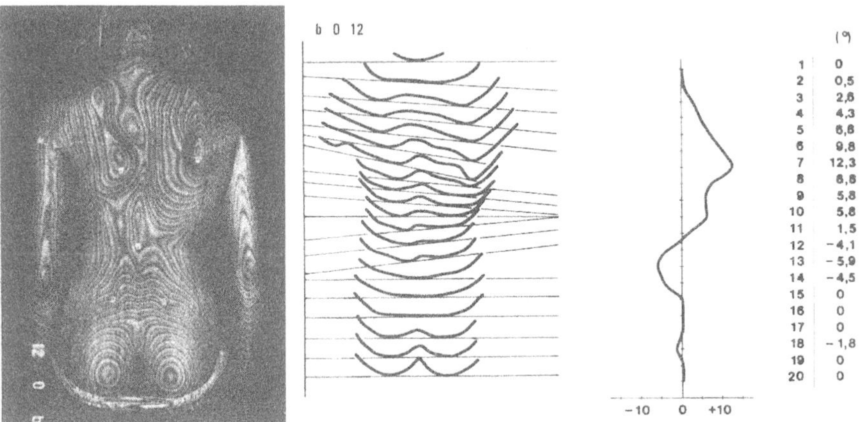

Fig.1. Moiré topogram, body cross-sections, and CMCT curve of a patient with idiopathic scoliosis

evalute the moiré images in such a way that the position or rotation of the object does not affect the result. Thus we have tried to produce a "corrected" moiré contourogram, which differs from the "optical" picture as it is not affected by rotation [3]. In order to obtain such a corrected picture several evaluation processes are required.

First of all, the distorsions caused by the "central projection" of the photographic image have to be eliminated. This is easy to obtain by means of a personal data computer.

Second, the corrected picture is subdivided into 20 horizontal sections in order to compare the contours of these sections. The top plane should be placed at the height of the vertebra prominens, that is, the seventh vertebra of the neck; this plane is called plane No.1. The subsequent planes are placed at identical distances. The last plane is located at the buttock points. The distance of the planes may differ on different topograms due to the height of the human body; however, the planes will always be situated in the same anatomic regions.

Third, we again use the computer, and the outcome is 20 contour lines. These contour lines correspond with the body contours. Their tangents are horizontal and parallel in a healthy individual. Any deviation of the tangents corresponds to a surface asymmetry usually due to vertebral rotation. This method of analysing moiré patterns is called C.M.C.T., which means corrected moiré contouro-tomogram (Fig. 1).

If we indicate all clockwise deviations of the tangents by a plus sign and all counter-clockwise deviations by a minus sign, the angle between the tangent and the horizontal line may be measured in degrees. A curve may be plotted on the basis of the plane positions and extent of deviation. This curve gives an indication of the trunc asymmetry and its relationship to the vertebral rotation. We call it the CMCT curve. A similar method was described in 1970 using the stereophotogrammetry [4]. The next problem was to find the relationship between the curve of scoliosis, the X-ray picture, and our new CMCT curve.

First, however, we wanted to know if the CMCT curve is reproducible. So we made topograms of one and the same girl with a time gap of one hour and on different days (Fig.,2).

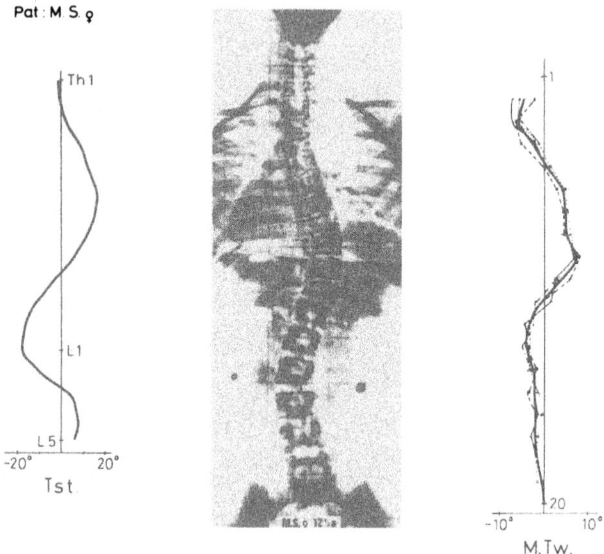

Fig.2. X-ray picture of a scoliotic patient with corresponding CMCT curves of four moiré topograms (M.Tw.). The left curve (Tst.) shows the evaluation of the spine curvature according to Tiderström

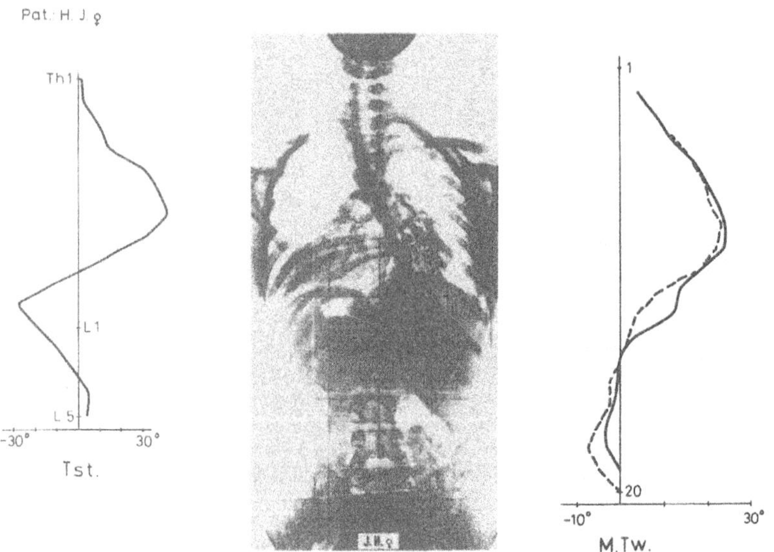

Fig.3. Corresponding CMCT curves (M.Tw.) and Tiderström curve demonstrating their similarity

The CMCT curve was almost identical. Figure 2 shows the mean variability of four curves. Secondly, we compared the X-ray picture with the CMCT curve.

There are some common methods to measure the angle of scoliosis in X-ray photographs. We use either the "middle-of-curve method" by Ferguson or the "end-of-curve method" by Cobb [5]. In the last few years these methods have been used by nearly all investigators. But there exists a third and special method by Tiderström, which shows the inclination of every vertebral body. The result of Cobb or Ferguson is only a single number, which characterizes a distinct angle. But both do not express anything about the form of the spine and its scoliosis.

Tiderström's method produces a curve for itself, but there does not seem to be any apparent similarity with the curve of scoliosis. The CMCT curve, however, seems to look much more like the curve of Tiderström (Fig. 3). But we suppose that it is not at all necessary to find or construct such a likeness. The Cobb, Ferguson, Tiderström results and the CMCT curve, each has its own value.

We are sure that the CMCT curve is not only a good supplement to the X-ray picture, but is a documentation with a special value, too. Starting to follow up a scoliosis, we order an X-ray picture and a moiré topogram at the same time; three months later for controlling we order moiré topogramming only. But, and this is the most important effect, we do not need a further X-ray photograph as long as the CMCT curve does not show a distinct change! Thus we are able to save many X-ray pictures. Therefore the radiation exposure for the child obviously decreases.

References

1 H. Takasaki: Moirê Topography. Appl. Opt. *9*, 1467 (1970)

2 G. Windischbauer: Ein Gitter zur Moirê-Topographie räumlicher Objekte mit Unterscheidung konkaver und konvexer Oberflächenbereiche". Optik *59*, 131 (1981)

3 H. Neugebauer, G. Windischbauer: "Effects of Rotating the Patient on Moirê Contourograms and Verification of Derotation in Scoliosis by Moirê Contourography", in *Moirê Fringe Topography and Spinal Deformity*, ed. by M.S. Moreland, M.H. Pope, G.W.D. Armstrong (Pergamon, Oxford 1981)

4 J.E. Hugg, P.R. Harrington, R.E. Herron: "Idiopathic Scoliosis in Two, Three and Four Dimensions", Technical papers from ASP-Univ. of Illinois Symp. on Close-Range Photogrammetry, Urbana, Jan. 1971, p.211

5 H. Neugebauer: Cobb oder Ferguson - Eine Analyse der beiden gebräuchlichen Röntgenmeßmethoden von Skoliosen. Z. Orthop. *110*, 342 (1972)

High-Precision Moiré Topography Based on Scanning Techniques

T. Yatagai, M. Idesawa, and S. Saito

The Institute of Physical and Chemical Research
2-1, Hirosawa, Wako, Saitama 351, Japan

1. Introduction

The ease of measuring diffuse surfaces has been significantly increased by the development of moiré topographic techniques [1]. Recently, we reported a promising automatic 3-D shape measuring method, called the scanning moiré method [2]. In this method, moiré fringes are generated by the electronic scanning and sampling of a deformed shadow grating, instead of superposition of a reference grating in the conventional moiré topography. The electronic scanning and sampling technique automatically eliminates the shape ambiguity problem, which is serious in the conventional moiré topography. A practical measuring system has been developed [3], especially for medical applications. An optical system of a moiré camera for medical application is used, and data from the photodiode array are processed by a microcomputer.

This paper describes the principle of the scanning moiré method, and then we discuss a technique of moiré fringe interpolation using the scanning moiré method.

2. Scanning Moiré Method

The schematic diagram of the scanning moiré topography is shown in Fig. 1. An optical system of a moiré camera consists of a projection system and an observation system. A grating is projected on an object to be measured. The casted shadow grating on the object is the input to an electronic scanning device. To make a stable and reliable measurement with the scanning moiré method, it is necessary that the image scanning device has high positional accuracy. Therefore, we employed a photodiode array on a precision stage. This device has 1728 photosensitive elements. A scanning servo mechanism moves the photodiode array in its transverse direction, so that all the image area can be covered. An A/D converter transforms the corrected signals to 8 bit digital signals, which are stored in the moiré processor memory.

By scanning and sampling signals of deformed gratings, we can generate moiré fringe signals. These procedures correspond to superposition of a reference grating in conventional projection-type moiré topography. When the sampling period is equal to the pitch of the projection grating, the moiré fringes obtained correspond to the contours of the object as in case of conventional moiré topography.

Merely from a moiré contour pattern we cannot distinguish between a depression and an elevation of an object, since the relative order or sign between two adjacent contours is unknown. Additional information is necessary to remove this ambiguity. In the case of the scanning moiré method, we can change the phase, the pitch, and the direction of the scanning lines, and therefore, various contours can be immediately generated. This means

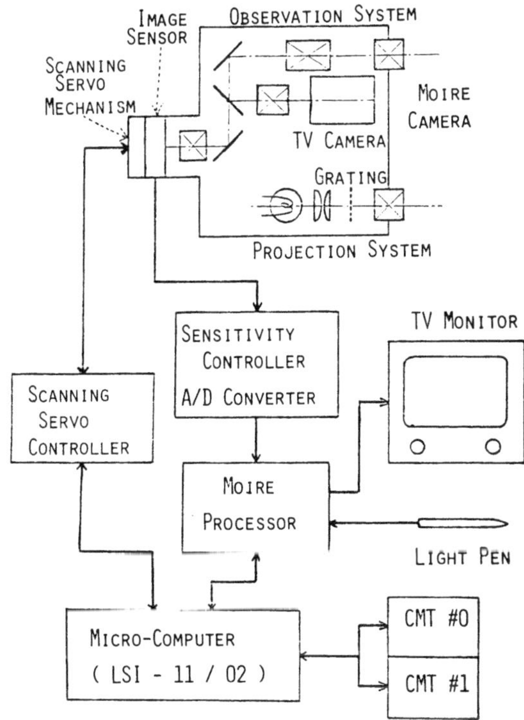

Fig.1. Block diagram of the automatic measurement system using the scanning moiré method

that sign determination and interpolation of moiré contours are performed. In practice, we have determined the sign of moiré fringes by sampling a shadow image of the projected grating with 3 different phases of sampling lines.

3. Fringe Interpolation by Synchronous Detection Technique

As is well known, there are many similarities between moiré topography and interferometric testing [4]. The profile of the moiré fringe is written in the same form of equations as that for the fringe pattern produced by a Twyman-Green interferometer with the object as the mirror in one arm. The only differences in these two equations, i.e., of the moiré system and the interferometer, are fringe contrast and the unit of the fringe sensitivity; one is in the order of the grating pitch and the other in that of the wave length. According to the scanning moiré method, a series of moiré fringes are generated in a computer by superimposing virtual reference gratings, whose phases are slightly different to each other. The digital AC interferometric technique [5] is employed to measure phase differences of generated moiré fringes. It is expected to attain accuracies of better than 1/100 of a fringe, because the electrical phase can be measured to less than a few degrees.

Figure 2(a) is a shadow grating image obtained by the photodiode array, and Fig. 2(b) shows moiré fringes produced by sampling a grating image as that shown in Fig. 2(a). The sampling period is the same as that for the reference grating but the phase is different. By the principle of the AC

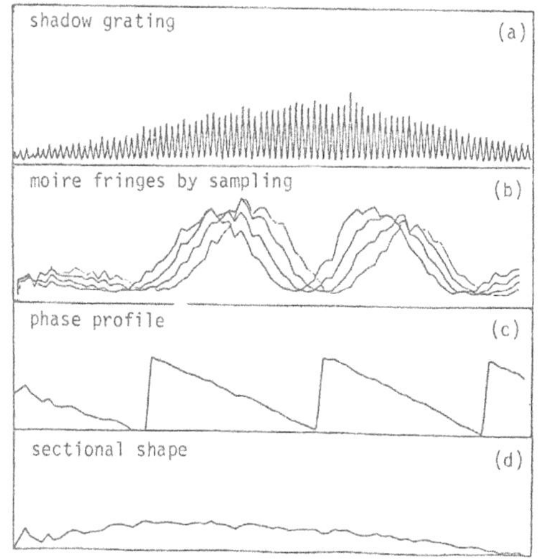

Fig.2. Synchronous detection of moiré fringes

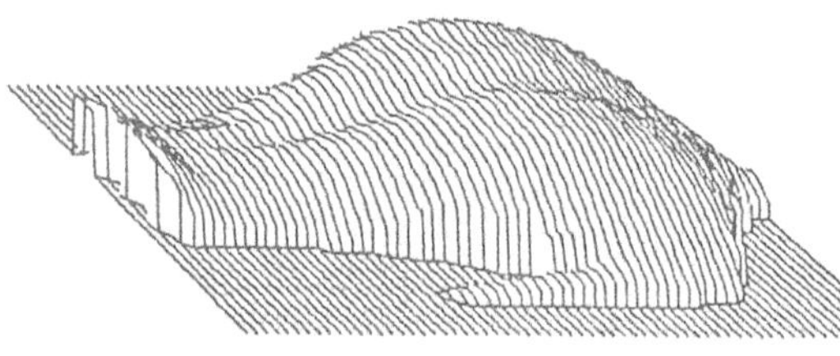

Fig.3. 3-D plot of a human back measured by the automatic system

interferometry, we calculate the phase distribution of the moiré fringe, as shown in Fig. 2(c). The phase distribution has discontinuities because a computer function subprogram gives a principal value ranging from π to π. These discontinuities can be corrected and subtracting the tilt term of the corrected phase distribution gives the object profile, as shown in Fig. 2(d).

Figure 3 shows an example of 2-D measurements. The object was a human body. By using the scanning moiré method and the interpolation technique, 64 sectional shapes were measured.

4. Conclusion

In this paper, we have presented a method of automating 3-D shape measurement by a new type of moiré topography. Sampling a shadow grating with electronic scanning techniques, instead of superimposing the reference grating in the conventional projection type moiré topography, produces various sets of moiré contours, so that we can realize automatic sign determination of moiré contours. On the base of the scanning moiré method, we have developed an interpolation method of moiré fringes, which is the extension of the synchronous detection technique of interferometric fringe reduction.

For medical applications, especially for quantitative analysis of scoliosis, several computer programs have been developed. Numerical parameters which indicate the extent of scoliotic and kyphotic deformities have been defined [6]. The parameters estimate spinal conditions well and can be used for automatic diagnosis of spinal deformities.

References

1 H. Takasaki: *Holography in Medicine and Biology*, ed. by G. von Bally, Springer Series in Optical Science, Vol.18 (Springer, Berlin, Heidelberg, New York 1979) p.45
2 M. Idesawa, T. Yatagai, T. Soma: Appl. Opt. *16*, 2152 (1977)
3 T. Yatagai, M. Idesawa: *Optics in Four Dimensions*, ed. by M.A. Machado (American Institute of Physics, New York 1981) p.579
4 A.J. MacGovern: Appl. Opt. *11*, 2972 (1972)
5 J.H. Bruning, D.R. Herriott, J.E. Gallagher, D.P. Rosenfeld, A.D. White, D.J. Brangaccio: Appl. Opt. *13*, 2693 (1974)
6 A. Shinoto, J. Ohtsuka, S. Inoue, T. Yatagai, M. Idesawa: Proc. 1st International Symposium on Moiré Contourography in Scoliosis, Sept. 22-24, 1980, Burlington, VT, USA

Some Problems in Analytical Reconstruction of Biological Shapes from Moiré Topograms

B. Drerup

Orthopädische Universitätsklinik Münster, Abteilung Biomechanik
D-4400 Münster, Fed. Rep. of Germany

1. Introduction

Among the different moiré techniques [1], those methods generating the topogram in a single process [2] are most frequently used for medical applications. This may be due to a somewhat lower proneness to misalignment compared, e.g. with projection methods. However, an important additional reason is the suggestive character of the finished topogram.

As there is no intrinsic coordinate system in patients, positioning is to certain extent arbitrary. Thus the same object is represented by a variety of different but equivalent contour patterns which may give rise to some confusions. On the other hand, medical interpretation requires a comprehended surface description which is independent of the particular positioning. The comprehension of form and the following morphometric comparisons are done analytically [3]. In the present study some problems associated with the extraction of a surface description from the topograms, which is appropriate to further analysis, are treated.

2. Point Reconstruction

The optical arrangement for shadow moiré topography is shown in Fig.1. The film plane is assumed to be parallel to the grating plane. A coordinate system is attached to the grating with the x,y axes parallel to the u,v axes which span the film plane. The origin of the film coordinate system is the principal point H: $H=(H_x,H_y,H_z)=(0,0,-l+f)$, with f being the focal length and $|l|$ the distance from the camera to the grating. Assuming P=(x,y,z) to be an object point and Q=(u,v) to be its image point, P can be reconstructed:

$$
\begin{aligned}
z &= l\cdot N/(N-d/g) \\
x &= u\cdot(l-z)/f \qquad (1)\\
y &= v\cdot(l-z)/f
\end{aligned}
$$

with d/g being the ratio of the grating constant to the distance between lens and light source. N is the ordinal number of the fringe. It equals the number of grating periods between the intersection points of the grating plane with the object ray and the image ray. If P happens to lie on a contour line, N will be an integer.

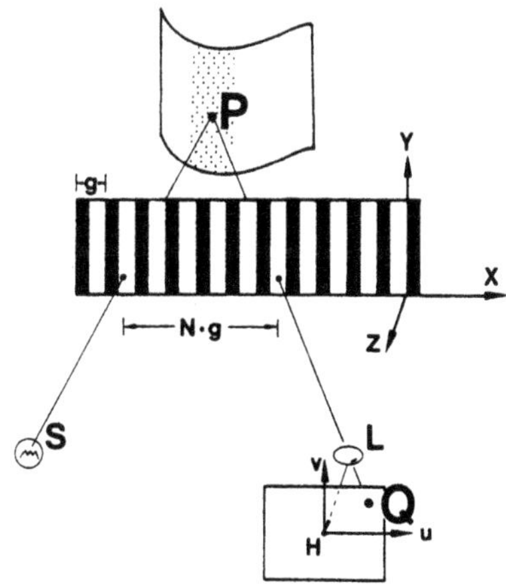

Fig.1 Principal arrangement for moire' topography of basic type

On biological surfaces contour lines generally form irregular patterns and the measuring points are therefore irregularly distributed too. However, an analytical shape analysis based on a representation of the spatial coordinates as explicit functions of u and v (parametric representation) requires a regular distribution of surface points. Commonly these points are the corners (u_i,v_j) of a regular rectangular grid with grid lines $u=u_i, i=1,2..$ and $v=v_j, j=1,2...$. If z and consequently N - cf. (1) - is determined as a function of u_i and v_j, (1) is transcribed into a parametric representation. N therefore has to be interpolated in the grid points.

3. Interpolation of Ordinal Number

A straightforward procedure for the interpolation of N in the corners of a rectangular grid has been described in [4]: the topogram is digitized along the u_i coordinate lines by picking up N in all intersection points with the contour lines. By this convention N is always an integer and differences between neighboring measurement points are limited to $\Delta N=0, \pm 1$. The maximum difference of any values in an interval between two measurement points is limited to $|\Delta N| \leq 1$.

A cubic spline $S(u_i,v_j)$ approximates the data points, allowing the ordinal number ν_{ij} in (u_i,v_j) to be reconstructed. These interpolated values no longer need to be integers. For control reasons, the same procedure is repeated. However, this time digitization is performed along the v_j coordinate lines, resulting in the values $\mu_{ij}(u_i,v_j)$. The differences $\tau_{ij}=|\nu_{ij} - \mu_{ij}|$ are a measure for the digitization error, which includes the error by the spline interpolation. Fig. 2 shows a classification of the differences for one particular topogram of which the even numbered contour lines are outlined. The variance of τ_{ij} amounts to about 0.22, which is equivalent to $\Delta z=35$mm for the topographic arrangement we used. It is obvious that discrepancies between the two interpolated point sets are greatest in those regions where the contour lines either run parallel to one of the coor-

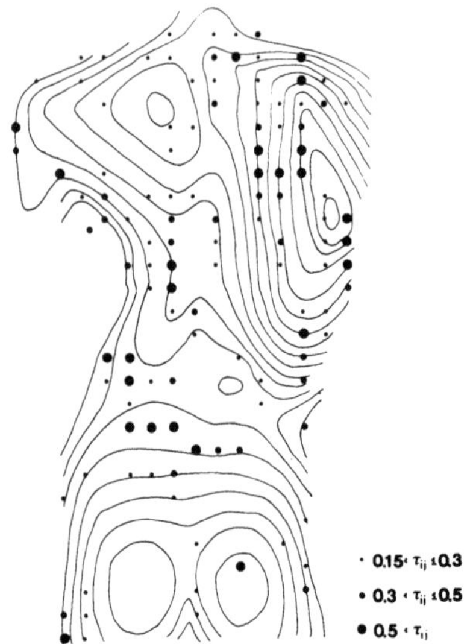

Fig.2 Distribution of digitization errors

dinate axes - resulting in long distances between data points on the respective coordinate lines - or where the line density in total is small; in both cases the spline is loosely bound and undulations may arise. Undulations are evident if in contrast to the measuring convetion an interpolated value exceeds the neighboring data point by more than $|\Delta N|=1$. In general unwanted undulations can be eliminated by the addition of half numbered data points. In the case studied, the additional measurement of a few extra points reduced the variance to $\Delta z=\pm 0.25$mm.

Another - well known - attempt to generate gridded data is based on polynomial least squares approximations [5]. In this case data points may be collected by digitizing the topogram either along coordinate lines or along contour lines. In each grid corner (u_i,v_j) a polynomial

$$\Phi_{ij} = \sum_{k=0}^{2} \sum_{l=0}^{k} a^{ij}_{l,(k-l)} \cdot \bar{u}_i^{l} \cdot \bar{v}_j^{(k-l)} \qquad (2)$$

with $\bar{u}_i=u-u_i$, $\bar{v}_j=v-v_j$, is computed by normal equations to give the best fit to the neighboring data points. The local character of this approximation is emphasized by weighting the data points by a weight function w

$$w_{ij}(u,v) = exp\{-(u_i^2 + v_j^2)/c^2\}$$

which is a function of the distance to the respective grid corner. The distance is measured in units of a constant c, which has to be selected appropriate to the grid spacing. The size of the surface segments around the grid corner can be adapted as to contain more than 6 data points, which are needed to calculate the 6 coefficients a^{ij}_{mn}. The first coefficient a^{ij}_{oo} delivers an interpolated value of N in (u_i,v_j), which turns out to be in good agreement with the mean value $(\nu_{ij} + \mu_{ij})/2$ of the above mentioned spline interpolations. The variance of the differences in all grid points was ≤ 0.1. The other coefficients can be interpreted as partial derivatives, i.e.

$$\frac{\partial^{m+n}}{\partial u^m \partial v^n} \Phi_{ij} = a_{mn}^{ij} .$$

4. Curvature

Calculation of curvature is done with the partial derivatives of first and second order of the point coordinates with respect to u and v [6]. If x,y and z are given in grid points, the derivatives can be determined from the coefficients of polynomials Ξ_{ij}, T_{ij} and Z_{ij} similarly to (2), which approximate the x,y and z values separately. Due to the regular arrangement along the grid, the least squares calculation can be simplified very much [7]. However, in our case a twofold approximation has been performed to obtain the derivatives from the data points. This twofold approximation can be avoided by computing the polynomials directly from the scattered data in the same way as discussed for Φ_{ij}. Derivatives of the coordinates may be determined just as well from a_{mn}^{ij} by application of the chain rule. This solution was found to be practicable, allowing on the one hand a high precision of computing, limiting on the other hand computational effort to a tolerable extent.

5. Summary

Moire topography is used as a diagnostic tool in medicine. For comparative studies, digitization of the topogram is necessary. Digitization can be performed by different methods. Two of these are discussed: 1) the topogram is digitized along a rectangular grid, 2) contour lines are digitized using scattered data points along these lines; in adaptive surface segments polynomials are fitted to the contour points by least squares methods. A check for rough reconstruction errors is given. Reconstruction errors mainly occur in surface regions oriented parallel to the reference plane, which therefore show few contour lines.

References

1. G. Windischbauer: "Survey on Applications of Moiré Techniques in Medicine and Biology, this volume
2. H. Takasaki: Moire' Topography, Appl. Optics 9 (1970)1467-1472
3. F.L. Bookstein: The Measurement of Biological Shape and Shape Change, Lecture Notes in Biomathematics, Vol. 24, Springer, Berlin, Heidelberg, New York, 1978
4. B. Drerup: A Procedure for the Numerical Analysis of Moire Topograms, Photogrammetria 36 (1981) 41-49
5. L. Schumaker: Fitting Surfaces to Scattered Data, in:Approximation Theory II, G. Lorentz et al. ed., Acad. Press London 1974
6. R. Courant: Differential and Integral Calculus II, Interscience Publications, New York, 1956
7. W. Frobin, E. Hierholzer: Rasterstereografie II, Krümmungsanalyse, Interner Bericht SFB 88/C1 ISSN 0173-6450 Nr. 21 Münster 1979

Automatic Measurement of Body Surfaces Using Rasterstereography

E. Hierholzer and W. Frobin

Abteilung für Biomechanik, Orthopädische Universitätsklinik
D-4400 Münster, Fed. Rep. of Germany

Rasterstereography is a photogrammetric method for three-dimensional measurement of body surfaces. It is similar to conventional stereophotography, except that one of the two cameras has been replaced by a projector with a raster diapositive (Fig.1). Thus, the diapositive and the camera image may be considered as a stereoscopic image pair and all the well known techniques of photogrammetric calibration and reconstruction may be applied with minimum modifications. However, in contrast to stereophotography, but similar to Moiré topography, the three-dimensional information is contained in a single image in the projected and distorted raster lines.

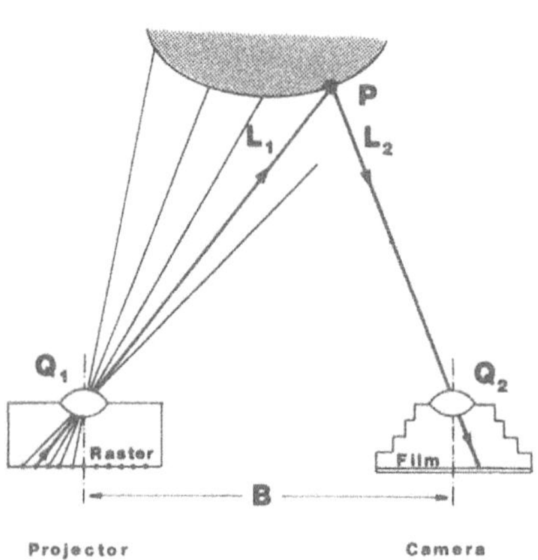

Fig.1: Basic principle of rasterstereography

For the purpose of automatic image processing a regular line raster is most favourable, although in principle any other raster type may be employed.

A rasterstereographic camera set consists of a camera and a projector mounted in a solid frame. In the rasterstereographic image the raster lines projected onto the surface to be measured are recorded (Fig.2). In addition, for calibration purposes a system of control points and planes is recorded from which the geometry of the whole system is determined. Details are reported in [1].

The basic principle of surface reconstruction is shown in Fig.3. Knowing the geometry of the camera and the projector, a surface point P can be reconstructed by spatial intersection of the pertinent projecting ray PQ_2 (camera system) with the plane defined by the appropriate raster line in the diapositive and the projection centre Q_1 of the projector system. In this way, a reconstruction is possible for any surface point lying on a projected raster line (intermediate points may, however, be obtained by interpolation). To reconstruct a surface point two different types of data must therefore be determined from the rasterstereograph:

i) the x and y coordinates of its camera image
ii) the position of the pertinent raster line in the diapositive (e.g. by using the line number).

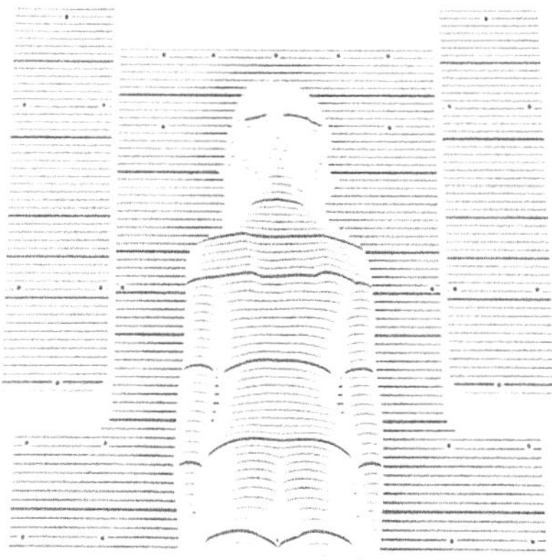

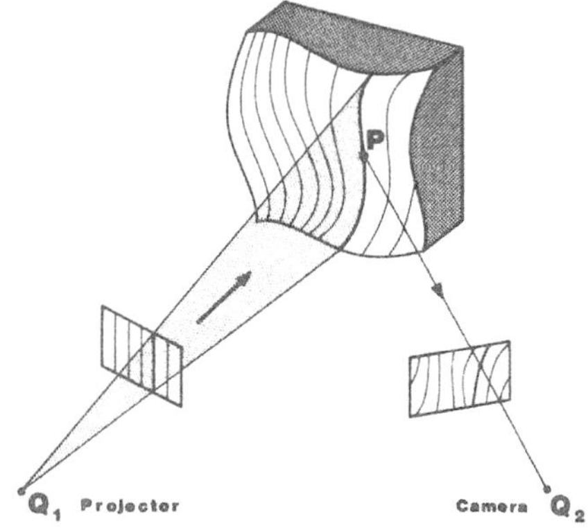

Fig.2: Rasterstereograph *Fig.3: Reconstruction*

In our automatic evaluation procedure the x and y image coordinates are measured by scanning the film with a linear sensor camera. The position (number) of any raster line in the diapositive is determined by analysis of the unique sequence of light and heavy lines (Fig.2). In the following, details of our method of image data acquisition and analysis are presented. The whole procedure consists of the image scan, line search, line sequence analysis, line numeration and reconstruction.

The *image scan* is performed with the apparatus shown schematically in Fig.4. The system consists of a linear sensor camera with 1024 elements and an x/y translator for film shifting. A projector is used as a light source. The video signal from the sensor is fed into the computer via a standard A/D interface. After each camera cycle the translator is advanced to the next scan position. The linear sensor is arranged approximately perpendicular to the raster lines. A number of 250 scan lines proved to be sufficient for a 6 cm x 6 cm film. For each scan line a raster line appears as a peak in the video signal. A peak is interpreted as a raster line only if certain conditions concerning height, width, positive and negative slope and peak area are met.

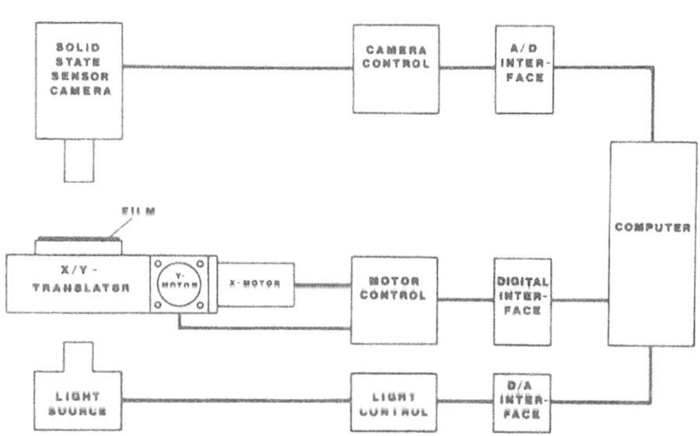

Fig.4: Scanning of rasterstereographs

Otherwise the peak will be discarded. Either the peak width or the peak area may be used for discriminating between light and heavy lines. By fitting an interpolating curve to the video signal the peak position may be calculated with an accuracy of a fraction of the pixel width. As a result of the image scan the raster lines are disassembled into series of peaks the data of which are stored on a magnetic disk.

By the *line search* procedure the peaks are reassembled to lines. This is shown schematically in Fig.5. According to a distance criterium adjacent peaks in consecutive scan lines are investigated, i.e. those peaks which belong to the same raster line. The peaks are classified as starting points (B), internal points (I) and end points (E) of a raster line, dependent on the existence of adjacent peaks in the preceding and following scan line. In addition, there may be isolated points (S) resulting e.g. from dust particles, and corner points (C) in the case of a sudden change in direction. As in most cases corner points originate from false connections of different raster lines, the line is dissected into two portions at such a point, and the adjacent peaks are classified as E and B respectively.

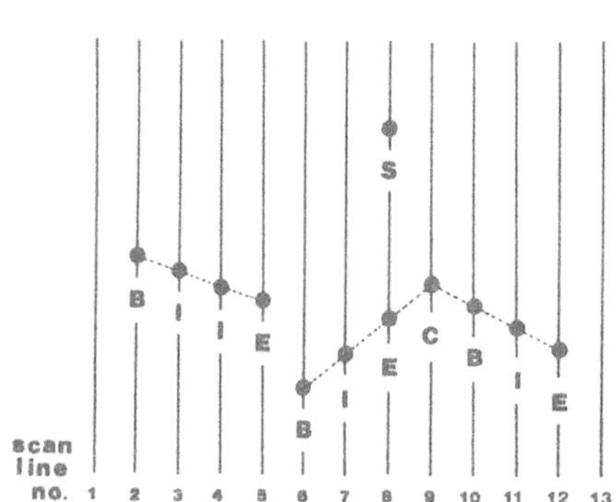

Fig.5: Peak classification

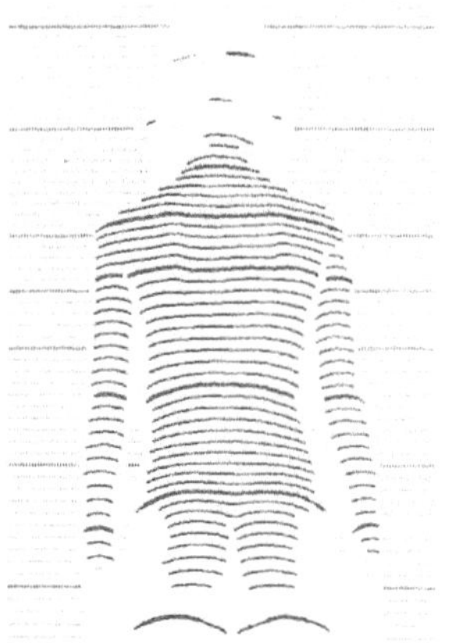

Fig.6: Line search

As a result of the line search any measured peak is supplied with a (preliminary) number of the raster line to which it belongs, and with a classification according to its position within that line. The result is visualized in Fig.6 where peaks belonging to the same raster line have been interconnected to a solid line.

In the *line sequence analysis* the array of measured raster lines is analysed with regard to neighboured lines. This is necessary for the subsequent line numeration, from which their position in the raster diapositive is determined. In the line sequence analysis any two measured raster lines are declared as neighbours if

i) there is no intermediate line between them
ii) their mean distance does not exceed a certain limit
iii) there is a certain overlap in the x (scanning) direction.

In regard to the surface to be measured two neighboured raster lines form a surface stripe. The whole surface may thus be divided into an ordered sequence of adjacent stripes. Any stripe may be classified as a 'starting', 'internal', 'end', 'forward node' or 'backward node' stripe according to the number of stripes adjacent to it in both directions (Fig.7). As a result of the line sequence analysis all surface stripes together with their bounding raster lines are brought into a unique relative order. If several nodes are present different line sequences with the same start and end may be possible. In this case the number of intermediate raster lines may be checked for compatibility of the different ways.

The *line numeration* is necessary for the determination of the raster line position in the diapositive. The relative ordinal number as determined in the line sequence analysis is converted into an absolute ordinal number by evaluating the sequence of line widths. To compensate for different projection scales in different portions of the rasterstereograph the ratio of the line widths of consecutive raster lines is used rather than the line width itself. The measured ratios are compared with the nominal ratios in a least squares procedure by shifting the relative ordinal numbers. If an appropriate pattern of light and heavy lines is employed a unique least squares minimum and hence a unique absolute line numeration is obtained.

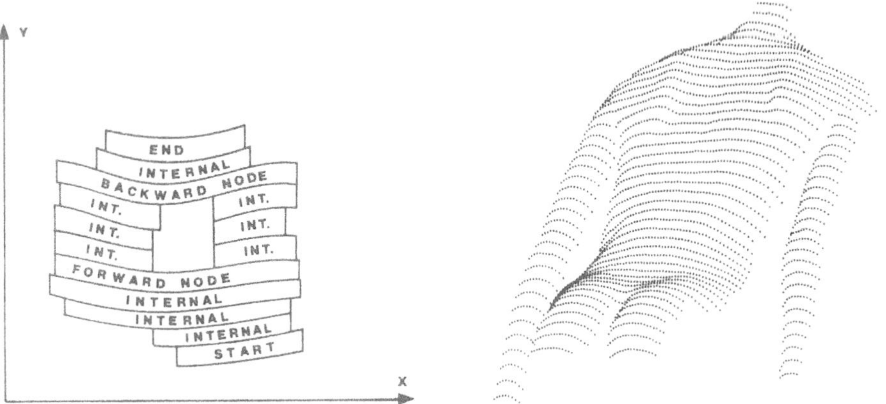

Fig.7: Line sequence analysis

Fig.8: Reconstructed surface

The *reconstruction* has already been described at the beginning. In Fig.8 the surface reconstructed from Fig.2 is shown. The computing time for the whole evaluation procedure (including calibration) was about 30 min on a PDP 11/45 computer. The depth resolution obtained is in the order of 0.5 mm.

Reference:

1 *Frobin, W., Hierholzer, E.: "Rasterstereography: A Photogrammetric Method for Measurement of Body Surfaces", Photogr. Eng. and Remote Sensing, in press*

Quantification of the Symmetry of the Nose Using an Analysis of Moiré Photographs of the Face

T. Katsuki, M. Goto, Y. Kawano, and H. Tashiro

Department of Oral Surgery, Faculty of Dentistry, Kyushu University
Fukuoka, 812 Japan

Correction of the symmetry of the nose is an important objective of the cleft lip operation. In the present study, we attempted to numerate the state of symmetry of the nose by using moire topography which provides a simple, one-step method for mapping the contour of the subject on a single photograph. Thus, a front view of the face of the subject was photographed with a moire camera, Fujinon FM 3012. Moire contour lines were subsequently processed by a coordinate measuring equipment GRADICON (Fig. 1), and the resulting values were applied to the equations listed below, in order to calculate the following four indices by a microcomputer, WANG 2200.

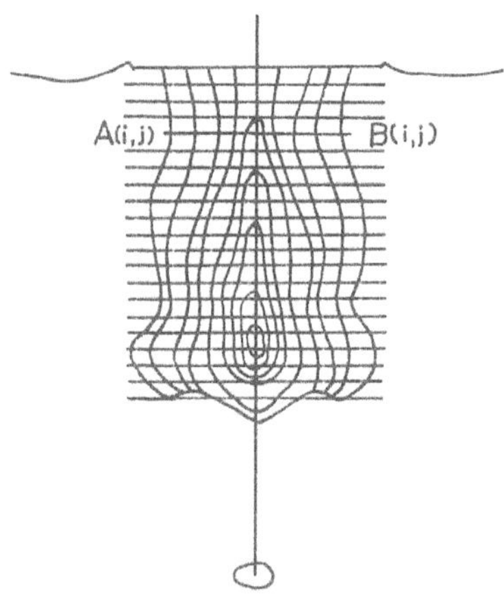

Fig. 1 Reading of the moire contour line

$$\text{Symmetry Index} = \frac{1}{N}\sum_{j=1}^{N}\frac{B(j)-A(j)}{B(j)+A(j)} \times 100 \quad (1)$$

$$\text{Total Symmetry Index} = \frac{1}{M}\sum_{i=1}^{M}\frac{1}{N}\sum_{j=1}^{N}\frac{B(i,j)-A(i,j)}{B(i,j)+A(i,j)} \times 100 \quad (2)$$

$$\text{Revised Symmetry Index} = \frac{1}{N}\sum_{j=1}^{N}\frac{B(j)-A(j)}{B(max)+A(max)} \times 100 \quad (3)$$

$$\text{Total Revised Symmetry Index} = \frac{1}{M}\sum_{i=1}^{M}\frac{1}{N}\sum_{j=1}^{N}\frac{B(i,j)-A(i,j)}{B(max)+A(max)} \times 100 \quad (4)$$

Twenty six normal individuals were examined in this way and the above four indices were calculated.

Table 1 Symmetry index of normal individuals

Level	Mean	S.D.
1	-3.18	8.63
2	-3.51	7.21
3	-6.58	9.79
4	-6.47	8.06
5	-7.62	8.73
6	-8.36	9.32
7	-8.00	8.15
8	-8.46	8.28
9	-9.98	10.43
10	-8.85	8.64
11	-8.66	8.23
12	-8.18	6.99
13	-8.49	7.69
14	-8.01	7.26
15	-7.15	7.29
16	-7.08	7.11
17	-6.63	7.25
18	-6.25	6.89
19	-5.93	8.87
20	-5.86	8.87

Table 2 Revised symmetry index of normal individuals

Level	Mean	S.D.
1	-0.67	1.65
2	-0.85	1.89
3	-1.55	2.35
4	-1.80	2.66
5	-2.18	2.75
6	-2.44	2.97
7	-2.65	3.05
8	-2.91	3.02
9	-3.02	3.12
10	-3.28	3.06
11	-3.38	3.24
12	-3.53	3.26
13	-3.58	3.43
14	-3.36	3.39
15	-3.28	3.67
16	-3.30	3.74
17	-3.21	3.81
18	-3.27	3.37
19	-3.02	3.66
20	-2.79	4.01

Table 3 Indices of normal individuals (n=26)

	Mean	S.D.
Total symmetry index	7.90	6.40
Total revised symmetry index	2.70	2.83

In order to determine the correlation between the symmetry index, revised symmetry index and subjective evaluation by examinors, 20 cleft lip nose patients were investigated. Calculation of the regression line (Fig. 2) and correlation coefficient (Fig. 3) clearly demonstrates that the symmetry indices and revised symmetry indices agree with the subjective evaluation by examinors.

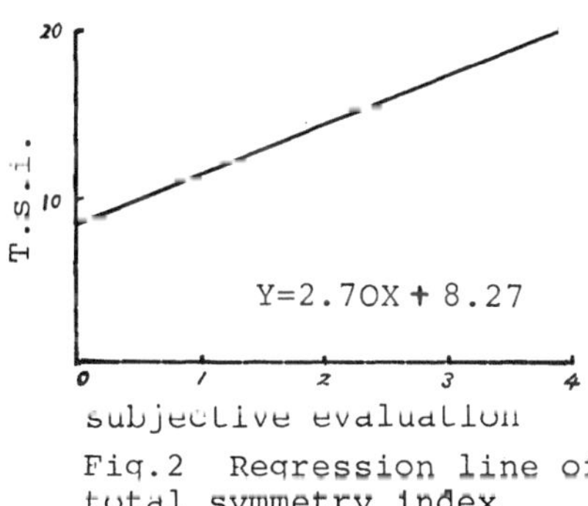

Fig.2 Regression line of total symmetry index

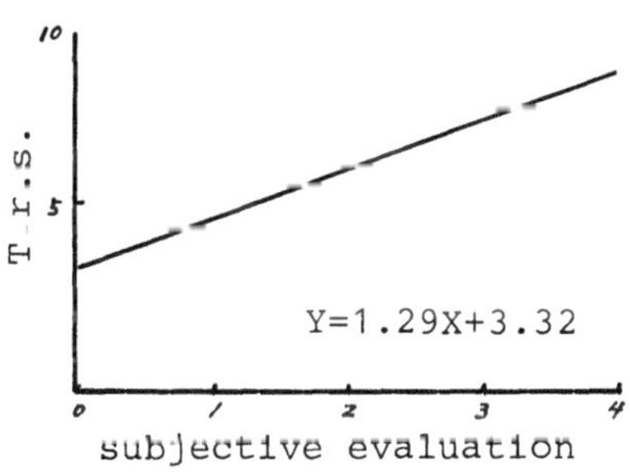

Fig.3 Regression line of revised total symmetry index

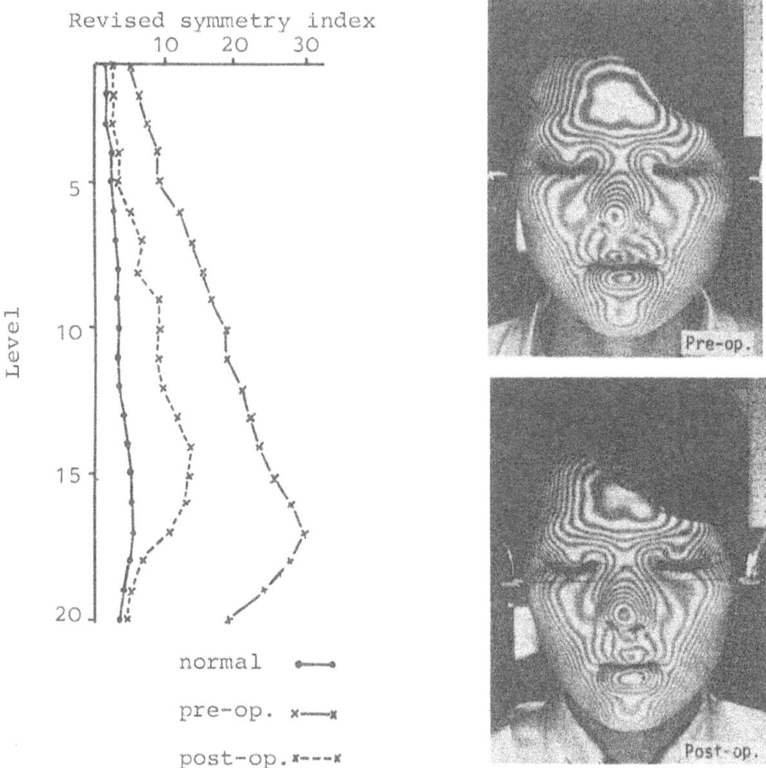

Fig.4 Change of moire contour lines and revised symmetry index by rhinoplasty

In order to compare the symmetry of the nose before and after rhinoplasty using this quantification procedure, symmetry index and revised symmetry index are presented graphically (Fig. 4 and 5).

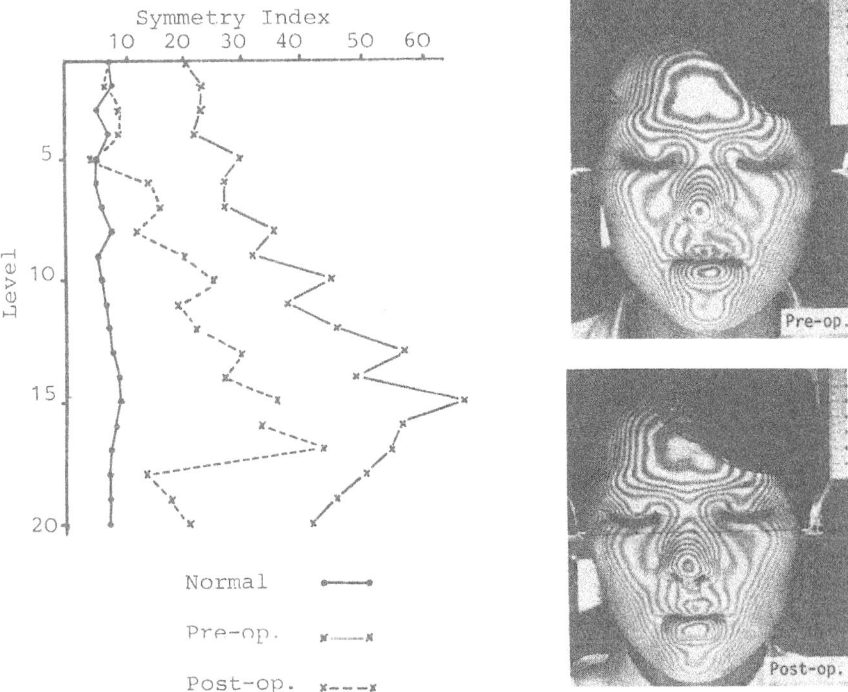

Fig.5 Change of moire contour lines and symmetry index by rhinoplasty

These indices may be also useful for the examination of the symmetry of other parts of the body such as face, back etc.

Part VII

Closing Remarks

Closing Remarks

N. Abramson

Production Engineering, Royal Institute of Technology
S-100 44 Stockholm, Sweden

As the time is short and most participants of this conference long for their homes I will be brief and concentrate my remarks on only those two lines of Fig. 1.

The upper line represents the high quality level of the presentations given in this conference while the lower line represents the level of understanding by me and many other participants. The upper level represents the excited state of the speakers, while the lower level represents the ground (or down-to-the-earth) state of the audiance.

As the upper level represents an excited state it also represents an inverted population the language of which is rather difficult to understand (as seen in Fig. 1) for those being in the ground state. Thus it is quite a difficult task for the speaker to fulfil his main purpose: to produce a <u>stimulating emission of information</u>.

To make still another analogy, let the upper line represent the high tension, or high voltage, of the highbrow physicists in the physical or medical laboratories while the lower line represents the down-to-the-earth people in the industrial or clinical workshops. The aim of a conference like ours is to work as sort of a transformer bringing the voltage down to a level that is useful in the workshops and clinics.

During this conference some of the speakers and contributers have not managed to bring their information all the way down to earth while others have extremely well succeeded in fulfilling this difficult mission.

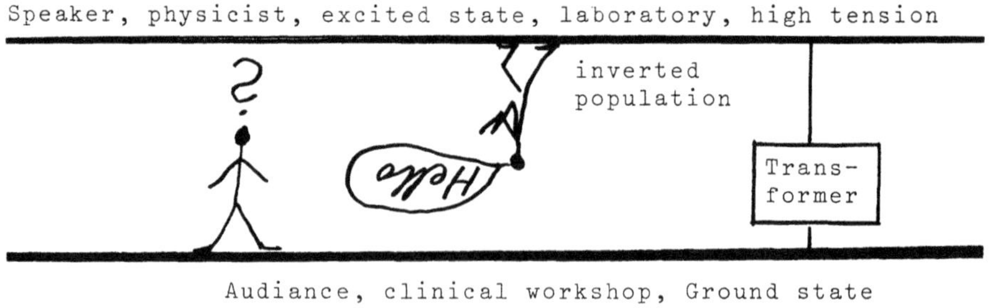

Speaker, physicist, excited state, laboratory, high tension

inverted
population

Trans-
former

Audiance, clinical workshop, Ground state

Fig. 1

Finally I want, in the name of all the participants, thank all those who have so well organized and arranged this conference. I thank Greguss and von Bally and all those nice people responsible for the great success of this conference. I hereby declare the closing of the ICO-12 Sattelite Meeting, Optics in Biomedical Sciences.

Index of Contributors

Topics in Applied Physics

Founded by H. K. V. Lotsch

A selection

Volume 6

Picture Processing and Digital Filtering

Editor: **T. S. Huang**
2nd corrected and updated edition. 1979.
113 figures, 7 tables. XIII, 297 pages
ISBN 3-540-09339-7

Contents:
T. S. Huang: Introduction. – *H. C. Andrews:* Two-Dimensional Transforms. – *J. G. Fiasconaro:* Two-Dimensional Nonrecursive Filters. – *R. R. Read, J. L. Shanks, S. Treitel:* Two-Dimensional Recursive Filtering. – *B. R. Frieden:* Image Enhancement and Restoration. – *F. C. Billingsley:* Noise Considerations in Digital Image Processing Hardware. – *T. S. Huang:* Recent Advances in Picture Processing and Digital Filtering. – Subject Index.

Volume 23

Optical Data Processing

Applications

Editor: **D. Casasent**
1978. 170 figures, 2 tables. XIII, 286 pages
ISBN 3-540-08453-3

Contents:
D. Casasent, H. J. Caulfield: Basic Concepts. – *B. J. Thompson:* Optical Transforms and Coherent Processing Systems – With Insights From Cristallography. – *P. S. Considine, R. A. Gonsalves:* Optical Image Enhancement and Image Restoration. – *E. N. Leith:* Synthetic Aperture Radar. – *N. Balasubramanian:* Optical Processing in Photogrammetry. – *N. Abramson:* Nondestructive Testing and Metrology. – *H. J. Caulfield:* Biomedical Applications of Coherent Optics. – *D. Casasent:* Optical Signal Processing.

Volume 42

Two-Dimensional Digital Signal Processing I

Linear Filters

Editor: **T. S. Huang**
1981. 77 figures. X, 210 pages. ISBN 3-540-10348-1

Contents:
T. S. Huang: Introduction. – *R. M. Mersereau:* Two-Dimensional Nonrecursive Filter Design. – *P. A. Ramamoorthy, L. T. Bruton:* Design of Two-Dimensional Recursive Filters. – *B. T. O'Connor, T. S. Huang:* Stability of General Two-Dimensional Recursive Filters. – *J. W. Woods:* Two-Dimensional Kalman Filtering.

Volume 43

Two-Dimensional Digital Signal Processing II

Transforms and Median Filters

Editor: **T. S. Huang**
1981. 49 figures. X, 222 pages. ISBN 3-540-10359-7

Contents:
T. S. Huang: Introduction. – *J.-D. Eklundh:* Efficient Matrix Transposition. – *H. J. Nussbaumer:* Two-Dimensional Convolution and DFT Computation. – *S. Zohar:* Winograd's Discrete Fourier Transform Algorithm. – *B. I. Justusson:* Median Filtering: Statistical Properties. – *S. G. Tyan:* Median Filtering: Deterministic Properties.

Volume 48

Optical Information Processing

Fundamentals

Editor: **S. H. Lee**

1981. 197 figures. XIII, 308 pages
ISBN 3-540-10522-0

Contents:
S. H. Lee: Basic Principles. – *S. H. Lee:* Coherent Optical Processing. – *W. T. Rhodes, A. A. Sawchuk:* Incoherent Optical Processing. – *G. R. Knight:* Interface Devices and Memory Materials. – *D. P. Casasent:* Hybrid Processors. – *J. W. Goodman:* Linear Space-Variant Optical Data Processing. – *S. H. Lee:* Nonlinear Optical Processing. – Subject Index.

Springer-Verlag Berlin Heidelberg New York

Topics in Applied Physics

Founded by H. K. V. Lotsch

A selection

Volume 9

Laser Speckle and Related Phenomena

Editor: **J. C. Dainty**
With contributions by J. C. Dainty, A. E. Ennos, M. Françon, J. W. Goodman, T. S. McKechnie, G. Parry
1975. 133 figures. XIII, 286 pages. ISBN 3-540-07498-8

Contents:
Introduction. – Statistical Properties of Laser Speckle Patterns. – Speckle Patterns in Partially Coherent Light. – Speckle Reduction. – Information Processing Using Speckle Patterns. – Speckle Interferometry. – Stellar Speckle Interferometry. – Additional References with Titles. – Subject Index.

Volume 13

High-Resolution Laser Spectroscopy

Editor: **K. Shimoda**
With contributions by N. Bloembergen, V. P. Chebotayev, J. L. Hall, S. Haroche, P. Jacquinot, V. S. Letokhov, M. D. Levenson, J. A. Magyar, K. Shimoda
1976. 132 figures. XIII, 378 pages. ISBN 3-540-07719-7

Contents:
Introduction. – Line Broadening and Narrowing Effects. – Atomic Beam Spectroscopy. – Saturation Spectroscopy. – High Resolution Saturated Absorption Studies of Methane and Some Methyl-Halides. – Three-Level Laser Spectroscopy. – Quantum Beats and Time-Resolved Fluorescence Spectroscopy. – Doppler-Free Two-Photon Absorption Spectroscopy.

Volume 18

Ultrashort Light Pulses

Picosecond Techniques and Applications

Editor: **S. L. Shapiro**
With contributions by D. H. Auston, D. J. Bradley, A. J. Campillo, K. B. Eisenthal, E. P. Ippen, D. v. der Linde, C. V. Shank, S. L. Shapiro
1977. 173 figures. XI, 389 pages. ISBN 3-540-08103-8

Contents:
Introduction – A Historical Overview. – Methods of Generation. – Techniques for Measurement. – Picosecond Nonlinear Optics. – Picosecond Interactions in Liquids and Solids. – Picosecond Relaxation Processes in Chemistry. – Picosecond Relaxation Measurements in Biology.

Springer-Verlag
Berlin
Heidelberg
New York

www.ingramcontent.com/pod-product-compliance
Ingram Content Group UK Ltd.
Pitfield, Milton Keynes, MK11 3LW, UK
UKHW021832190726
13853UKWH00003B/1283

* 9 7 8 3 6 6 2 1 3 5 2 4 2 *